I0061881

B. P. Pratten

Investigation of Diseases of Swine

And infectious and contagious diseases incident to other classes of domesticated

animals

B. P. Pratten

Investigation of Diseases of Swine
And infectious and contagious diseases incident to other classes of domesticated animals

ISBN/EAN: 9783337241209

Printed in Europe, USA, Canada, Australia, Japan

Cover: Foto ©berggeist007 / pixelio.de

More available books at **www.hansebooks.com**

DEPARTMENT OF AGRICULTURE.

SPECIAL REPORT—No. 12.

INVESTIGATION

OF

DISEASES OF SWINE,

AND

INFECTIOUS AND CONTAGIOUS DISEASES

INCIDENT TO

OTHER CLASSES OF DOMESTICATED ANIMALS.

WASHINGTON:
GOVERNMENT PRINTING OFFICE.
1879.

TABLE OF CONTENTS.

LIST OF ILLUSTRATIONS.

INVESTIGATION OF SWINE PLAGUE.

INTRODUCTORY.

Congress having previously appropriated the sum of $10,000 for defraying the expenses of a commission to investigate and determine the causes producing, and, if possible, discover remedies for, some of the more contagious and destructive diseases incident to domesticated animals, early in August last the Commissioner of Agriculture appointed examiners in the States of New York, Indiana, Illinois, Iowa, Kansas, Missouri, and North Carolina, to conduct such investigation. Still later in the season, on receiving information that not only diseases among swine were prevailing to an alarming extent in Virginia, but that a fatal disease resembling pleuro-pneumonia or contagious lung fever was destroying a good many valuable dairy cattle in some localities of that State, an additional examiner was appointed and instructed to investigate and report upon all the facts connected with the condition of both classes of animals in the infected districts of this State.

In the preliminary report of the Commissioner of Agriculture on the subject of diseases of domesticated animals, a tabular statement gives the total value of farm animals lost in the United States during the year 1877, principally from infectious and contagious diseases, at $16,653,428. These losses were based upon as accurate returns as could be obtained in the absence of an absolute census, but as they included data from but eleven hundred and twenty-five counties (about one-half the whole number of counties in the United States), the above sum falls far below the aggregate losses for that year. About two-thirds of this sum was occasioned by the loss of swine by diseases presumed to be of an infectious and contagious character. Notwithstanding these maladies had their origin near a quarter of a century ago, and had rapidly spread from one State and one county to another, there was great diversity of opinion as to their contagious or non-contagious character. Many intelligent farmers and stock-growers insisted that they were not transmissible from one animal to another, while perhaps equally as large a number contended that the diseases were of a highly infectious and contagious nature. As this was regarded as one among the most important facts to be determined by the investigation, two of the examiners devoted most of their time to experiments looking to a solution of this problem.

As the number and value of the annual losses among swine were much heavier than among all other classes of domesticated animals combined, the Commissioner deemed it best to devote the greater portion of the limited sum placed at his disposal to an investigation of the fatal diseases affecting this class of farm animals.

The preliminary investigation instituted and conducted under the supervision of this department, in the fall and winter of 1877-'78, established the fact that diseases prevail among these animals much more extensively during the late summer and early fall months than at other seasons of the year, and for this reason the examiners selected to conduct the investigation were employed for periods ranging from one to three months. It was assumed, and the subsequent history of the disease proved the assumption to be well founded, that the reduced temperature of the late fall and early winter months would cause an abatement of the disease, and in a measure deprive the examiners of subjects with which to continue their experiments. While, therefore, the very severe weather of the past winter caused a great reduction in the number of animals affected, the disease was not eradicated, nor did its fatality seem to be lessened. The spread of the infection from one herd to another was greatly diminished; but, in infected herds, where the malady was still prevailing when cold weather set in, there appeared but little difference in the rapidity of the transmission of the disease, from one animal to another, in the same herd. Dr. H. J. Detmers, V. S., of Chicago, who conducted his investigations and made his experiments in one of the worst infected of the many large hog-growing districts in Illinois, writing under date of January 7 last, speaks as follows of the effects of severe frosts on the spread of the disease:

Since my last letter the weather has continued extremely cold. Where I now am, in Lee County, some five or six miles west of Dixon, the thermometer indicated at seven o'clock on the morning of January 2, 28° below zero, and on the next morning 24° below zero. At present—to-day, yesterday and day before—the weather is a little milder. To-day it tried to snow a little; otherwise the sky has been clear every day. The wind is, and has been, west, except yesterday afternoon, when it was almost due south. Swine-plague during this cold weather does not seem to spread either so readily or so rapidly from one farm to another as a few months ago; but as to its spreading from one animal to another in the same herd in which it previously existed no difference can be observed. It seems to be just as fatal as in August, and its course, on the whole, is probably more acute, as severe affections of the lungs and of the heart are more frequent, a fact easily explained in the habits of swine crowding together and lying on top of each other in their sleeping places when the temperature is very low.

Dr. James Law, of Ithaca, N. Y., whose investigations have been solely confined to experiments intended to further establish the contagious and infectious character of the disease, the period of its incubation, &c., confirms the statement of Dr. Detmers, i. e., that the severe frosts of winter do not destroy the germs of the malady but simply retard their conveyance from one herd to another. In a letter of recent date, forwarded since his report was completed, Dr. Law says:

I have demonstrated that the freezing of the virulent matter does not destroy its activity, and that the virus loses nothing in potency by preservation for one or two months closely packed in dry bran. The same may be inferred of all other situations when it is closely packed and where the air has imperfect access. These two last points are of immense importance as bearing on the question of the preservation of the poison in infected pens and yards alike in winter and in summer, to say nothing of its possible conveyance in fodder, &c. The different modes in which the disease may be conveyed in the wet and dry condition, and in the bodies of rabbits, and probably sheep and other animals, speak in the strongest terms against keeping up the production of the poison by preserving sick animals, unless where they can be secluded in thoroughly disinfected buildings in which even the air shall be constantly charged with disinfectants.

In most of the States in which investigations have been made, the examiners have found the symptoms and *post-mortem* appearances of the disease the same, and hence agree as to the propriety of designating the affection under the head of a general disorder. Dr. Detmers has, therefore, given the disease the name of "Swine-plague," and Dr. Law has named it "Hog-fever." While either designation would seem to be eminently proper, that of "Swine-plague" will no doubt be generally adopted.

As in almost all general disorders, a certain variety of organs were found affected and diseased. Marked changes and extravasations in various parts of the body were observed, and inflammation of the lungs and large intestines was usually present. The heart, the pleura, the eyes, the epidermis, and many other important organs showed either slight or more serious affections, and in almost every case tested with the thermometer the temperature was found to be above normal heat before any other symptom of the disease was in the least apparent. In every herd where the disease had prevailed to any considerable extent, no case was found where death had occurred from a local malady, but all the lesions and appearances unmistakably indicated the existence of the general disorder. In but few cases was death found to have resulted from the affection of any single organ, but on the contrary seemed to have been the result of the various organic changes observed.

Dr. Detmers says that the morbid process, although in all cases essentially the same, is not restricted to a single part or organ, or to a set of organs, but can have its seat almost anywhere—in the tissue of the lungs; in the pleura and pericardium; in the heart; in the lymphatic system; in the peritoneum; in all mucous membranes, especially in those of the intestines; in the liver; in the spleen, and even in the skin. Only the pulmonal tissue and lymphatic glands are invariably affected.

The most constant and unvarying symptom of the disease is observed in the increased temperature of the body. Indeed, one of the examiners regards it as highly probable that a high temperature may exist several weeks before other symptoms are manifested, and that the disease may in some cases even be confined to and run its course in the blood without a localization in any other organ or organs. A few isolated cases are noted where this symptom was lacking, but it may have been present in a mild form before other symptoms were observed. The external symptoms of the disease, which were found to be almost identical in all the widely-separated localities in which examinations were made, were a dullness of the eyes, the lids of which are kept nearer closed than in health, with an accumulation of secretion in the corners. There is hanging of the head, with lopped ears, and an inclination to hide in the litter and to lie on the belly and keep quiet. As the disease advances, the animal manifests more or less thirst, some cough, and a pink blush or rose-colored spots, and papular eruption appears on the skin, particularly along the belly, inside of the thighs and fore legs, and about the ears. There is accelerated respiration and circulation, increased action of the flanks in breathing, tucked-up abdomen, arched back, swelling of the vulva in the female as in heat; occasionally, also, of the sheath of the male, loss of appetite, and tenderness of the abdomen, sometimes persistent diarrhea, but generally obstinate constipation. In some cases large abraded spots are observed at the projecting points of the body, caused by separation and loss of the epidermis. In such cases a slight blow or friction on the skin is sufficient to produce such abrasions. In many cases the eruption, blush, and spots are entirely absent; petechia are formed in only about one-third of the cases. In some cases there is considerable inflammation of and discharge from the eyes. Some animals emit a very offensive odor even before death. In large herds, where the disease prevails extensively, this offensive effluvia can be detected for a great distance to windward. In nearly all cases there is a weakness or partial paralysis of the posterior extremities, and occasionally this paralysis is so complete in the first stages of the disease as to prevent walking or standing.

As symptoms of special diagnostic value, which are scarcely ever absent in any case, the following are mentioned: Drooping of the ears and of the head; more or less coughing; dull look of the eyes; staring appearance of the coat of hair; partial or total want of appetite for food; vitiated appetite for excrements; rapid emaciation; great debility; weak and undecided, and frequently staggering, gait; great indifference to surroundings; tendency to lie down in a dark corner, and to hide the nose and even the whole head in the bedding; the specific offensive smell, and the peculiar color of the excrements. This last symptom is always present, at least in an advanced stage of the disease, no matter whether constipation or diarrhea is exisiting. Among other characteristic symptoms, which are not present in every animal, may be mentioned frequent sneezing; bleeding from the nose; swelling of the eyelids; accumulation of mucus in the inner canthi of the eyes; attempts to vomit, or real vomiting; accelerated and difficult breathing; thumping or spasmodic contraction of the abdominal muscles (flanks), and a peculiar, faint, and hoarse voice in the last stages of the disease.

The duration of the disease varies according to the violence and seat of the attack and the age and constitution of the patient. Where the attack is violent, and its principal seat is located in one of the vital organs—such as the heart—the disease frequently terminates fatally in a few days, and sometimes even within twenty-four hours; but when the attack is of a mild character, and the heart is not seriously affected, and the animal is naturally strong and vigorous, one or two weeks usually intervene before death ensues. If the termination is not fatal, convalescence requires an equal and not unfrequently a much longer time. A perfect recovery seldom occurs; in most cases some lasting disorder remains behind and more or less interferes with the growth and fattening of the animal. Those that do recover make but very poor returns for the food consumed; hence from a pecuniary standpoint it makes but little difference to the owner whether the animal recovers or not. The attack is always more violent and fatal when large numbers of animals are closely confined together in small and dirty inclosures or in illy ventilated and filthy pens.

The disease can have its seat in many different organs or parts of the body, and therefore produces a great variety of morbid changes. This accounts for its different aspect in different animals. In some cases the principal seat of the disease may be in the organs of respiration and circulation, and in others in the intestinal canal and organs of digestion. Death may therefore be the result of different causes in different cases. In some cases it results from a cessation of the functions of the heart, the lungs, &c., and in others it is in consequence of the inability of entirely different organs to perform their allotted functions. This being the case, the *post-mortem* appearances would necessarily greatly vary, but in all animals similarly affected the lesions and morbid changes were found identical.

Perhaps the most important point to be determined by this investigation was the contagious or non-contagious character of the disease. In order to do this a series of experiments were instituted and conducted solely with this end in view, by Dr. Detmers, of Illinois, and Dr. Law, of Cornell University, New York. These experiments resulted in determining the fact that the disease is both infectious and contagious, and that it is not confined alone to swine, but that other animals may contract it in a mild form and retransmit it to swine in its most virulent and malignant character.

On the 6th day of September, Dr. Detmers fed a portion of the stomach, the cæcum, and the spleen of a pig that had died on that day to two healthy pigs. On the 19th of the same month they showed signs of illness, and the symptoms continued to grow in intensity until the 23d, when, finding that the animal must die in a few hours, one of them was killed by bleeding. The other pig was found dead in the pen on the morning of September 30. The symptoms and *post mortem* appearances were those of swine-plague, as they revealed the same lesions as those observed in an examination of the pig from which the diseased products had been taken for the purpose of infection. On the 24th day of September, the day following the death of the first pig, a healthy pig of mixed Poland-China and Berkshire was confined in the same pen with the sick pig that died on the 30th of that month. It showed no signs of sickness until the 2d day of October, when the first symptoms of the disease were observed. It continued to grow rapidly worse, and was found dead in its pen on the morning of the 11th, nine days after the first symptoms were observed.

Experiments were made with a large number of other animals to test the infectious and contagious character of the plague. These experiments included the confinement of healthy with sick animals, and the inoculation of healthy animals with the diseased products of those suffering with the fever. In almost every case, as will be seen from his detailed report, Dr. Detmers was successful in transmitting the disease from sick to healthy animals.

The microscopic investigations of Dr. Detmers also revealed some important facts. His discovery of a new order of *bacteria* or *bacillus*, which he names *bacillus suis*. as it is common only to this disease of swine, and his failure to inoculate healthy animals with virus from which these germs had been removed by filtration and otherwise, would lead to the conclusion that these microphytes are the true seeds of the hog fever.

Dr. Detmers invariably found these germs, in one form or another, in all fluids. So constantly were they observed in the blood, urine, mucus, fluid exudations, &c., and in the excrements and in all morbidly affected tissues of diseased animals, that he regards them as the true infectious principle. They would seem to undergo several changes, and to require a certain length of time for further propagation; therefore, if introduced into the animal organism, a period of incubation or colonization must elapse before the morbid symptoms make their appearance. These germs were generally found in immense numbers in the fluids, but more especially in the blood and in the exudations of the diseased animals. With the proper temperature and the presence of a sufficient amount of oxygen they soon develop and grow lengthwise by a kind of budding process. A globular germ, constantly observed under the microscope, budded and grew under a temperature of 70° F. twice the original length in exactly two hours. and changed gradually to rod-bacteria or *bacilli*. Under favorable circumstances these *bacilli* continue to grow in length until, when magnified 850 diameters, they appear from one to six inches long. A knee or angle is first formed where a separation is to take place, and then a complete separation is effected by a swinging motion of both ends. After the division. which requires but a minute or two after this swinging motion commences, the ends thus separated move apart in different directions. These long bacteria seem pregnant with new germs; their external envelope disappears or is dissolved, and then the numerous bacillus germs become free, and in this way effect propagation. Some of the *bacilli* or rod-bacteria move very rapidly, while others are apparently motionless. A certain degree of heat would seem to be necessary for their propagation, as, under the microscope, the motion increases and becomes more lively if the rays of the light, thrown upon the slide by the mirror, are sufficiently concentrated to increase the temperature of the object. Another change observed by Dr. Detmers, but the cause of which he was not able to determine, was observed in the fact that the globular bacteria or bacillus germs commence to bud or grow, when, very suddenly, their further development ceases, and partially developed *bacilli* and simple and budding germs congregate to colonies, agglutinate to each other, and form longer or smaller irregularly-shaped and apparently viscous clusters. These clusters are frequently found in the blood and in other fluids, and invariably in the exudations of the lungs; and in the lymphatic glands in pulmonal exudation and in blood serum this formation can be observed under the microscope if the object remains unchanged for an hour or two. In the ulcerous tumors on the intestinal mucous membrane but few of these clusters will be found, but the fully-developed *bacilli*, many of which appear very lively, are always exceedingly numerous. These tumors or morbid growths in the intestines seem to afford the most favorable conditions for the growth and development of the *bacilli* and their germs. The presence of such immense numbers of these microphytes and their germs in the excrements and other morbid products of swine leads Dr. Detmers to regard them, beyond doubt, as the principal disseminators of the plague. Whether these colonies or viscous clusters are instrumental in bringing about the extensive embolism of the lungs and other tissues by merely closing the capillary vessels in a mechanical way, or whether the presence, growth, development, and propagation of the *bacilli* and their germs produce peculiar chemical changes in the composition of the blood, thereby disqualifying it from passing with facility through the capillaries, or which cause a clotting and retention of the same in the capillary system, Dr. Detmers is not able positively to decide. He is of the opinion, however, that these colonies or viscous clusters of bacillus germs and partially developed *bacilli* cause sufficient obstruction of the capillaries to produce fatal embolism.

The vitality of the *bacilli* and bacillus-germs is not very great, except where preserved in a substance or fluid not easily subject to decomposition; for instance, in water which contains a slight admixture of organic substances. Where contained in such a fluid and preserved in a vial with a glass stopper, they will remain for at least five or six weeks in nearly the same condition, or develop very slowly, according to the amount of oxygen and degree of temperature maintained. In an open vessel the development is a more rapid one. If oxygen is excluded, or the amount available is exhausted, no further change takes place. In the water of streamlets, brooks, ditches, ponds, &c., their vitality is retained or preserved for some time. In fluids and substances subject to putrefaction, they lose their vitality and are destroyed in a comparatively brief period; at least they disappear as soon as those fluids and substances undergo decomposition. In the blood they disappear as soon as the blood-corpuscles commence to decompose or putrefy. They are also destroyed if brought in contact with or acted upon by alcohol, carbolic acid, thymol, iodine, &c. The destruction of these germs by decomposition would seem to account for the harmless nature of thoroughly putrid products when consumed by healthy animals. (See drawings, *bacilli* and *bacillus-germs*.)

Dr. Law also discovered bacteria in the blood of pigs suffering with the disease, and in one case, on the second day before death, he found the blood swarming with them, all showing very active movements. (See drawings, Plate xiii, Fig. 3.) The blood from another pig, which had been inoculated from this one, showed the same living, actively-moving germs in equal quantity. They were further found in the blood of a rabbit and of a sheep inoculated from the first-mentioned pig. In an abscess of a puppy, which had also been inoculated, the germs were abundant. In the examination of blood from healthy pigs the microscope failed to reveal the presence of these organisms. Dr. Law states that in his experiments the greatest precautions were taken to avoid the introduction of extraneous germs. The caustic potash employed was first fused, then placed with reboiled distilled water in a stoppered bottle which had been heated to red heat. The glass slides and cover-glasses were cleaned and burned, the skin of the animal cleaned and incised with a knife that had just been heated in the flame of a lamp. The caustic solution and the distilled water for the immersion-lens were reboiled on each occasion before using, and finally the glass rods employed to lift the latter were superheated before being dipped in them. On different occasions, when the animal was being killed, the blood from the flowing vessels was received beneath the skin into a capillary tube which had just been purified by burning in the flame of a lamp. With these precautions Dr. Law thinks it might have been possible for one or two bacteria to get in from the atmosphere, but this would not account for the swarms found as soon as the blood was placed under the microscope.

The most scrupulous care was observed by Dr. Law in his experiments in inoculation. The isolated and non-infected locality where the experiments were conducted offered special advantages for a series of experiments of this character, as there were no large herds of diseased and exposed swine, and, consequently, no danger of accidental infection from other sources than the experimental pens. The number of animals subjected to experiment was limited by the necessity for the most perfect isolation of the healthy and diseased, for the employment of separate attendants for each, and for the disinfection of instruments used for scientific observations, and of the persons and clothes of those necessarily in attendance. The experimental pens were constructed on high ground in an open field, with nothing to impede the free circulation of air. They were large and roomy, with abundant ventilation from back and front, with perfectly close walls, floors, and roofs, and in cases

where two or more existed in the same building, the intervening walls
were constructed of a double thickness of matched boards, with building pasteboard between, so that no communication could possibly take
place except through the open air of the fields. When deemed necessary, disinfectants were placed at the ventilating orifices. On showing
the first signs of illness, infected pigs were at once turned over to the
care of attendants delegated to take charge of these alone. The food,
utensils, &c., for the healthy and diseased animals were kept most carefully apart. When passing from one to the other for scientific observations, the healthy were first attended, and afterward the diseased, as
far as possible in the order of severity. Disinfection was then resorted
to, and no visit was paid to the healthy pigs until after a lapse of six or
eight hours, with free exposure to the air in the interval. In the pens
the most scrupulous cleanliness was maintained, and deodorizing agents
used in sufficient quantities to keep them perfectly sweet.

The experiments of Dr. Law have shown the period of incubation to
vary greatly, though in a majority of cases it terminated in from three
to seven days after inoculation. One animal sickened and died on the
first day, three on the third, two on the fourth, one on the fifth, two on
the sixth, four on the seventh, and one each on the eighth and thirteenth
days respectively. Referring to experiments of others for determining
the period of incubation, Dr. Law says that Dr. Sutton, observing the
result of contact alone in autumn, sets the period at from thirteen to
fourteen days; his own observations in Scotland, in summer, indicated
from seven to fourteen days; Professor Axe, in summer, in London, concluded on from five to eight days; Dr. Budd, in summer, from four to
five days; and Professor Osler, in autumn, at from four to six days.
Dr. Detmers gives the period of incubation from five to fifteen days, or
an average of about seven days. A comparison of these results would
seem to indicate that both extremes have been reached.

In experimenting in this direction, Dr. Law first sought to ascertain
the tenacity of life of the dried virus. Some years ago Professor Axe
had successfully inoculated a pig with virus that had remained dried
upon ivory points for twenty-six days. In order to carry this experiment still further, Dr. Law inoculated three pigs with virulent products
that had been dried on quills for one day, one with virus dried on a quill for
four days, one for five days, and one for six days. These quills had been
sent from North Carolina and New Jersey, wrapped in a simple paper
covering, and were in no way specially protected against the action of
the air. Of the six inoculations, four took effect. In the two exceptional
cases the quills had been treated with disinfectants before inoculation,
so that the failure was anticipated.

Three pigs were inoculated with diseased intestine which had been
dried for three and four days respectively. The intestine was dried in
the free air and sun, and the process was necessarily slower than in the
case of the quills, where the virus was in a very thin layer, hence there

was more time allowed for septic changes. In all three cases the inoculation proved successful. This experiment would prove that the morbid products, even in comparatively thick layers, may dry spontaneously, and retain their vitality sufficiently to transmit the disease to the most distant States.

Another pig was inoculated with a portion of moist diseased intestine sent from Illinois in a closely-corked bottle. The material had been three days from the pig, and smelt slightly putrid. The disease developed on the sixth day. A second pig was inoculated with blood from a diseased pig that had been kept for eleven days at 100° F. in an isolation apparatus, the outlets of which were plugged with cotton wool. Illness supervened in twenty-four hours.

A solitary experiment of Dr. Klein's having appeared to support the idea that the blood was non-virulent, Dr. Law tested the matter by inoculating two pigs with the blood of one that had been sick for nine days. They sickened on the seventh and eighth days respectively, and from one of these the disease was still further propagated by inoculating with the blood three other animals. Notwithstanding the success of these three experiments, Dr. Law is still doubtful of the blood being virulent at all stages of the disease.

But one or two experiments were instituted by Dr. Law to test the question of infection through the air alone. A healthy pig placed in a pen between two infected ones, and with the ventilating orifices within a foot of each other, front and back, had an elevated temperature on the ninth, tenth, and eleventh days, with lameness in the right shoulder, evidently of a rheumatic character. On the twenty-fourth day the temperature rose two degrees, and remained 104° F. and upward for six days, when it slowly declined to the natural standard.

A healthy pig was placed in a pen from which a sick one had been removed thirteen days before. The pen had been simply swept out, but subjected to no disinfection other than the free circulation of air, and as the pig was placed in the pen on December 19, all moist objects had been frozen during the time the apartment had stood empty. The pig died on the fifteenth day, without having shown any rise of temperature, but with *post-mortem* lesions that showed the operation of the poison. Dr. Law refers to this case as an example of the rapidly fatal action of the disease, the poison having fallen with prostrating effect on vital organs—the lungs and brain—and cut life short before there was time for the full development of all the other lesions. It fully demonstrates the preservation of the poison in a covered building at a temperature below the freezing point.

Perhaps the most important experiments conducted by Dr. Law were those relating to the inoculation of other animals than swine with the virus and morbid products of pigs suffering with the plague, and the transmission of the disease from these animals back to healthy hogs. A merino wether, a tame rabbit, and a Newfoundland puppy were inoculated with blood and pleural fluid containing numerous actively · moving bacteria, taken from the right ventricle and pleuræ of a pig that had died of the fever the same morning. Next day the temperature of all three was elevated. In the puppy it became normal on the third day, but on the eighth day a large abscess formed in the seat of inoculation and burst. The rabbit had elevated temperature for eight days, lost appetite, became weak and purged, and its blood contained myriads of the characteristic bacteria. The wether had his temperature raised for an equal length of time, and had bacteria in his blood, though not so abundantly as in that of the rabbit. The sheep and rabbit had each been unsuccessfully inoculated on two former occasions with the blood of sick pigs, in which no moving bacteria had been detected. Subsequently, after two inoculations with questionable results, made with the blood of sick pigs in which no microzymes had been observed, Dr. Law succeeded in inoculating a rabbit with the pleural effusion of a pig that had died the night before, and in which were numerous actively moving bacteria. Next day the rabbit was very feverish and quite ill, and continued so for twenty-two days, when it was killed and showed lesions in many respects resembling those of the sick pigs. The blood of the rabbit contained active microzymes like those of the pig. On the fourth day of sickness the blood of the rabbit containing bacteria was inoculated on a healthy pig, but for fifteen days the pig showed no signs of illness. It was then reinoculated, but this time with the discharge from an open sore which had formed over an engorgement in the groin of the rabbit. Illness set in on the third day thereafter and continued for ten days, when the pig was destroyed and found to present the lesions of the disease in a moderate degree. A second pig, inoculated with frozen matter which had been taken from the open sore on the rabbit's groin, sickened on the thirteenth day thereafter, and remained ill for six days, when an imminent death was anticipated by destroying the animal. During life and after death it presented the phenomena of the plague in a very violent form.

The results of these experiments have convinced Dr. Law, as they must convince others, that the rabbit is itself a victim of this disease, and that the poison can be reproduced and multiplied in the body of this rodent and conveyed back with undiminished virulence to the pig. Dr. Klein had previously demonstrated the susceptibility of mice and guinea pigs to the disease. The rabbit, and still more the mouse, is a frequent visitor of hog pens and yards. The latter eats from the same feeding troughs with the pig, hides under the same litter, and runs constant risk of infection. Once infected, they may carry the disease to long distances. During the progress of severe attacks of the disease, their weakness and inability to escape will make them an easy prey to the omnivorous hog; and thus sick and dead alike will be devoured by the doomed swine.

Dr. Law says that the infection of these rodents creates the strongest presumption that other genera of the same family may also contract the disease, and by virtue of an even closer relation to the pigs, may succeed in conveying the malady to distant herds. The rat is suggested as being almost ubiquitous in piggeries, and more likely than any other rodent to contract and transmit the disease to distant farms. In order to test its susceptibility to the poison, Dr. Law inoculated a rat with the virus from a sick pig, but unfortunately the subject died on the second day thereafter. The body showed slight suspicious lesions, such as congested lungs with considerable interlobular exudation, congested small intestines, dried-up contents of the large intestines, and sanguinous discoloration of the tail from the seat of inoculation to the tip. With the fresh congested small intestine of the rat he inoculated one pig, and with the frozen intestine one day later he inoculated a second. The first showed no rise of temperature, loss of appetite, or digestive disorder; but on the sixth day pink and violet eruptions, the size of a pin's head and upwards, appeared on the teats and belly; and on the tenth day there was a manifest enlargement of the inguinal glands. In the second pig inoculated, the symptoms were too obscure to be of any real value. Dr. Law will continue his experiments with this rodent.

In addition to the above, Dr. Law experimented on two sheep of different ages, an adult merino wether and a cross-breed lamb, and in both cases succeeded in transmitting the disease. With the mucus from the anus of the wether he inoculated a healthy pig, which showed a slight elevation of temperature for five days, but without any other marked symptoms of illness. Eleven days later it was reinoculated with scab from the ear of the lamb, and again three days later with anal mucus from the sheep. The day preceding the last inoculation it was noticed that the inguinal glands were much enlarged, and in six days thereafter the temperature was elevated and purple spots appeared on the belly. At the time that Dr. Law closed his report this fever had lasted but a few days, but he regards the symptoms, taken in connection with the violent rash and the enlarged lymphatic glands, as satisfactory evidence of the presence of the disease. It can, therefore, be affirmed of the sheep as of the rabbit, that not only is it subject to this disease, but that it can multiply the poison in its system and transmit it back to the pig.

Among the later experiments by Dr. Law was one inaugurated with the view of testing the vitality of frozen products of the disease. This point was briefly alluded to above, but its importance would seem to call for further attention. In two cases healthy pigs were inoculated with virulent products which had been frozen hard for one and two days respectively. In both instances the resulting disease was of a very violent type, and would have proved fatal had it been left to run its course. The freezing had failed to impair the virulence of the product; on the contrary, it had only sealed it up to be opened and given free course on the recurrence of warm weather. Once frozen no change could take place until it was again thawed out, and if it was preserved for one night unchanged in its potency, it would be equally unaffected after the lapse of many months, provided its liquids had remained in the same crystalline condition throughout. It is in this way, no doubt, that the virus is often preserved through the winter in pens and yards, as well as in cars and other conveyances, to break out anew on returning spring. The importance of this discovery, as applied to preventive measures, cannot be overestimated. Infected yards and other open and uncovered places may not be considered safe until after two months' vacation in summer, and not then if sufficient rain has not fallen during the interval to insure the soaking and putrid decomposition of all organic matter near the surface. This will be made more apparent by reference to an experiment which resulted in the successful inoculation of pigs with virus that had been kept for a month in dry wheat bran. In winter, on the other hand, the yard or other open and infected place may prove non-infecting for weeks and even months and yet retain the virus in readiness for a new and deadly course as soon as mild weather sets in. Safety under such circumstances is contingent on a disuse of the premises so long as the frost continues, and for at least one month or more thereafter. Even during the continuance of frost such places are dangerous, as the heat of the animal's body or of the rays of the sun at midday may suffice to set the virus free.

Several of the examiners treat at length of hygienic and sanitary measures, and the attention of the reader is directed to their detailed reports, which will be found below, without further comment.

2 SW

INVESTIGATION OF SWINE-PLAGUE.

REPORT OF DR. H. J. DETMERS, V. S.

Hon. WM. G. LeDUC,
Commissioner of Agriculture:

SIR: Having been appointed by you as one of the inspectors to make an investigation of the diseases prevailing among swine, I forwarded to you my written acceptance, immediately after I received my appointment, on July 29, 1878, and took at once the necessary steps to obtain reliable information as to the localities where the disease of swine, known to the farmers as "hog-cholera," was at that time prevailing. I made also such other preparations as I deemed necessary to successful investigation, and provided myself with a good Hartnack microscope, divers chemicals and medicines, a clinical thermometer, &c. Among all the places and localities at which the disease, as reported, was very frequent, I selected Champaign, Champaign County, Illinois, as affording the greatest facilities for the intended investigation, or the most suitable basis for my operations, and repaired to that place on the second day of August. I found what I expected, *i. e.*, numerous cases of disease in the vicinity of Champaign and Urbana, and offers of assistance by F. W. Prentice, M. D., and M. R. C. V. S., who is lecturer on veterinary science in the Illinois Industrial University at Urbana, and of Prof. T. J. Burrill, M. A., who is professor of botany and microscopist in the same institution. Dr. Prentice had even the kindness of offering to me, for experimental purposes, the free use of his veterinary infirmary buildings. That offer, of course, was accepted. Besides that, Dr. Prentice, who is a very able and well-educated veterinary surgeon, has assisted me otherwise very essentially in my work, and took charge of my experimental animals whenever I was obliged to be absent for a short time. I am, therefore, very much indebted to him for his valuable help and kind assistance. Professor Burrill has assisted me in my microscopical examinations.

Arrived at Champaign I made my plans as to the manner in which to proceed with my investigation. Knowing that an enemy can only be conquered by being well known, I determined to ascertain first the real nature of the disease I had to deal with. That accomplished, I proposed to direct my attention exclusively to investigating and ascertaining the causes, reasoning that, if the causes are known, it cannot be very difficult to devise proper and efficient means of prevention, and, perhaps, remedies that will effect a cure. At any rate, a knowledge of the causes of a disease affords not only a sound, but in fact the only basis of successful prevention and rational treatment. This plan I have executed as far as circumstances and the time granted have permitted me to do.

In order to become thoroughly acquainted with the nature of the so-called "hog-cholera," or more appropriately "swine-plague," called also typhoid, pig-typhoid, enteric fever, pneumo-enteric fever, hog or swine disease, &c., I have made during the time from August 2 to

November 1, 54 visits to 26 different herds of diseased swine, and 53 *post-mortem* examinations, and have examined microscopically the blood, diverse other fluids, morbid products, and tissues of 42 sick or dead animals.

For the purpose of ascertaining the cause or causes of the disease, I have also made numerous experiments, a detailed account of most of which will be found in this report. After having inquired into the causes, I have made other experiments in regard to prevention and treatment.

The following may be considered as the result of my investigations:

1. DESCRIPTION OF SWINE-PLAGUE.

The disease, commonly known as "hog-cholera" to the farmers, but which may, more appropriately, be called swine-plague—a name which I shall use exclusively hereafter—is a disease *sui generis*, peculiar to swine, is neither cholera nor anthrax; it somewhat resembles the enteric fever, or dothinenteria, of man, but is not identical with the same; is communicated from one animal to another by direct and indirect infection; has usually a subacute course; is extremely fatal, especially among young animals; and exempts neither sex, age, nor breed, but seems to prefer. in its attacks, for reasons hereafter to be explained, large herds, and is always most fatal in such sties, pens, and yards in which many animals are crowded together. Some individual animals seem to have more predisposition to the disease than others. The morbid process, although in all cases essentially the same, is not restricted to a single part or organ, or to a set of organs, but can have its seat almost everywhere—in the tissue of the lungs, in the pleura and pericardium, in the heart, in the lymphatic system, in the peritoneum, in all mucous membranes, especially in those of the intestines, in the liver, in the spleen, and even in the skin. Only the pulmonal tissue and the lymphatic glands are invariably affected.

2. THE SYMPTOMS.

The symptoms, although presenting certain characteristics, observed more or less in the affected animals, vary considerably in different cases, even in one and the same herd, but still more so in different herds, and in different seasons and localities. The causes of these differences will hereafter be fully explained.

To convey a better idea of the features of swine-plague, as presented in the living animal, I shall first give an outline of all the symptoms observed in a large number of hogs and pigs, and shall append, in order to show what combinations may occur in an individual animal, a description of the symptoms presented by some of my experimental pigs.

Swine-plague announces its presence very often by a cold shivering, lasting from a few minutes to several hours, frequent sneezing, and more or less coughing. The symptoms of shivering and sneezing are generally noticed. At the beginning of the disease the temperature of the body seems to be increased. The thermometer indicated from 104° to 106° F. Still, not much reliance can be placed on the temperature, as indicated by the thermometer. In some cases it was found to be very high—in one case as high as 111° F.—and in others below normal. It was always more or less variable, and has been found decreasing at the very height of the disease. I have come to the conclusion that in diseases of swine thermometry is of a very doubtful practical value, because to ascertain the temperature of a hog, that is not extremely low or in a dying condition, by introducing a thermometer into the rectum, requires the use of force, because a hog or pig can very seldom be persuaded to submit to that operation without struggling and without being held; and struggling, according to my observation, increases the temperature of such an irritable animal immediately. The general appearance of the animal, if correctly analyzed, is of much more diagnostic and prognostic value than the differences of temperature as indicated by the thermometer. In diseases of swine the latter is, at best, a nice and interesting plaything in the hands of the inexperienced.

The first symptoms are usually followed within a short time by a partial, and afterwards by a total loss of appetite; a rough and somewhat staring appearance of the coat of hair; a drooping of the ears (characteristic); loss of vivacity; attempts to vomit (in some cases); a tendency to root in the bedding, and to lie down in a dark and quiet corner; a dull look of the eyes, which not seldom become dim and injected; swelling of the head (observed in several cases); eruptions on the ears and on other parts of the body (quite frequent); bleeding from the nose (in a few cases); swelling of the eyelids, and partial or total blindness (in five or six cases); dizziness or apparent pressure upon the brain; accelerated and frequently laborious breathing; more or less constipation, or, in some cases, diarrhea; a gaunt appearance of the flanks; a pumping motion of the same at each breath; rapid emaciation; a vitiated appetite for dung, dirt, and saline substances; increased thirst (sometimes); accumulation of mucus in the canthi of the eyes (very often at an early stage of the disease); more or less copious discharges from the nose, &c. The peculiar offensive and fetid smell of the exhalations and of the excrements may be considered as characteristic of the disease. This odor is so penetrating as to announce the presence of the disease, especially if the herd of swine is a large one, at a distance of half a mile or even farther, provided the wind is favorable. If the animals are inclined to be costive, the dung is usually grayish or brownish black, and hard; if diarrhea is present the feces are semi-fluid, and of a grayish-green color, and contain, in some cases, an admixture of blood. In a large number of cases the more tender portions of the skin on the lower surface of the body, between the hind legs, behind the ears, and even on the nose and on the neck, exhibit numerous larger or smaller red spots, or (sometimes) a uniform redness (Red Soldier of the English). Toward a fatal termination of the disease this redness changes frequently to purple. A physical exploration of the thorax reveals, if pleuritis is existing, frequently a plain rubbing sound. As the morbid process progresses the movements, of the sick animal become weaker and slower; the gait becomes staggering and undecided; the steps made are short, as if the animal was unable to advance its legs without pain; sometimes lameness, especially in a hind leg (not very often), and sometimes great weakness in the hind quarters, or partial paralysis (oftener) make their appearance. The head, if the animal is on its legs, seems to be too heavy to be carried, and is kept in a drooping position with the nose almost touching the ground; but as a general rule the diseased animals are usually found lying down in a dark and quiet corner with the nose hid in the bedding. If a fatal termination is approaching, a very fetid diarrhea (usually one or two days before death) takes the place of the previous costiveness; the voice becomes very peculiar, grows very faint and hoarse; the sick animal manifests a great indifference to its surroundings, and to what is going on; emaciation and general debility increase very fast; the skin (especially if the disease has been of long duration) becomes wrinkled, hard, dry, parchment-like, and very unclean; a cold clammy sweat breaks out (observed several times, once as early as forty-eight hours before death), and death ensues either under convulsions (comparatively rare), or gradually and without any struggle. A peculiar symptom, which, however, has been observed only once, in a litter of nine pigs, about a week old, at the beginning, or in the first stage of the disease, may here be mentioned. It consisted in a peculiar and constant twitching of all voluntary muscles. All nine pigs died, and I am sorry that I had no opportunity to make any *post mortem* examination.

In some cases numerous eruptions (ulcerous nodules) appeared on the tender skin on the lower surface of the body between the legs and behind the ears, and in a few cases whole pieces of skin (in one case as large as a man's hand) were destroyed by the morbid process, sloughed off, and left behind a raw, ulcerous surface. In another case a part of the lower lip, of the gums, and of the lower jaw-bone had undergone ulcerous destruction.

Wherever pigs or hogs had been ringed, the wounds thus made showed a great tendency to ulcerate. In several cases the morbid process had caused sufficient ulcerous destruction to form an opening directly into the nasal cavities large enough to enable the animal to breathe through, instead of through the nostrils, which had become nearly closed by swelling and by exudations and morbid products adhering to their borders.

In those few cases in which the disease has not a fatal termination .the symptoms gradually disappear, coughing becomes more frequent and easier; the discharges from the nose, for a day or two, become copious, but soon diminish, and finally cease altogether: appetite returns, and becomes normal; the offensive smell of the excrements disappears; sores or ulcers that may happen to exist show a tendency to heal; the animal becomes more lively, and gains, though slowly, in flesh and strength; but some difficulty of breathing, and a short, somewhat hoarse, hacking cough remains for a long time.

Symptoms of special cases.—Experimental pigs Nos. 5 and 6, both of the same litter, and about fifteen weeks old, were fed on the sixth day of September with the stomach, cut in pieces, the cæcum, and the spleen of experimental pig No. 2, which had died the same day.

September 7.—Pig No. 5 coughs a little, but eats well; pig No. 6 has a slight catarrh; some yellow mucus in inner canthus of one eye.

September 8.—Both pigs the same as yesterday.

September 9.—Both pigs have very good appetite.

September 10.—Both pigs seem to be as well as possible; consume all their food greedily.

September 11.—Both pigs apparently healthy: neither one shows any symptoms of disease.

September 12.—Both pigs evidently sick: they are tardy in their movements; their ears are drooping; their appetite diminished. Pig No. 5 made attempts to vomit.

September 13.—Both pigs, but especially pig No. 5, are very sick: take scarcely any food; show a tendency to hide themselves in a corner; .coat of hair looks rough and staring; flanks are thin; accumulation of mucus in the inner canthi of the eyes. No. 6 has discharges from the nose, especially from the right nostril.

September 14.—Pig No. 5, both eyes nearly closed: is weak, though not very; emaciates rapidly; appetite is poor. No. 6 has its eyes yet open; otherwise about the same as No. 5.

September 15.—Pig No. 5, eyes closed; is very loath to move, and shows plain symptoms of pneumonia. Pig No. 6, too, shows symptoms of pneumonia, but they are less pronounced; is without appetite, and just as much emaciated as No. 5. The skin of both animals is hard and dry; and their coat of hair rough and staring; their bowels are costive; but little dung is voided. Both animals betray plain indications of pain and suffering; neither one seems to be very thirsty.

September 16.—Pig No. 5 very weak, breathes one hundred times per minute; its flanks are working forcibly; slight lameness in left hind leg. Pig No. 6 is also very weak, but is yet able to run; passed a large quantity of urine of a bright yellow color. The appetite of both pigs for food is reduced to nothing, but both exhibit a vitiated appetite, and eat each other's dung, or their own, as soon almost as it drops. The skin is very hard to the touch, parchment-like, and seems to stick to the bones. In the evening pig No. 5 is extremely weak; is scarcely able to move; its breathing is difficult and distressing. No. 6 is about the same as in the morning.

September 17.—Pig No. 5 shows symptoms of dropsy in the chest, and breathes with great difficulty, about one hundred times per minute. In the evening the pumping motion of the flanks is increased, but the respiration is slower—about fifty-six breaths per minute. Pig No. 6 is a little more lively than No. 5, but also very sick, and has no appetite. Both pigs failed to void any dung from 8 o'clock a. m. to 6 o'clock p. m.

September 18.—Pig No. 5 exceedingly emaciated, some rattling noise in the respiratory passages. Pig No. 6 about the same as yesterday.

September 19.—Pig No. 5 emaciated to the utmost, but otherwise apparently not worse. Pig No. 6 shows apparent improvement; is a little livelier than before; has some appetite; consumed one ear of corn during the last twenty-four hours. In the evening pig No. 5 breathes with the greatest difficulty, one hundred and four times per minute. No. 6 unchanged.

September 20.—Pig No. 5 very sick; breathes with great difficulty. No. 6 apparently improving

September 21.—Pig No. 5 just alive. Both pigs have been lying nearly all day in one corner of their sty, their noses buried in the bedding. In the evening pig No. 5 is perspiring; sweat cold and clammy.

September 22.—Pig No. 5 breathes sixty-four times per minute, with jerking motions of the flanks, and so far has been more or less constipated, but now has diarrhea; feces grayish-green, semi-fluid, and exceedingly fetid. Pig No. 6 is less emaciated than No. 5, has no diarrhea, and eats a little. Urine of No. 5, examined under the microscope, contained innumerable bacillus-germs (micrococci of Hellier), and a few *bacilli suis.** (See drawing I, fig. 1.)

September 23.—Pig No. 5 a mere skeleton, and extremely weak; breathes only forty-eight times per minute. Pig No. 6 not quite so low; breathes only thirty-six times per minute. In afternoon pig No. 5 too weak to stand on its legs; breathes fifty-two times per minute; is sweating; the sweat cold and clammy. Seeing that the animal could not possibly live till next morning, and desiring to make the *post-mortem* examination before putrefaction should set in, I killed pig No. 5 by bleeding at 6 o'clock p. m. (As to result of *post-mortem* examination, see chapter on Morbid Changes.)

September 24.—Pig No. 6 very sick; eats scarcely anything.

___

* I have chosen the name "*bacillus suis*" because the *bacilli*, as will appear hereafter, seem to be peculiar to swine-plague, and have not been before named as far as I have been able to learn.

September 25.—Pig No. 6 shows slightly increased appetite, and fully as much, if not more, liveliness than on any day last week. It almost seems as if some real improvement is going on, notwithstanding very serious morbid changes must have taken place.

September 26.—Pig No. 6 eats some in the morning, but does not seem to care for any food at noon; appears to be a trifle bloated; droops its head, and holds its nose to the ground.

September 27.—Pig No. 6 decidedly worse; breathes seventy-two times per minute; head drooping; nose to the ground; back arched; skin very dry and hard to the touch; no appetite whatever.

September 28.—Pig No. 6, which was very low last night, has somewhat recuperated, and is moving again; consumed some water, and also a little food.

September 29.—Pig No. 6 exceedingly emaciated and very weak; breathes thirty-eight times per minute; holds its nose persistently to the ground, and has no appetite whatever.

September 30.—Found pig No. 6, at 7 o'clock a. m., lying dead in a corner of its sty. (See chapter on Morbid Changes as to result of *post mortem* examination.)

It may be well to add a brief account of the symptoms and the progress of the disease, as observed in experimental pig B, a sow pig, about fourteen weeks old, and of mixed Poland China and Berkshire stock. Pig B was put in pen No. 3, together with pig No. 6, on September 24. The same was and remained perfectly healthy until October 2, when the first symptoms of disease made their appearance. I find in my diary the following notes:

October 2.—Pig B shows symptoms of sickness: sneezes; has an eruption on both ears; is not quite as lively as it used to be; appetite is diminished: curl is out of its tail.

October 3.—Pig B has but little appetite: is decidedly sick. In afternoon shows unmistakable symptoms of sickness; ears are drooping; no appetite: great tendency to lie down in a corner; hides its nose in the bedding.

October 4.—Pig B about the same as yesterday; has eaten a little.

October 5.—Pig B hides its nose in the bedding; has no appetite whatever: emaciation has taken place. B, although a week ago a better and heavier pig than C, a full sister, and of the same litter, is now considerably lighter.

October 7.—Pig B very sick; still, seems to have a desire to eat, but takes hold of an ear of corn so feebly as to make it appear that it has not sufficient strength in the jaws to shell the corn: gave it, therefore, shelled.

October 8.—Pig B very sick; hides in its corner: ears are cold: other parts of the body warm: no appetite: great indifference to surroundings: emaciation rapid.

October 9.—Pig B about the same as yesterday.

October 10.—Pig B is getting worse: does not eat anything.

October 11.—Pig B found dead in its pen in the morning.

These three cases show that the symptoms vary in different cases, and that those which are constant can scarcely be considered as very characteristic. Still, if the various symptoms presented by an individual animal are taken as a whole, a diagnostic mistake is scarcely possible.

The diagnosis is very easy, especially if swine-plague is known to be prevailing in the neighborhood, or has already made its appearance in the herd, and if the anamnesis, and the fact that many animals are attacked at once, or within a short time and in rapid succession, are taken into consideration. As symptoms of special diagnostic value, scarcely ever absent in any case, may be mentioned the drooping of the ears and of the head; more or less coughing; the dull look of the eyes; the staring appearance of the coat of hair; the partial or total want of appetite for food; the vitiated appetite for excrements; the rapid emaciation; the great debility; the weak and undecided, frequently staggering, gait; the great indifference to surroundings; the tendency to lie down in a dark corner, and to hide the nose, or even the whole head in the bedding, and particularly the specific, offensive smell, and the peculiar color of the excrements. This symptom is always present, at least in an advanced stage of the disease, no matter whether constipation or diarrhea is existing. As other characteristic symptoms, though not present in every animal, deserve to be mentioned frequent sneezing; bleeding from the nose; swelling of the eyelids; accumulation of mucus in the inner canthi of the eyes; attempts to vomit, or real vomiting; accelerated and difficult breathing; thumping or spasmodic contraction of the abdominal muscles (flanks) at each breath, and a peculiar, faint and hoarse voice in the last stages of the disease.

3. THE PROGNOSIS AND TERMINATION.

The prognosis is decidedly unfavorable, but is the more so the younger the animals or the larger the herd. Among pigs less than three months old the mortality may be set down as from 90 to 100 per cent.; among animals from three to six or seven months old the same is from 75 to 90 per cent.; while among older animals that have been well kept and are in good condition, and naturally strong and vigorous, the mortality sometimes may not exceed 25 per cent., but may, on an average, reach 40 to 50 per cent. The prognosis is comparatively favorable only in those few cases in which the morbid process is not very violent; in which the seat of the disease is confined to the respiratory organs and to the skin; in which any thumping or pumping motion of the flanks is absent; and in which the patient is, naturally, a strong, vigorous animal, not too young and in a good condition; in which, further, where but a few, not more than two or three, animals are kept in the same pen or sty, and receive nothing but clean uncontaminated food and pure water for drinking, and in which a frequent and thorough cleaning of the sty or pen prevents any consumption of excrements.

The duration of the disease varies according to the violence and the seat of the morbid process, the age and the constitution of the patient, and the treatment and keeping in general. Where the morbid process is violent, where its principal seat is in one of the most vital organs— in the heart, for instance—where a large number of animals are kept together in one sty or pen, where sties and pens are very dirty, or where the sick animals are very young, the disease frequently becomes fatal in a day or two, and sometimes even within twenty-four hours. On the other hand, where the morbid process is not very violent or extensive, where the heart, for instance, is not seriously affected, and where the patients are naturally strong and vigorous, and well kept in every respect, it usually takes from one to three weeks to cause death. If the termination is not a fatal one, the convalescence, at any rate, requires an equal and probably a much longer time. A perfect recovery seldom occurs; in most cases some lasting disorders—morbid changes which can be repaired but slowly or not at all—remain behind, and interfere more or less with the growth and fattening of the animal.

From a pecuniary standpoint, it makes but little difference to the owner whether a pig affected with this plague recovers or dies, because those which do survive usually make very poor returns for the food consumed, unless the attack has been a very mild one.

4. MORBID CHANGES.

The morbid process, although everywhere essentially the same (see chapter on Contagion, Causes, and Nature of Morbid Process), can have its seat in many different organs or parts of the body, and produces, therefore, a great variety of morbid changes. The disease, in consequence, very often presents a somewhat different aspect in different animals. In some cases the principal seat of the morbid process is in one organ or set of organs (organs of respiration and circulation, for instance), and in others in entirely different parts (intestinal canal and organs of digestion, &c.) Death, therefore, has very often a different cause in different cases; in some cases it results from a cessation of the functions of the heart, the lungs, &c., and in others it is the consequence of an inability of entirely different organs to perform their functions, of the digestive apparatus, for instance.

But few morbid changes have ever been found entirely absent at any of the fifty-three *post-mortem* examinations made since August 2. and may, therefore, be considered as a constant occurrence. All others have been found absent a larger or smaller number of times. These constant morbid changes consist—

1. In a more or less perfect hepatization of a larger or smaller portion of the lungs, or a more or less extensive accumulation of blood, blood-serum and exudation in the pulmonal tissue. In some cases the morbid changes (hepatization) found in the lungs are so extensive as to cause the latter, if thrown into water, to sink like a rock, but in other cases the hepatization is limited to about one-sixth or one-eighth of the whole pulmonal tissue. In some cases, especially those in which the morbid changes were of a recent origin, no real hepatization, fully developed, had yet been effected; the lungs were merely gorged with exudation or blood-serum; the texture was not yet destroyed or seriously changed, but innumerable small red spots or specks, indicating incipient embolism, were plainly visible to the naked eye. (See photograph, Plate I, half-size lungs, right side of experimental pig No. VII, and photograph, Plate II, enlarged section of same lungs.) In other cases a part of the exudation had changed, organized, or become a part of the tissue, and had caused the latter to become more or less perfectly impermeable to air. In some lungs hepatization was found only in certain insulated places, while in others the hepatization extended uninterruptedly over whole portions. In all these cases in which the hepatization was very limited, it was found principally in the anterior lobes. In some animals (that is in those which had been sick for some time), old or so-called gray, more recent or brown, and very new or red hepatization were frequently found side by side, or in more or less distinctly limited patches, showing plainly that the morbid changes had not been produced at once, but at several intervals. In others, usually the upper parts of the same lungs, the exudation or blood-serum was but recently deposited, and was yet in a fluid condition. The blood-serum, examined under the microscope, invariably contained. besides blood-corpuscles, numerous *bacilli suis*, some moving and some without motion, and innumerable bacillus-germs, of which some had budded,

A Hoen & Co Lithocaustic Baltimore

Half size of right lung of experimental pig, No.VII.

some were budding, and others had conglomerated. (See drawing II, figs. 3 and 4, and drawing III, fig. 1.)

2. The lymphatic and mesenteric glands were found invariably more or less enlarged. In some cases they presented even a brownish or blackish color, and contained not only deleterious matter, but even effusions of blood in sufficient quantities to push aside the normal glandular tissue. Whether neoplastic formations (a proliferous growth of cells) had taken place I have not ascertained, but have not the least doubt that it had. Under the microscope, particles of lymph and glandular substance, taken from the interior of the lymphatic gland, presented, besides normal tissue and lymph-corpuscles, a few blood-corpuscles, some granular detritus, and innumerable *bacilli* and bacillus-germs. (See drawings III and IV, figs. 5 and 6.) As lymphatic glands always most conspicuously enlarged and morbidly changed, may be mentioned the superficial and deep inguinal and the axillary glands, the bronchial and mediastinum glands in the chest, and the mesenteric, gastric, gastroepiploic, and hepatic glands in the abdominal cavity.

3. The trachea and the bronchial tubes contained in all cases more or less of a frothy mucus—in some cases the bronchial tubes were full of it—which consisted, examined under the microscope, of broken-down epithelium-cells, and contained a large number of bacillus-germs and *bacilli*. (See drawing III, fig. 2.) The mucous membrane of the trachea and of the bronchial tubes appeared to be congested, and more or less swelled in every case.

4. The pulmonal and costal pleura, the mediastinum, and the pericardium presented almost invariably some morbid changes ; only in a few cases no visible morbid changes could be found. In some animals those membranes appeared to be smooth, but either the thoracic cavity or the pericardium, usually both, contained a smaller or larger quantity (from one ounce to one pint or more) of straw-colored serum. In a great many cases one or more, and sometimes all, of those membranes were coated to some extent with plastic exudation. In several cases a more or less firm adhesion between costal and pulmonal pleura and mediastinum, between pulmonal pleura and diaphragm, or between pulmonal pleura and pericardium, had been effected. In a few cases the whole surface of the lungs appeared more or less firmly united with the walls of the thorax. In one case the whole external surface of the heart was firmly, and in another one partially, coalesced with the inner surface of the pericardium. The pig (a fine animal about four months old), in which the pericardium adhered with its whole interior surface firmly and inseparably to the external surface of the heart, had severe convulsions during life. It was killed in my presence by a professional butcher, who stuck it in the usual way and severed the trunk of the carotides ; only a few drops of blood issued, but the pig died immediately. The other morbid changes consisted in hepatization in the lungs, enlargement of the lymphatic glands, and the presence of large and numerous morbid growths in the cæcum and colon.

5. In nearly every animal the heart itself has been found more or less affected in one way or another. In some animals it was flabby and dilated, but in most cases it was more or less congested. The capillary vessels, especially of the auricles, were, in a large number of cases, gorged with blood to such an extent as to give them a brownish-black appearance, almost similar to gangrene. On closer inspection, however, it could be seen very plainly that the brownish-black color was caused exclusively by an accumulation of blood in the capillary vessels.

6. In forty-eight cases out of fifty-three, characteristic morbid changes

have been found in the cæcum and colon. The same consist in peculiar
morbid growths or ulcerous tumors on the mucous membrane of those
intestines. They are of various sizes, nearly round or (sometimes) irregular in shape, more or less elevated above the surface of the mucous
membrane, and frequently, especially the older and larger ones, dark-
pigmented on their surface. Their size varies from that of a pin's head
(incipient tumors or nodules) to that of a quarter of a dollar. The
smaller ones are usually of an ocher color, and but slightly projecting
above the surface of the mucous membrane (see photograph, Plate III),
but the larger ones are of a grayish-black-brown (see photograph, Plate
IV.) or blackish color; project considerably above the surface of the
membrane, in some cases fully half an inch; have usually a slight concavity in the center, and frequently a plain neck or thick pedicle. (See
photographs, Plates V, VI, and VII.) Under the microscope these morbid growths or excrescences appear to be composed, on their surface,
of a granular detritus and morbid epithelium cells, and contain innumerable *bacilli suis*, some of which have a very rapid motion. (See
drawing V, fig. 1.) The stroma of these morbid growths consists mainly
of a dense connective tissue. In some cases these morbid growths, especially the smaller ones, or those of a recent origin (see photograph,
Plate III), are situated merely on the surface of the mucous membrane,
and are easily scraped off with the back of the scalpel. Thus removed
they leave behind an uneven, excoriated surface, not dissimilar to granulation. The older and larger tumors, however, extend deeper into the
membranes of the intestine; they usually penetrate the mucous membrane, and extend into the muscular coat, and even penetrate the latter,
and extend into the external or serous membrane. In some cases all
three membranes of the cæcum or colon have been found degenerated
and destroyed beneath such a morbid growth, so as to show perforation
on the removal of the latter. The immediate surrounding of such a
deep-seated degeneration presented some, but not very much, inflammation. These morbid growths, usually, were found most developed near
the ileo-cæcal valve in the cæcum, but also in larger or smaller numbers,
and of various sizes, large and small, in all parts of the cæcum and
colon.

7. The same, or very similar morbid growths, occurred also, though
not so often, in other intestines. In one case (experimental pig No.
VII) a diffuse, decaying morbid growth coated the whole interior surface of the jejunum for a length of several feet. Examined under the
microscope it was found to consist of broken-down epithelium cells and
a granular detritus, and contained numerous *bacilli* and bacillus-germs.
(See drawing VI. fig. 1.)

In another case one ulcerous tumor was found on the mucous membrane of the gall-bladder. In three cases the same, or at least very
similar morbid changes, presented themselves on the mucous membrane
of the stomach. (See photograph, Plate VIII.) In a few cases some
ulcerous tumors were found in the duodenum, and in one case even in
the right horn of the uterus. In a few cases similar morbid changes—
small, knotty, tubercle-like, yellowish, or ocher-colored excrescences of
the size of a small pea—were found on the surface of the spleen. In
one case similar small excrescences were also found on the external surface of the vena cava posterior. In two cases the liver was found to be
degenerated by an hypertrophic condition of the connective tissue, a
morbid change which may or may not constitute a product of the morbid process of swine-plague.

8. Morbid changes in the serous membranes of the abdominal cavity.

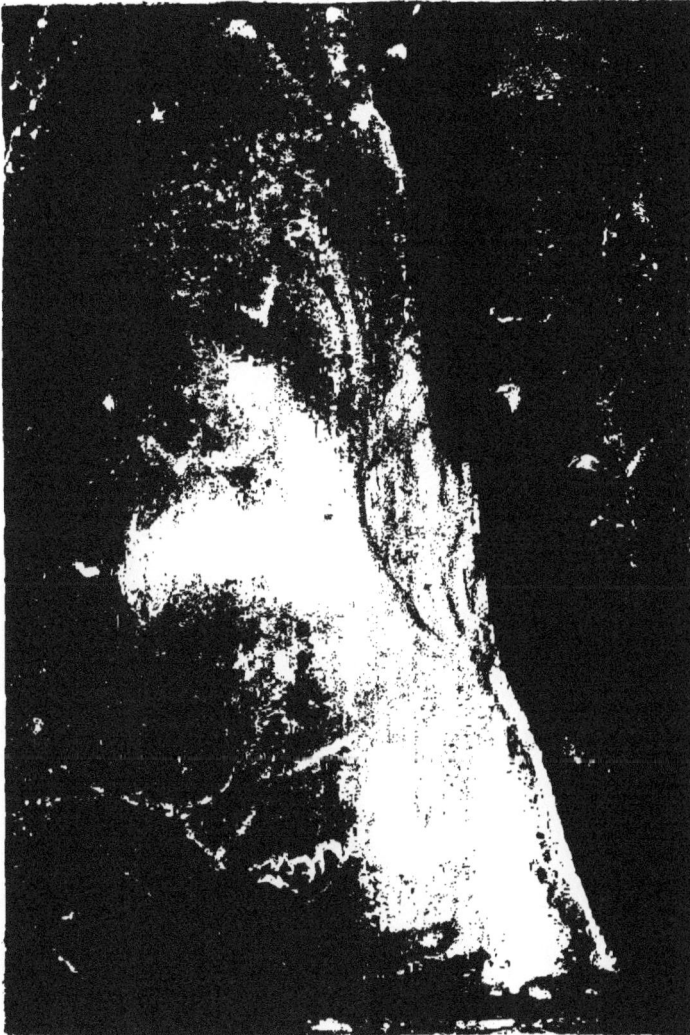

A. Hoen & Co. lithocaustic Baltimore

Enlarged section of right lung of experimental pig No. VII

In some cases the peritoneum and the serous membranes of the intestines appeared to be perfectly smooth, but a larger or smaller quantity of straw-colored serum, from two ounces to one quart or more, was found in the abdominal cavity. In others, adhesions between the intestines and the peritoneum, between the intestines themselves, or with other organs, had been effected. More or less coalescence between cæcum and colon, between cæcum and ilium, or between the convolutes of the colon, sometimes not separable except by means of the knife, presented itself in almost every case, and in which the ulcerous tumors or morbid growths in the cæcum and colon were extensive, large, and sufficiently deep-seated to affect the serous membrane.

9. The contents of the gall-bladder in a large number of cases were found to consist of a semi-solid, granular, and dirty brownish-colored substance. In most of those cases, however, the ductus choledochus appeared to be thickened, and its membranes swelled; and so it may be that the semi-solid condition of the bile was due to some extent to the partially or totally obstructed passage.

10. In one case a morbid enlargement or hypertrophy of the pancreas presented itself, and slight changes (congestion) were found in a few cases in the kidneys.

11. Morbid changes, similar in every respect to those occurring on the mucous membrane of the cæcum and colon, presented themselves in two cases on the conjunctiva, or mucous membrane of the eye. But as the conjunctiva is exposed more or less to the influence of the atmosphere, the morbid growth was not projecting in the same way as in the cæcum and colon; over the surface of the membrane the decay was more complete, and, perhaps, more rapid, so that instead of an excessive growth loss of tissue could be noticed. In both cases the eyes themselves appeared congested, and the animals seemed to be perfectly blind.

12. In one case the gums of the lower jaw presented similar changes, but in these, too, considerable loss of tissue had taken place. The morbid process extended into the lower jaw-bone, and enough of it had been decayed and destroyed to expose the roots of the incisors, and to cause some of them to drop out.

13. Morbid changes, ulceration, and decay have been observed twice in one of the spermatic chords of pigs which had been castrated a short time before the disease was contracted. In both pigs an abscess was found in the scrotum, being the only instances in which real matter or pus was observed.

14. In nearly all those hogs and pigs which had been ringed to prevent them from rooting, the parts thus wounded presented more or less decay, in about a dozen cases to such an extent as to cause a formation of large holes directly from the superior surface of the nose into the nasal cavities. These holes presented very ragged or corroded borders, coated with a dirty-yellowish detritus, and were, in several instances, sufficiently large to enable the animals to breathe through instead of through the nostrils.

15. Morbid changes in the skin, but of a different character, were found to be of frequent occurrence. In three or four cases numerous small morbid growths (eruptions) extending but slightly into the cutis, but causing a complete degeneration of the epidermis, and leaving behind, if removed, an uneven, raw, or excoriated surface, in appearance not unlike granulation, were found on the comparatively fine skin on the lower surface of the body, between the legs and behind the ears. In two other cases whole pieces of degenerated and decayed skin had sloughed off and fallen out. The corroded borders and the bottom of

the ulcers thus produced were coated with a dirty-yellowish looking granular detritus.

In a great many cases, that is, in nearly half of the whole number examined, red or purple spots and patches, and even continuous or confluent redness, of a purple hue, presented themselves in the skin on the lower surface of the body, between the legs, behind the ears, &c. At the autopsy the skin and the subcutaneous tissue appeared to be congested, the capillary vessels were gorged with blood, and more or less exudation and small extravasations of blood were found to have taken place. In one case a large piece of skin on the lower surface of the body was mortified.

16. In two cases quite extensive extravasations of blood presented themselves in the mucous membrane of the stomach and intestines.

17. The blood presented some quantitative and qualitative changes in every case. Its quantity appeared to be diminished in every animal, in some cases to such an extent that not more than, say, four or five ounces could have been collected if the animal had been killed by bleeding. Still, the actual want of blood was never as great as it appeared to be, because a considerable quantity was locked up in the tissues, especially in the lungs, and had become stagnant in the capillary vessels. The blood was dark-colored in all cases in which death had been caused by extensive morbid changes in the lungs, or in which, on account of those changes, respiration had been very imperfect; but it presented a normal color, and was perhaps a little lighter colored and thinner or more watery than in a healthy hog, in all cases in which death had been caused by other morbid changes, or in which the affection of the lungs was comparatively unimportant. It invariably coagulated as soon as it became exposed to the influence of the atmosphere, to a loose and spongy clot, containing a considerable quantity of serum. Hence, it must be supposed that it was rich in fibrinogen, but probably poor in fibrin, a condition due, unquestionably, to the fact that during the disease the process of waste had been largely in excess of that of repair.

Under the microscope the blood-corpuscles of fresh blood appeared sometimes nearly all normal or round, and sometimes more or less angular and star-shaped, but after a while they all became more or less angular and of an irregular shape, and showed more or less tendency to congregate in rows and clusters. The fresh blood contained numerous bacillus-germs, many of them simple, small, round bodies, some in process of budding, others budded or double, and still others congregated into, apparently, viscous clusters. (See drawing II, fig. 1; drawing IV, fig. 4; drawing VII, figs. 1 and 4; drawings VIII, IX, and X, fig. 1.) In a few cases fully developed *bacilli suis* were found in the fresh blood, but they were, comparatively, few in number. In blood which had been kept twenty-four hours or longer in well-closed vials, *bacilli* were always more numerous, and sometimes were found in large numbers. As soon, however, as putrefaction or decomposition had set in, the *bacilli* disappeared. White blood-corpuscles, a few in number, were found only in three or four cases.

18. A microscopic examination of the blood-serum or exudations, deposited in the pulmonal tissue, invariably revealed, besides some angular red blood-corpuscles, an immense number of *bacilli suis*, and of bacillus-germs in all stages of development, single, budding, budded, or double, and congregated into clusters. (See drawing III, fig. 1, and drawing II, figs. 3 and 4.)

That every one of these morbid changes does not occur in one and the

same animal, and that sometimes some and sometimes others are more developed and constitute the immediate cause of death, has already been indicated, and does not need any further explanation. To convey, however, a clearer idea of the morbid features and changes presented after death, I will copy from my notes the result of the *post-mortem* examinations of a few of my experimental pigs. Of pigs Nos. 5 and 6 the symptoms, observed during life, have already been noted.

Post-mortem examination of pig No. 5.—On opening the chest, the ribs, usually tough in a young animal, broke very easily, and seemed to be deficient in organic substances. No serum in the chest; pulmonal pleura rough, partially coated with plastic exudation; lower half of both lobes of lungs hepatized; no serum in the pericardium, but apex of heart firmly coalesced with the inner surface of the pericardium; thick, white, and frothy mucus, but no *strongili paradoxi* in trachea and bronchial tubes. Cæcum and colon firmly agglutinated to each other with their external surfaces; adhesion separable only by means of the knife. Numerous large and small ulcerous tumors or morbid growths in both cæcum and colon. (See photograph, Plate V, which shows the cæcum, and Plate VI, which shows the colon, natural size of pig No. 5.) Lymphatic and mesenteric glands enlarged. Ulcerous decay in mucous membrane of the stomach. (See photograph, Plate VIII, which presents the interior surface of the stomach of pig No. 5, natural size.) Besides those essential changes mentioned, one large nematoid was found in the ductus choledochus, extending from the duodenum through the choledochus and the gall-bladder into an hepatic duct. Another worm of the same kind was found in the cæcum.

Autopsy of pig No. 6.—An abscess in right side of the scrotum, about seven-eighths of an inch in diameter, and connected with ulceration in right spermatic chord. Inguinal and axillary lymphatic glands considerably enlarged. One-fourth of right and one-fifth of left lobe of lungs hepatized; the rest gorged with blood-serum or exudation. Cæcum and colon agglutinated to each other; cæcum also adhering to peritoneum. Mesenteric glands very much enlarged; right spermatic chord ulcerated. (Pig had been castrated a few weeks before it contracted the disease.) Extensive morbid growth, in process of decay, in cæcum, and also a large number in colon. Some exudation on lower surface of spleen. Ulcerous decay in mucous membrane of anterior portion of stomach, and wine-colored infiltration and extravasations of blood in mucous membrane of pyloric portion of same intestine.

Autopsy of pig B.—Some redness between hind legs and on lower surface of the body; greenish mucus oozing from the nose; axillary and inguinal glands very much enlarged; ribs deficient in organic substances, at any rate very brittle; both lungs spotted all over, indicating plainly capillary embolism in early stage of development; hepatization limited, just commencing; lymphatic glands in chest very much enlarged; the heart, but especially the auricles, very much congested; auricles almost black; small quantity of straw-colored serum (not exceeding two ounces) in thoracic cavity, and still less in pericardium. In the abdominal cavity mucous membrane of anterior part of stomach wine-colored; some diffuse morbid growth, in process of decay, in posterior (pyloric) portion of same membrane. No food whatever in stomach and intestines; bile thickened, semi-solid; no ulceration nor any morbid growth whatever in cæcum, colon, or any other intestine.

Results of post-mortem examination of experimental pig No. VI.—Decaying blotches or nodules of the size of a five-cent piece and smaller on skin of lower surface of body and between the legs; right spermatic chord ulcerated, and an abscess the size of a hen's egg in right side of scrotum. Internally all lymphatic and mesenteric glands enlarged; anterior portion of both lungs everywhere, with their whole external surface, and posterior portion at some places adhering (coalesced) to the costal pleura; numerous smaller and larger embolic tubercles, presenting the appearance of incipient abscesses, in anterior portion of both lobes of the lungs, but more numerous and more developed in right lobe than in the left; remainder—posterior parts of both lobes—gorged with exudation; small quantity of straw-colored serum in the chest and in the pericardium. In abdominal cavity, liver rather hard (sclerotic), its connective tissue apparently hypertrophied. One small tape-worm, not over one and a half inches long, in jejunum, and numerous small, incipient morbid growths or ocher-colored decaying nodules in cæcum. (See photograph, Plate III.) No other morbid changes.

Besides these numerous morbid changes, which must be looked upon as products of the morbid process of swine-plague, some species of entozoa, a few of which have already been mentioned, have occasionally been met with; but as their presence is merely accidental, that is, has nothing whatever to do with the disease in question, a brief mention of this occurrence will be sufficient. *Strongilus paradoxus* has been found in small numbers in the bronchial tubes of a few pigs in one herd only— Mr. Bassett's. *Trichocephalus crenatus* has been found in small numbers in the blind end of the cæcum of four animals, belonging to two different herds. A small tape-worm was once found in the jejunum, as has been stated, and a few other entozoa (nematoids) were found in four or five instances in the choledochus, gall-bladder, and hepatic ducts (in one case as many as twelve worms), and twice in other intestines.

What I have so far related was comparatively easily ascertained. Numerous examinations of diseased animals, frequent visits to affected herds, and fifty-three *post-mortem* examinations revealed the facts, and all that was necessary was to observe and take notes. But the principal object of the investigation was to devise means to prevent the immense losses caused every year by that most fatal disease, swine-plague. (I have adopted that name, because the disease, if anything, is a real plague; and the name is sufficiently comprehensive to cover the whole morbid process, and so simple that I have no doubt it will soon supercede, even among farmers, that very improper name of hog cholera.)

To devise such means, a more reliable basis than a mere knowledge of the various features of the disease had to be gained. The real nature of the morbid process, and the true cause or causes, had to be ascertained. Above all, it had to be decided as to whether swine-plague is a contagious disease or not; and if contagious, the means by which the contagion is conveyed from one place and from one animal to another; the manner in which it enters the animal organism, and, if possible, the nature of the same. This could not be done by simply visiting diseased herds and examining sick and dead animals; it was necessary to make experiments and to observe and to record the results. This I have done, and before I proceed any further it may be best to give, first, a condensed account of the experiments which I have made for the purpose of settling those points, so as to give others an opportunity to form an opinion as to the correctness of the conclusions I have arrived at. I will mention again, that in making those experiments, in noting the results, and in making the necessary and very numerous microscopical examinations, I have been ably assisted by my friends, Dr. F. W. Prentice and Prof. T. J. Burrill, of the Illinois Industrial University. I commenced those experiments after I had gained con-

siderable information as to the various features of the disease during life and after death, and as to the conditions and surroundings under which the same makes its appearance. The first series of experiments has been made for the purpose of settling the question as to the contagiousness or non-contagiousness of Swine-Plague. This was the more necessary from the fact that those who had suffered severe losses were decidedly divided on that question.

FIRST SERIES OF EXPERIMENTS.

After encountering considerable difficulty in finding indubitably healthy pigs, belonging to a perfectly healthy herd, which had never been in contact with diseased animals, I succeeded finally, on the 20th of August, in buying of Mr. Harris, south of Champaign, three Berkshire sow pigs about three and a half months old, perfectly healthy, and without any lesions whatever. I designated them as pigs Nos. 1, 2, and 3. Dr. Prentice, at the same time, had the kindness of placing at my disposition two box-stalls in his veterinary hospital, a new building which had never been entered by any hog or pig. About one hundred and fifty yards east of the veterinary hospital building, on a piece of ground never trodden by hogs, as far as known, I built of new lumber a pen eight feet square. This pen I designated pen No. 1, and the box-stalls, which are twelve feet square, as pens Nos. 2 and 3 respectively. Pig No. 1 was put in pen No. 1, and pigs Nos. 2 and 3 together in pen No. 2.

It may be well to state here that pen No. 1 having no floor, but resting on the ground, was moved to another place (each time its own width) every other day, usually at noon, in order to preserve cleanliness, and pens Nos. 2 and 3 were cleaned and swept once a day, except where stated otherwise in the following pages. The food of all experimental pigs was the same, and consisted of corn in the ear, and occasionally a little green clover and purslane at noon or in the evening. The water for drinking was drawn three times a day from a well.

1. *Account of pig No. 1.*—On August 21 I procured from Mr. Bassett, four miles north of Champaign, a diseased Chester white pig, four months old (pig No. 4), which I put with pig No. 1 in pen No. 1. This diseased pig which arrived at 10.30 o'clock, a. m., exhibited plain and unmistakable symptoms of swine-plague; its temperature was $106\frac{1}{2}°$ F., and its skin, on lower surface of the body, between the legs, &c., was considerably reddened. The temperature of pig No. 1, which objected to being examined and struggled hard, was $104\frac{1}{2}°$ F.

August 22.—Pig No. 1 all right; has vigorous appetite. Pig No. 4 at 8 o'clock a. m. very sick: has a peculiar, short, abrupt cough; at 1 o'clock, p. m., dead.

Post-mortem examination.—Capillary redness in the skin on lower surface of body and between the legs; considerable enlargement of lymphatic glands; more than two-thirds of the lungs hepatized and gorged with blood-serum; some straw-colored serum in thoracic cavity and pericardium; and morbid growths in process of decay (ulcerous tumors), in cæcum and colon.

Received at 1 o'clock, p. m., three more pigs, each about three months old (cross of Berkshire and scrub), of Mr. Schumacher, a butcher in Champaign, who had bought the same of a farmer ten miles southeast of Champaign. I designated the same as pigs Nos. 5, 6, and 7. Pigs Nos. 5 and 6 appeared to be perfectly healthy, and were put together in pen No. 3. Pig No. 7 was apparently indisposed: it had been transported ten miles, crowded together with twenty others, most of them larger and older, and exposed for several hours to the burning rays of the sun, in an open farm-wagon on a very hot day. It was panting for breath, and showed symptoms of congestion of the lungs. It was put in pen No. 1 with pig No. 1, before dead pig No. 4 had been removed.

August 23.—Pig No. 1 perfectly healthly. Pig No. 7 very sick; breathes ninety-two times per minute; shows plain symptoms of pleurites; has no appetite, but is attentive and moves quickly when disturbed. It died at 8 o'clock p. m. *Post-mortem* examination revealed pleurites and pericarditis; the whole surface of the lungs was loosely agglutinated to costal pleura, and the substance of the same was gorged with exudation. *No other morbid changes whatever.* Whether this was a case of swine-plague or not, I leave to my readers to decide for themselves. I am decidedly of the opinion it was not, because none of the other twenty pigs, except Nos. 5 and 6 (see account of them) have, up to date, contracted the disease, as I have learned from a reliable source. It is true two other pigs of the same lot showed some indisposition on the 24th, 25th, and 26th days of August, but were all right again the next day, and are healthy yet.

August 24.—Pig No. 1 perfectly healthy; vigorous appetite.

August 25.—No change.

August 26.—No change.

August 27.—No change.

August 28.—Weather very hot and sultry; in afternoon severe thunder-storm and rain, which effected a sudden cooling of the atmosphere. Pig No. 1 in perfect health.

August 29.—Pig No. 1 coughed once; being exposed in an open pen to the changes of weather and temperature, it has possibly taken cold.

August 30.—Pig No. 1 perfectly healthy; is very lively, and has vigorous appetite.

August 31.—The same.

September 1.—The same.

September 2.—The same.

September 3.—The same.

September 4.—The same. At 6.30 o'clock, p. m., diseased experimental pig No. 2 (see account of the same further down) was put in pen No. 1 with pig No. 1.

September 5.—Pig No. 1 perfectly healthy. Pig No. 2 eats nothing; shows plain symptoms of pneumonia.

September 6.—Pig No. 1 perfectly healthy. Pig No. 2 died at 6 o'clock, p. m. (For *post-mortem* examination, which was made immediately, see account of pig No. 2.)

September 7.—Pig No. 1 perfectly healthy, and has remained so up to date. Has always first-rate appetite, has never refused a meal, and is to-day a strong, vigorous, and thriving pig. (Made use of the same for another experiment on *November* 13.)

2. *Account of pigs Nos. 2 and 3.*—*August* 21.—Both pigs are perfectly healthy; have good appetite, and are active and lively.

August 22.—Both pigs perfectly healthy. Inoculated both in right ear at 1.30 o'clock. p. m., with blood-serum from the lungs of pig No. 4, which had died at 1 o'clock, p. m. The operation was performed by means of a small inoculation-needle, made for the purpose of inoculating sheep with the virus of sheep pox. Each pig received two slight punctures on the external surface of the ear; the serum inoculated was less than one-fourth of a drop per animal. The blood-serum used was of a faint reddish color, and almost limpid. Examined under the

Plate V.

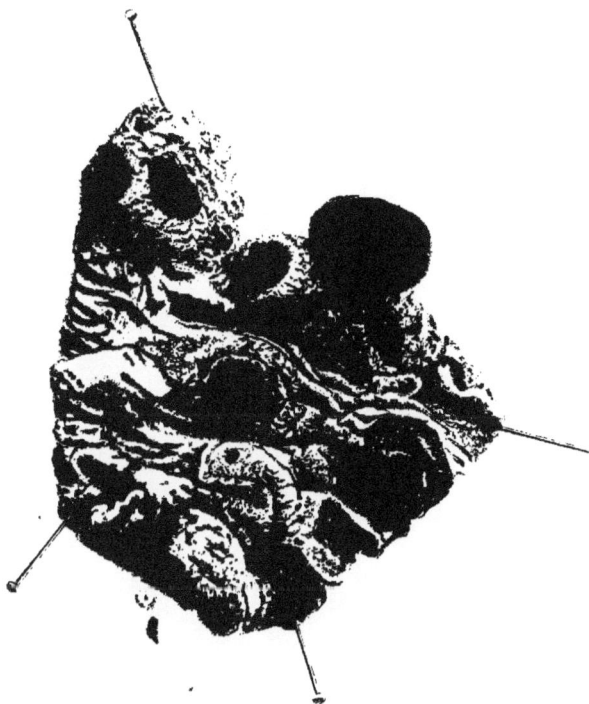

A. Hoen & Co. Lithocaustic Baltimore

Ulcerous tumors on mucous membrane of intestines, showing concavity in center.

microscope it contained a few red blood-corpuscles, numerous bacillus-germs, and some developed *bacilli suis.*

August 23.—Pigs Nos. 2 and 3 perfectly healthy. No visible reaction.

August 24.—Both pigs perfectly healthy; have very good appetite.

August 25.—No change.

August 26.—No change.

August 27.—Pig No. 2 appears to be slightly indisposed. Pig No. 3 apparently healthy.

August 28.—Both seem to be healthy; eat well.

August 29.—Pig No. 2 not quite as lively as a healthy pig; does not seem to have very good appetite. Pig No. 3 shows no symptoms of disease. Temperature of pig No. 2, 105.4° F., and of No. 3, 104½° F. Both pigs struggled very much while being examined.

August 30.—Pig No. 2 not very lively, and shows a tendency to lie down; does not eat as well as formerly; temperature, 104½° F. At feeding time in the evening it did not arise, nor did it seem to care for its food. Pig No. 3 apparently all right.

August 31.—Pig No. 2 shows plain symptoms of sickness; arches its back, and moves with short undecided steps. Pig No. 3 appears to be less lively.

September 1.—Both pigs, Nos. 2 and 3, show plain symptoms of swine-plague.

September 2.—Pig No. 3 seems to be worse than pig No. 2. In afternoon the eyes of pig No. 3 appeared congested, and the conjunctiva infiltrated with blood. Appetite of both animals rather poor. Both are thirsty.

September 3.—Pigs Nos. 2 and 3 do not eat anything; are evidently very sick; show great indifference to surroundings, and do not like to come out of their corner. Both are very weak, and look as if they suffer from pressure upon the brain.

September 4.—Pigs Nos. 2 and 3 have not touched any food; they huddle together in their corner, lie down, and will not get up unless compelled to do so. Both show increasing muscular weakness and emaciation. At 6.30 o'clock, p. m., pig No. 2 was removed to pen No. 1. (See account of pig No. 1.)

September 5.—Pig No. 2 (now in pen No. 1) eats nothing; has plain symptoms of pneumonia. Pig No. 3 (in pen No. 2) is getting very weak; at 7 o'clock, p. m., is lying flat, and in a dying condition.

September 6.—Pig No. 2 (in pen No. 1) very sick. Pig No. 3 (in pen No. 2) dead in the morning, with well-marked *rigor mortis.*

Post-mortem examination.—Skin normal; lymphatic glands enlarged; left lobe of lungs partially hepatized; right lobe the same, but hepatization more extensive; no serum in thoracic cavity; about two drachms in pericardium; heart normal; spleen enlarged; partially coalesced with peritoneum of abdominal wall, which shows traces of inflammation; some small ulcerous tumors on surface of spleen, and adhesion between the latter and the colon; mesenteric glands considerably enlarged; morbid growths or ulcerous tumors, and a few worms (*trichocephalus crenatus*), the latter partially embedded in the smaller cæcal mucous membrane in cæcum; blood extravasations, and capillary congestion in mucous membrane of cæcum, colon, ilium, and stomach; liver somewhat enlarged; kidneys normal. The blood, examined under the microscope, contained, besides red blood-corpuscles with ragged, irregular or star-shaped outlines, a few white blood-corpuscles (from one to five in the field), numerous bacillus-germs in various stages of development, and a few developed *bacilli suis.*

Pig No. 2 died at 6 o'clock. p. m.   (See account of pig No. 1.)

Post-mortem examination.—Skin normal; lungs partially hepatized; hepatization most marked in anterior lobes; small quantity of serum in pericardium; liver enlarged; one nematoid in choledochus; abdominal cavity free from serum: ecchymosis on the external surface of colon and cæcum; capillary hyperæmia and swelling in cæcal mucous membrane; several small ulcerous tumors in cæcum, especially near the ilio-cæcal valve; swelling, capillary congestion, and extravasations of blood in mucous membrane of colon and ilium; kidneys normal; bladder empty; mucous membrane of stomach similar in appearance to that of cæcum, colon, and ilium.

Account of pigs Nos. 5 and 6.—Pigs Nos. 5 and 6, which arrived, as has been stated before, August 22. at 1 o'clock, p. m., were put in pen No. 3, and at 1.30 o'clock, p. m., the colon, the heart, and a piece of the diseased lungs of pig No. 4 were given to them. They, however, touched neither colon, heart, nor piece of lung.

August 23.—Both pigs, Nos. 5 and 6, in good health, and eat their food greedily, but have not touched the colon, heart, and piece of lung. The colon, having become very putrid, had to be removed; heart and piece of lung were thrown into the feed-trough.

August 24.—Both pigs healthy. Heart and piece of lung have disappeared, but whether they have been consumed by the pigs or by rats I am not able to decide.

August 25.—Both pigs healthy: have good appetite, and eat greedily.

August 26.—The same.

August 27.—The same.

August 28.—The same. August 28th was a very hot day, but a severe thunder-storm in the afternoon effected a sudden cooling of the atmosphere.

August 29.—Both pigs, Nos. 5 and 6, seem to have a slight catarrh, probably in consequence of the sudden reduction of temperature and change of weather. Both cough some.

August 30.—Both pigs, to all appearances, all right, except that occasionally a slight cough can be heard. Both have first-rate appetites.

August 31.—Both pigs apparently in perfect health; appetite good.

September 1.—Both pigs all right.

September 2.—The same.

September 3.—The same. Pig No. 5 coughed once or twice, but has excellent appetite.

September 4.—Pig No. 5 coughs again a few times, but is lively, and has very good appetite. No. 6 is all right in every respect.

September 5.—Both pigs all right.

September 6.—Both pigs have good appetite, are very lively, and seem to enjoy good health. At 10.30 o'clock, a. m., the entire stomach, cut up into five pieces, the cæcum, and the spleen of pig No. 3 were given to them, and consumed immediately in the presence of Dr. Prentice.

September 7.—Both pigs, Nos. 5 and 6, have very good appetite. No. 5 has a slight cough, and a slight accumulation of mucus in the inner canthi of the eyes.   (For further particulars see the accounts given of pigs Nos. 5 and 6 in the chapter on Symptoms and Morbid Changes.)

Having thus ascertained by experiments, just related, that swine-plague is infectious, and can be communicated by inoculation, and also through the digestive canal by a consumption of morbid tissues, I considered it to be of great importance to ascertain, if possible, the nature of the infectious principle; that is, to decide by experiments whether it consists in something corporeal, endowed with life and power of propa-

SWINE FEVER.

Plate VI.

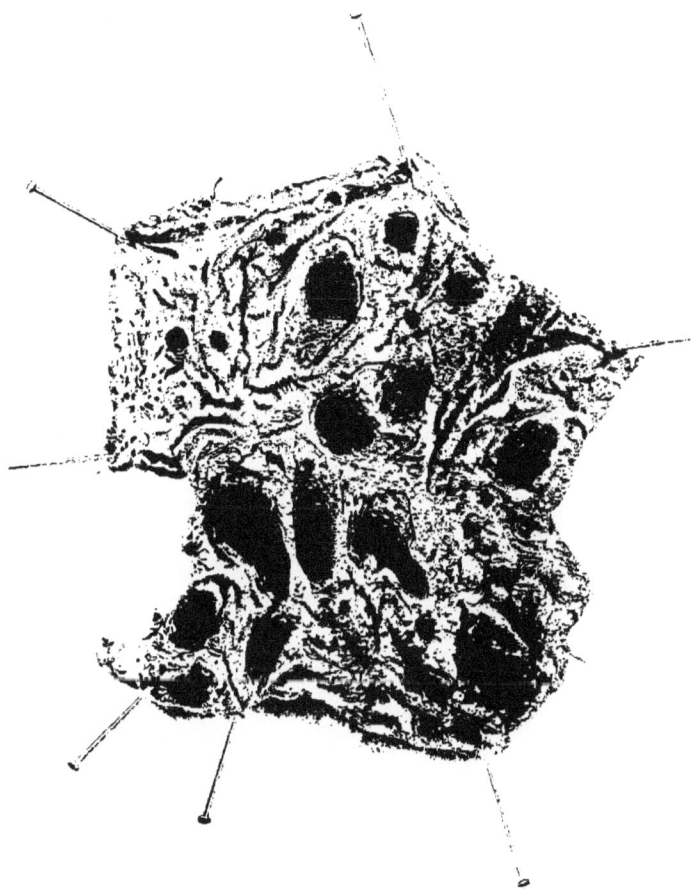

A.Hoen & Co.Lithocaustic Baltimore

Ulcerous tumors on mucous membrane of intestines, showing different view.

gation, or in some invisible chemical agency or mysterious fluid per-
meating, as has been supposed, the whole animal organism, and contained in, or clinging to, all those substances which possess infectious
properties, or constitute the bearers or vehicles of the contagion. As
all microscopical examinations of the blood, morbid tissues, and morbid
products of forty-two animals, which had been affected with swine-
plague and had died of that disease or been killed by bleeding, and
repeated microscopical examinations of the excretions (urine and excrements) of diseased animals, have revealed in every case the presence of
numerous bacillus-germs (micrococci of Hallier) and developed *bacilli
suis*, I deemed it necessary to ascertain first, if possible, the relation
which these extremely small microscopic bodies may have to the morbid process and to the infectious principle. For that purpose I commenced another series of experiments, and bought again, on September
24th, three very nice, perfectly healthy pigs, each a little over three
months old, of Mr. Burton, residing four miles southeast of Champaign.
I designated one of them, a nearly full-bred Berkshire barrow, as pig A;
another one, a Poland-China sow, as pig B; and the third one, also a
Poland-China sow, as pig C.

Account of pigs A, B, and C.—The same arrived at 10 o'clock, a. m.
Pig A was put in pen No. 1 with pig No. 1; pig B in pen No. 3 with pig
No. 6; and pig C by itself in the thoroughly cleaned and disinfected pen
No. 2, formerly occupied by pigs Nos. 2 and 3. Pen No. 2 had been
clean and empty since September 6th, and was again disinfected with
carbolic acid before pig C was put in.

September 25.—All three pigs, A, B, and C. perfectly healthy.

September 26.—All three pigs perfectly healthy; have good appetite.

September 27.—The same; inoculated pig C with cultivated *bacilli*
and bacillus-germs. On September 23d, Professor Burrill charged two
drachms of fresh cow-milk with a mere speck, smaller than a pin's head,
of a decaying morbid growth, or ulcerous tumor of the cæcum of pig
No. 5, and kept the vial well closed, at a temperature of 92° F. On
the evening of September 26th the milk was examined under the microscope, and was found to contain numerous *bacilli suis* and bacillus-
germs (see drawing III, figs. 3 and 4), the same as found in the blood-
serum, or exudation of diseased lungs, and in the decaying substance
of the intestinal morbid growths. The inoculation with this milk was
executed in the same way as the inoculations of pigs Nos. 2 and 3; two
punctures were made on the external surface of the left ear.

September 28.—All three pigs perfectly healthy. The inoculation-
punctures on the ear of C slightly swelled.

September 29.—Pigs A, B, and C, all right.

September 30.—All three pigs perfectly healthy; no symptoms of
disease.

October 1.—The same.

October 2.—Pig A perfectly healthy : pig B shows symptoms of sickness, sneezes, has eruption on the ears, diminished appetite, and is not
as lively as formerly. As a full account of pig B has already been given
in the chapter on symptoms and morbid changes, it will not be necessary
to repeat what has been said there, and pig B may be dropped. Pig C
apparently all right in the morning. At noon, pig C, too, commences to
sneeze; sneezes a good deal, and shivers like a man suffering from ague,
but has good appetite.

October 3.—Pig A perfectly healthy. Pig C shows slightly diminished
appetite and other plain symptoms of indisposition; is less lively, and
has a tendency to lie down; the sneezing continues.

October 4.—Pig A in first-rate health. Pig C a little more lively; has fair appetite, but is not as greedy as formerly.

October 5.—Pig A in fine condition, and all right in every respect. Pig C shivers, and sneezes again a good deal, but does not show any other perceptible symptoms of disease, except some eruptions behind the ears, and on the external surface of the same.

October 6.—Pig A all right in every respect. Pig C about the same as preceding day.

October 7.—Pig A perfectly healthy. Pig C has good appetite, and with the exception of its coat of hair being a little rougher than usual, does not show any plain symptoms of disease.

Made two *post-mortem* examinations of pigs which had died of swine-plague at Mr. Hossack's place, five miles southwest of Champaign. In the evening I examined microscopically the blood-serum or exudations of the diseased lungs of one of Mr. Hossack's pigs, and found normal red blood-corpuscles, numerous bacillus-germs in all stages of development—single, budding, budded, or double, and aggregated into clusters—and some developed *bacilli suis.*

October 8.—Pig A all right. Pig C shivering again. In the forenoon I filtered some of the blood-serum of the diseased lungs of Mr. Hossack's pig through eight filters—the very finest used in chemical laboratory of the I. I. University—for the purpose of freeing the serum from the *bacilli* and bacillus-germs; but notwithstanding that I have taken all possible precautions, the filtrate, which was almost limpid, still contained, as examined under the microscope, a great many bacillus-germs. I preserved it in a vial with a tight-fitting ground-glass stop.

October 9.—Pig A healthy. Pig C has fair appetite, but is not greedy. I filtrated the filtrate once more through two filters, and obtained a limpid fluid, which, however, at a microscopic examination, was found to still contain some bacillus-germs. Preserved the filtrate again in a clean vial, with a perfectly-fitting ground-glass stop.

October 10.—Pig A healthy. Pig C eats its food, but is rather slow at it.

October 11.—Pig A healthy. Pig C about the same as on preceding day.

October 12.—Pig A healthy: pig C, no perceptible change.

October 13.—Pig A all right in every respect: pig C does not show any plain symptoms of disease in the morning, but is sneezing again in the evening.

October 14.—Pig A in perfect health; pig C sneezes a good deal, but has fair appetite. Took up again the filtrated blood-serum, and finding, on examination under the microscope, that the bacillus-germs had changed to *bacilli* (see drawing XI. figs. 1 and 2), I filtrated the same again through four papers. Dr. Prentice and myself examined the filtrate obtained under the microscope (850 diameters), and neither of us being able to discover any bacillus-germs, I inoculated pig A on the left ear with the filtrate in the same manner in which the other pigs had been inoculated. Made two punctures, but used a needle a trifle larger than the one used before.

October 15.—Pig A all right: no reaction whatever. Pig C sneezing, but fair appetite.

October 16.—Pig A perfectly healthy, and has remained so up to date (November 11th). It has never refused a meal, and has been always very active and lively. It is now a very fine pig and in a first-rate condition. (Made use of the same for another experiment on November 13th.) Pig

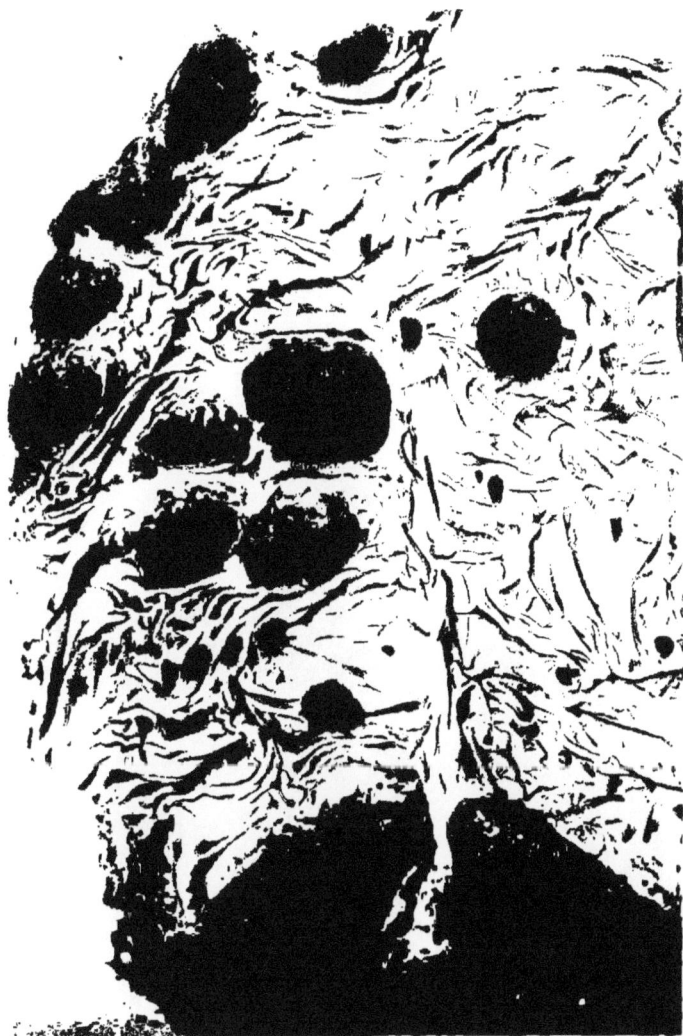

A. Hoen & Co.Lithocaustic Baltimore

Ulcerous tumors on mucous membrane of intestines, showing different view.

C shows plain symptoms of disease: its appetite is poor, and some emaciation has gradually taken place; at least C has not improved like A, and weighs about half as much as the latter, notwithstanding A is in an open pen, exposed to the inclemencies of the weather, and C in a good, new building, with a shingled roof, in which it is amply protected against the changes of the weather.

October 17.—Pig C rather poor appetite; breathing a little accelerated, and coat of hair somewhat rough and staring.

October 18.—Pig C exhibits plain symptoms of swine-plague; its breathing is accelerated; it sneezes a good deal, and its appetite is poor. Eats some in the evening.

October 19.—Pig C improving; has better appetite.

October 20.—Pig C much improved; eats its food again, but is not greedy.

October 21.—No change.

October 22.—Pig C is lively again, and eats well—at any rate, seems to care more for its food. The sores on the ears are healing and disappearing.

October 23.—Pig C must be considered as fully recovered from its slight attack.

Up to date pig C has presented the appearance of a perfectly healthy pig. Its ears have healed, and are now (November 11th) perfectly smooth. It is lively and greedy for its food, but has grown very little, and weighs to-day about half as much as pig A. It can be seen very plainly that pig C has been sick. When I received A, B, and C, A was slightly the best pig. B came next, and C was the smallest, but the difference was only a trifling one.

The experiments just related show that the *bacilli* and their germs must have a causal connection with the morbid process of swine-plague, because an inoculation with *bacilli* and bacillus-germs, cultivated in such an innocent and harmless fluid as milk, produced the disease, while an inoculation with blood-serum from diseased lungs—a highly infectious fluid, if not deprived of its *bacilli* and bacillus-germs—remained without the slightest effect after it had been freed from its *bacilli* and bacillus-germs. I know very well that the result obtained can hardly be considered as conclusive, and that some more experiments of the same kind are needed to confirm the conclusions arrived at.

5. THE CONTAGION, THE CAUSES, AND THE NATURE OF THE MORBID PROCESS.

That swine-plague is an infectious disease, which can be communicated to heathy animals, has been demonstrated by my experiments. It has further been proven that an exceedingly small quantity of an infectious or contagious substance (blood-serum or exudation, for instance) if inoculated, or directly absorbed by the vascular system, is sufficient to produce the disease. It has also been proven that morbid tissues and morbid products, if consumed by healthy pigs, will cause them to become affected with the plague. Consequently, two ways of infection have been ascertained with certainty. Further, if the results of the *post-mortem* examinations are inquired into more closely, it will be found that the principal morbid changes have occurred in the digestive canal, but especially in the cæcum and colon, in all those cases in which the disease had been communicated by way of the digestive apparatus; and that, on the other hand, the principal seat of the morbid process has been in the organs of respiration and circulation, or in the organs situated in the thorax if the contagion had been inoculated or been introduced into the system through wounds and absorbed by the veins and lymphatics.

Whether an inhalation of the contagious or infectious principle into the respiratory passage or into the lungs is sufficient to produce the disease is doubtful. One pig (pig No. 1), an animal free from any lesions or wounds whatever, has been exposed twice and has not contracted the disease; but while exposed and immediately after its pen was moved once a day, and as the pen was thus kept clean, and as dry earth is a good disinfectant, it must be supposed that the animal was never obliged to consume the contagious principle clinging to the excrements of the diseased animals, neither with its food nor with its water for drinking. Its trough was cleaned three times a day, and always before fresh water was poured in. Pig B, however, was exposed only once, by being kept together with pig No. 6, and contracted the disease in due time. But the conditions were entirely different. Pen No. 3, in which both pigs were kept, contains a wooden floor: pig B was put in soon after pig No. 5 had died, and the pen, otherwise always cleaned once a day, had been left dirty (uncleaned) on purpose. So it happened that the ears of corn, thrown on the floor for food, became soiled, though perhaps only slightly, with the dung and the urine of dead pig No. 5 and diseased pig No. 6. Further, both pigs (B and No. 6) tramped through the excrements and soiled their feet, and, as pigs will do, went with their dirty feet into the trough which contained the water for drinking. So it is but fair to suppose that pig B contracted the disease, not by inhaling the contagion, but by consuming the same with its food and water for drinking. Hence I have come to the conclusion that swine-plague is probably not communicated through the lungs by an inhalation of the atmosphere surrounding diseased animals or by simple contact, but that, in order to effect a communication of the disease, the contagion or infectious principle must be introduced directly into a wound within the reach of the veins and lymphatics or be taken up by the digestive apparatus. This conclusion of mine has been corroborated by several facts, some of which I had an opportunity to observe myself, and some of which have been related to me by reliable persons. To mention a few will suffice: Mr. Henry Yothy, who lives four miles north of Urbana, informed me that his neighbor, Mr. Stickgrath, who lives only one hundred yards south of him, lost every hog but one on his place: that he, Yothy, had nineteen head of swine shut up in a yard, and has not lost a single animal, notwithstanding Stickgrath's diseased animals have been running at large, have tramped all around Yothy's pens, and come every day close to the fence: but that his, Yothy's, hogs have no lesions or wounds whatever, and having remained separated from Stickgrath's hogs by a fence, had no opportunity to consume food or water soiled with the excrements or urine of the latter, and to become infected in that way.

Mr. L. Harris, a few miles north of Champaign, kept his shoats and pigs separate from his older hogs. Among the former, swine plague made its appearance, and proved to be very fatal. They were kept in a yard west of the house, and had access to a pasture to the west and an orchard to the south. The peculiar, offensive smell emanating from that yard was so marked that I perceived it several times very plainly when passing by, at a distance of half a mile or more, so it is to be supposed that considerable contagion must have been floating in the air. The yard in which Mr. Harris kept his old hogs (they were intended to be fattened and were not allowed to run out into a pasture) was not over fifty yards south or southeast of the yard occupied by the diseased and dying shoats

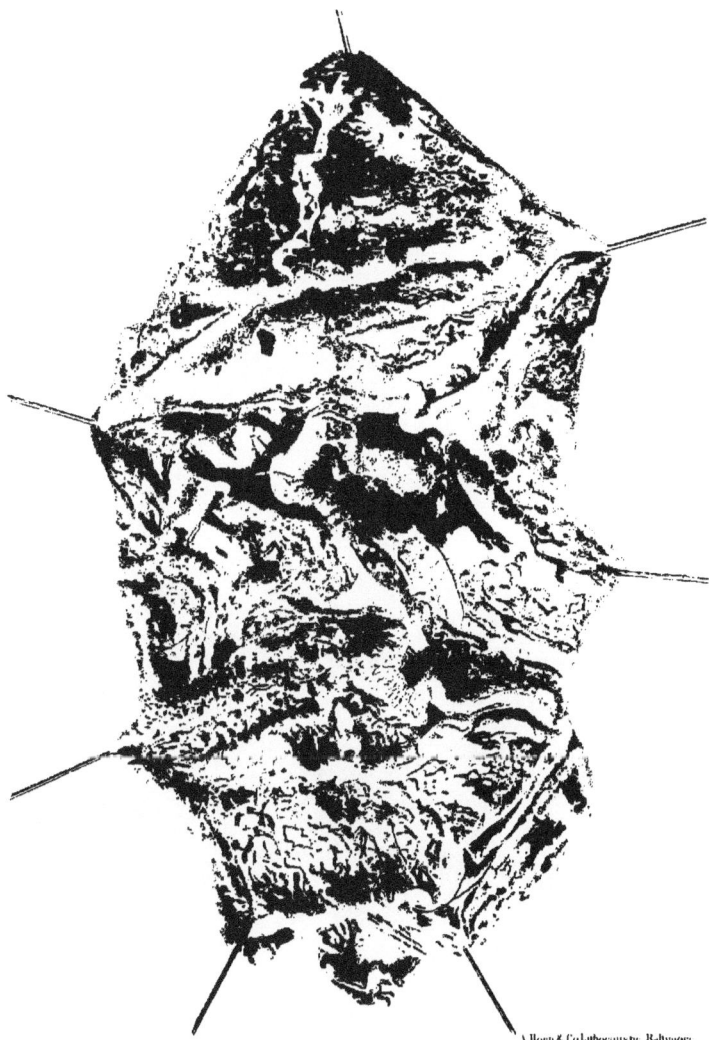

A Hoen & Co Lithocaustic Baltimore

Ulcerous tumors on mucous membrane of the stomach.

and pigs, consequently the wind, usually in the south, carried the effluvia and the foul atmosphere of the former almost constantly into the
yard occupied by the old hogs. The latter, notwithstanding, remained
exempted. It may yet be stated that the old hogs were fed exclusively
with corn, and received nothing but well-water for drinking. On the other
hand, I have not been able to learn of any herd remaining exempted
after the disease had once made its appearance in the immediate neighborhood, unless the animals constituting the herd were free from any external lesions, were watered from a well, fed with clean food, and shut up
during the night and in the morning till the dew had disappeared from
the grass, in a bare yard not containing any old straw-stacks, or in sties
or in pens. Animals allowed to run out on a pasture or on grass, clover,
or stubble fields at all times of the day, and animals that had external
sores or wounds, contracted the disease sooner or later in every instance
where the plague made its appearance in the neighborhood. Further,
the plague, at least during the summer or while south wind was prevailing, seemed to have a special tendency to spread from south to north. If
the history of swine-plague is inquired into it will probably be found
that that tendency has been prevailing every year. This year, for
instance, the disease made its appearance, as I have been informed, for
the first time, in Wisconsin. These facts, of course, could not fail to be
suggestive. So I conceived the idea that the contagious or infectious
principle, abundant in the excretions of the diseased animals, might rise in
the air in daytime, be carried off a certain distance by winds, and come
down again during the night with the dew. That such might be the
case appeared to be possible, because the excrements of hogs, if exposed
to the influence of sunlight, heat, rain, and wind, are soon ground to
powder (partially at least), which is fine enough to be raised into the
air and to be carried off by winds. Moreover, as the bacillus-germs,
which, I have no doubt, must be looked upon as the infectious principle,
are so exceedingly small, it appears to be possible and even probable
that they are carried up into the air by the aqueous vapors arising from
evaporating urine and moisture contained in the excrements, and from
other evaporating fluids (small pools of water), which may have become
polluted with the excretions of sick hogs. To ascertain the facts, I collected dew from the herbage of a hog-lot occupied by diseased animals,
and also from the grass of an adjoining pasture, and on examining the
same under the microscope I found the identical *bacilli* and bacillus-
germs invariably found in the blood, other fluids, and morbid tissues
of swine affected with the plague. (See drawing VII, fig. 5.) Consequently I have come to the conclusion that the bacillus-germs rise into
the air during the day, are carried from one place to another by the
wind, and are introduced into the organism of the animal either by eating herbage (grass, clover, &c.), or old straw covered with dew, or by
entering wounds and being absorbed by the veins and lymphatics.
There is, however, still another way by which the contagious or infectious principle is conveyed from one place to another. It is by means
of running water. It has been observed that wherever swine-plague
prevailed among hogs that had access to running water (as small
creeks, streamlets, &c.), that all the hogs and pigs which had access
to the creek or streamlet below contracted the disease, usually within
a short time, while all the animals which had access above remained
exempted, unless they became infected by other means. I could cite a
large number of instances, but as this observation has been made everywhere, probably nobody who is at all acquainted with swine-plague will
ask for any further proof.

As to the distance which the infectious principle can be conveyed through the air, I cannot make any accurate statements, but have reasons to believe that swine located a distance of one mile from any diseased herd will be safe. To decide this point, which is of very great importance, requires careful experiments.

The nature of the infectious or contagious principle.—The experiments with pigs A and C, though not conclusive and needing repetition, indicate very strongly, as has already been mentioned, that the *bacilli* and their germs formed invariably in the blood, in the morbidly changed tissues, and in the excretions of the diseased swine, must constitute the infectious or contagious principle of swine-plague. I, for my part, am convinced that such is the case. Still I should hesitate to express this opinion if it was supported only by those experiments and not by other facts, such as the peculiarities in the spreading of the disease, the manner in which the infectious principle is acting and is communicated to healthy animals, and the workings of the morbid process. (See next chapter.) At any rate, if the *bacilli* and bacillus-germs constitute the infectious principle, all the strange features of swine-plague find a satisfactory explanation; but if the infectious principle consists in an unknown and mysterious chemical something, the peculiarities of the disease are, to say the least, enveloped in mystery and cannot be explained. What Professor Beale calls bioplasm could not be discovered under the microscope.

In want of a better name I have called the *bacilli* "*bacilli suis*," because the same, as far as I have been able to learn, are peculiar to and characteristic of swine-plague. The bacillus-germs are small round bodies of—as near as I can figure without the aid of a micrometer—about 0.0007 millimeter diameter, and reflect the light very strongly. The *bacilli suis* are small, almost straight, cylindrical bodies of about 0.003 to 0.005 millimeter in length, and 0.0007 to 0.0008 millimeter in thickness, sometimes moving and sometimes without motion, and in certain stages of development slightly moniliform, but in others apparently not. (See drawings.)

The causes.—Whether the disease is caused exclusively by infection—by the *bacilli* and their germs being conveyed directly or indirectly from diseased animals to healthy ones—or whether those *bacilli suis* and their germs can be produced independently from, and outside of, the organism of swine; whether, in other words, swine-plague is a pure contagion, caused exclusively by means of the infectious or contagious principle, or can develop spontaneously, is a very important question, which can be solved only by protracted experiments, and may not be solved at all until the question as to whether a "*generatio equivoca*" is possible or actually taking place or not has found a definite solution. If the *bacilli suis* and their germs constitute the sole cause of swine-plague, as they undoubtedly do, the disease must be considered as a pure contagion, like many other contagious or infectious diseases, not capable of a protopathic or spontaneous development, as long as the possibility of a "*generatio equivoca*" is denied, but if the latter is admitted, or proved to be taking place, a spontaneous development must be considered not only as possible but also as very probable.

If the conclusions I have arrived at concerning the cause of the disease are correct, and I have scarcely any doubt they are, the question as to the causes has been solved. Still, as a *positive* knowledge of the true cause or causes is of the greatest importance, and as my experiments are not numerous enough to be absolutely conclusive, further investigations and more experiments of the same or of similar kind will be very desirable, and, indeed, necessary, in order to obtain *absolute certainty* as to the true nature of the cause or causes.

One thing I am sure of, and that is that an exclusive corn diet, as has been asserted by several agricultural writers, wallowing in dirt and nastiness, starvation, in and in breeding, &c., although by no means calculated to promote health or to invigorate the animal organism, cannot constitute the cause and cannot produce a solitary case of swine-plague, unless the infectious principles (the *bacilli* and their germs) are present. If they are, then, of course, dirt and nastiness, consumption of unclean food and of dirty water, facilitate an infection, and warmth and moisture, pregnant with organic substances, or organic substances in a state of decay, are undoubtedly well calculated to preserve the bacillus-germs and to develop the *bacilli.*

Whether the disease can be communicated to other animals besides swine or not, is a question I am trying at present to decide. Some time ago I had an occasion to throw away some morbid tissues (parts of diseased lungs) of a diseased hog, which I had used for microscopical examination. I threw them—very carelessly, I admit—into an empty lot full of rank weeds, across the road. About a week after several chickens (four or five) died in the neighborhood, of so-called "chicken-cholera." Although there was no proof whatever that these chickens had consumed the morbid tissues, there was a possibility that they had. I bought two healthy chickens, kept them separate, each in a coop, and fed them with the morbidly changed colon of a diseased pig. They consumed the same in my presence, but up to date (November 12th) no results have made their appearance. Further, as no case of an infection of any other animals besides swine has come to my knowledge, it would seem that swine-plague is a disease peculiar to swine like pleuro-pneumonia to cattle.

6. THE MORBID PROCESS.

Concerning the nature of the morbid process, or the manner in which the morbid changes are brought about, the microscope has made some important revelations.

In all those *post-mortem* examinations (fifty-three in number) which I have made since August 3rd, and in all those I had an opportunity of making before that time, I found the lungs more or less affected. The same were partially hepatized, and still partially filled with fluid exudation or blood-serum. Besides that, where the morbid changes in the lungs were of recent origin, innumerable small red specks, caused by capillary hyperaemia, or, rather, a stagnation of the blood, or embolism in the capillaries, could be observed. In several other cases—four or five in number—where the morbid changes in the lungs were not of a recent origin, or older than, say, two weeks, innumerable small, round, and larger confluent tuberculous-looking centers of beginning suppuration or decay (incipient abscesses) presented themselves, especially in the lower and anterior portions of the lungs, and usually more pronounced in the right lobe than in the left one. My friend, Dr. Prentice, who is not only a veterinary surgeon, but also a practicing physician, pronounced the lungs of Mr. Bassett's boar (two years old, and three weeks sick), thus changed, similar or identical in appearance to the consumptive or tuberculous lungs of a human being. Close investigation, however, soon revealed the fact that all the morbid changes found in the lungs of different animals—innumerable small red specks, accumulation of blood-serum or exudation, hepatization, red, brown, and gray, and incipient abscesses—are the products or the consequences of extensive capillary embolism. The other morbid changes, usually found in the thoracic cavity, such as pleuritis, pericarditis, accumulation of straw-colored serum, and the morbid changes found sometimes in the heart, but especially in the auricles, in which, in numerous cases, the capillary vessels have been found to be gorged with blood, tend also to show that embolism constitutes the cause, or at least the main cause, of all those changes. The microscope very fortunately has revealed how this embolism is effected. The capillaries of the lungs, as is well known, are narrower than those in other parts of the body. The blood of the diseased animals, and especially the blood-serum deposited in the affected pulmonal tissue, contain invariably large numbers of bacillus-germs and *bacilli*. These bacillus-germs, as I have observed with the microscope, and as Hallier, who calls them micrococci, nine years ago found, bud and develop to *bacilli*, and show, at a certain period of their development, a great tendency to agglutinate to each other, and to form in that way larger or smaller, irregular-shaped, and apparently somewhat viscous clusters. (See drawing II, fig. 1; drawing IX, fig. 1 *a*.) These clusters, or some of them, are large enough to close or to obstruct the finer capillaries, and to stop in that way the capillary circulation. As a necessary consequence, the serum of the blood transudes through the walls of the capillary vessels, and is deposited in the tissue of the lungs, in the thoracic cavity, and in the pericardium. In some cases, and at some places, the tender walls of the finer capillaries yield to the pressure and rupture, and then extravasations of blood, such as have been observed in several cases, are the consequence. The capillary redness, and the red and purple spots observed in certain comparatively fine portions of the skin, and in the subcutaneous tissues, I have no doubt are also a product of the same process, and are caused by capillary embolism. If the animals would only live long enough, gangrene or mortification of parts of the skin would be met with quite often, but as other morbid changes cause death, and thus terminate the morbid process usually before the stagnation of the blood in the skin becomes perfect, gangrene or mortification has been found only once in the skin on the lower surface of the body. Certain morbid changes in the abdominal cavity, such as abdominal dropsy, and the blood extravasations found repeatedly in various organs, such as stomach and intestines, are due to the same cause. The clusters of bacillus-germs also constitute probably the cause of the swelling of the lymphatic glands. Microscopic examinations of the interior of those glands (see drawing IV, fig. 3) revealed invariably, besides some lymph-corpuscles, immense numbers of *bacilli* and bacillus-germs in different stages of development, some budding, some agglutinated to each other, and some in process of agglutination, &c. These clusters of bacillus-germs, it seems, not only close the capillary blood-vessels, but probably also the finer lymphatics ramifying in the glands; a swelling of the latter, therefore, is a natural consequence.

The production of the morbid growths (swine-plague tumors would be a good name), which are found in nearly every case on the mucous membrane of the cæcum and colon, and sometimes, though not so often, on the mucous membrane of other intestines, such as ilium, jejunum, duodenum, stomach, gall-bladder, and uterus, and even on the conjunctiva and the gums, is not so easily explained. It seems that a proliferous process is taking place; new epithelium-cells and connective-tissue corpuscles are formed rapidly, but decay before fully developed. These new morbid and rapidly decaying cells are imbedded in a stroma of a dense connective tissue which, too, is a morbid product, and formed rapidly. In the older and larger morbid growths or tumors in the cæcum and colon this connective tissue is usually very abundant, especially in the frequently pedicle-shaped foot or basis. The proliferous morbid growths which occur in the small intestines are almost destitute of it. If these morbid growths or tumors are examined under the microscope, immense numbers of *bacilli suis*, some of them moving very rapidly and others at rest (sometimes some other bacteria), and comparatively few bacillus-germs will be seen. (See drawing III, fig. 5; drawing VI, fig. 1; drawing V, fig. 2; drawing IV, fig. 2; drawing VII, fig. 2, and drawing X, fig. 2.)

It appears to be probable that the excessive proliferous growth of the epithelium-cells and connective-tissue corpuscles is caused by a constant irritation of the mucous membrane, or of the *membrana intermedia* (basement or limitary membrane, Fleming) produced by the *bacilli*. This is the more probable, as those morbid growths occur especially in such parts of the alimentary canal in which the food is known to tarry the longest, in the cæcum and in the colon. The morbid changes (ulcerations) found occasionally in the skin, where they sometimes cause whole portions to become mortified or decayed and to slough off, occur, it seems, only in parts where a wound or lesion has been existing into which the infectious principle, the *bacilli* or their germs, have been introduced; so, for instance, in the teats of brood-sows wounded by pigs, and in the nose of hogs and pigs that have been ringed. These morbid changes in the skin, it would seem, are produced in a similar way as the morbid growths in the intestines, with only this difference, that instead of an excrescence loss of substance makes its appearance. The skin is constantly exposed to the atmospheric air, and to a much lower and more changeable temperature than the mucous membrane of the intestines, and in consequence of that the process of decay may be more rapid and may exceed the probably slower process of production.

7. PERIOD OF INCUBATION.

The period of incubation—perhaps more correctly "stage of colonization," Klebs—or the time passing between an infection and the first outbreak of the disease, I have found to be from five to fifteen days, or on an average of about seven days. Still, I have no doubt that in single cases an outbreak may take place a day or two sooner, and in others, though rarely, a day or two later.

8. MEASURES OF PREVENTION.

As swine-plague is a contagious or infectious disease, which spreads everywhere by means of direct and indirect infection, and as a spontaneous development is problematic, or has not yet been proven, the principal means of prevention must consist in preventing a dissemination of the contagious or infectious principle, and in an immediate, prompt, and thorough destruction of the same wherever it may be found. To prevent successfully a dissemination of the contagion and to secure a prompt destruction of the same, Congressional legislation will be necessary. State legislation, for reasons to be mentioned hereafter, will never be sufficiently effective. As it is, the contagion or the infectious principle is, and has been, disseminated through the whole country in a wholesale manner, as I shall show immediately. During the first month of my presence in Champaign I stopped at the Doane House, a hotel belonging to the Illinois Central Railroad Company, and constituting also the railroad depot. Every night car-loads of diseased hogs destined for Chicago passed my window. Only a very short time ago, on one of the last days of October, a farmer, J. T. M., living near Tolono, sold sixty-seven hogs (some, if not all of them, diseased and a few of them already in a dying condition) for two cents a pound, to be shipped to Chicago. I could cite numerous instances, but I think it is not necessary, because these facts are known to every one where swine-plague is prevailing. Besides, in nearly every little town in the neighborhood of which cases of swine-plague are of frequent occurrence, is a rendering establishment to which dead hogs are brought. These establishments pay one cent a pound, and the farmers haul their dead hogs, sometimes ten or fifteen miles, in open wagons, past farms, barns, and hog-lots, and disseminate thereby the germs of the disease through the whole country. The transportation of dead hogs by wagon, I admit, might be stopped by State laws, but the latter prove usually to be ineffective where railroad companies (inter-State and international traffic) are concerned. I include international traffic, because swine-plague is or has been prevailing in Europe. Besides that, there are other contagious diseases which spread exclusively by means of their contagion—I will mention only glanders, foot and mouth disease, aphthæ, and pleuro-pneumonia of cattle—and can be stamped out and be prevented from spreading only by efficient Congressional legislation. Pleuro-pneumonia particularly deserves special attention. It has already gained a firm foothold in the East, and would undoubtedly invade the West very soon, or would have done so long ago, if the traffic in cattle were from East to West instead of from West to East. It may, however, at any time be carried to the West by shipments of blooded cattle from the East the same as it was imported from Holland to New York, and having once entered any of the Western States or Territories it will soon find ample means to spread toward the East again and to sweep the whole country. If it comes to that it will prove to be much more disastrous to the live-stock interest of the United States than swine-plague or any other contagious disease.'

If any transportation of, or traffic in, diseased and dead swine is effectually prohibited by proper laws, a spreading of the swine-plague on a large scale will be impossible, and its ravages will remain limited to localities where the disease-germs have not been destroyed, and been preserved till the same find sufficient food again. In order to prevent such a local spreading, two remedies may be resorted to. The one is a radical one, and consists in destroying every sick hog or pig immediately, wherever the disease makes its appearance, and in disinfecting the infected premises by such means as are the most effective and the most practicable. If this is done, and if healthy hogs are kept away from such a locality, say for one month after the diseased animals have been destroyed, and the sties, pens, &c., disinfected with chloride of lime or carbolic acid, and the yards plowed, &c., the disease will be stamped out. I know that this is a violent way of dealing with the plague, but in the end it may prove to be by far the cheapest. The other remedy is more of a palliative character, and may be substituted if swine-plague, as is now the case, is prevailing almost everywhere, or in cases in which the radical measures are considered as too severe and too sweeping. It consists in a perfect isolation of every diseased herd, not only during the actual existence of the plague but for some time, say one month, after the occurrence of the last case of sickness, and after the sties and pens have been thoroughly cleaned and disinfected with carbolic acid or other disinfectants of equal efficiency, and the yards, &c., plowed. Old straw-stacks, &c., must be burned, or rapidly converted into manure. It is also very essential that diseased animals are not allowed any access to running water, streamlets, or creeks accessible to other healthy swine. Those healthy hogs and pigs which are within the possible influence of the contagious or infectious principle, perhaps on the same farm or in the immediate neighborhood of a diseased herd, must be protected by special means. For these, I think, it will be best to make movable pens, say eight feet square, of common fence-boards (eleven fence-boards will make a pen); put two animals in each pen; place the latter, if possible, on high and dry ground, but by no means in an old hog-lot, on a manure-heap, or near a slough, and move each pen every noon to a new place, until after all danger has passed. If this is done the animals will not be compelled to eat their food soiled with excrements, and as dry earth is a good disinfectant, an infection, very likely, will not take place. Besides this, the troughs must always be cleaned before water or food is put in, and the water for drinking must be fresh and pure, or be drawn from a good well immediately before it is poured into the troughs. Water from ponds, or that which has been exposed in any way or manner to a contamination with the infectious principle, must not be used. If all this is complied with, and the disease notwithstanding should make its appearance and attack one or another of the animals thus kept, very likely it will remain confined to that one pen.

If the hogs or pigs cannot be treated in that way, it will be advisable to keep every one shut up in its pen, or in a bare yard, from sundown until the dew next morning has disappeared from the grass, and to allow neither sick hogs nor pigs, nor other animals, nor even persons, who have been near or in contact with animals affected with swine-plague, to come near the animals intended to be protected. That good ventilation and general cleanliness constitute valuable auxiliary measures of prevention may not need any mentioning. The worst thing that possibly can be done, if swine-plague is prevailing in the neighborhood, is to shelter the hogs and pigs under or in an old straw or hay stack, because nothing is more apt to absorb the contagious or infectious principle, and to preserve it longer or more effectively than old straw, hay, or manure-heaps composed mostly of hay or straw. It is even probable that the contagion of swine-plague, like that of some other contagious diseases, if absorbed by or clinging to old straw or hay, &c., will remain effective and a source of spreading the disease for months, and maybe for a year.

Therapeutically but little can be done to prevent an outbreak of swine-plague. Where it is sufficient to destroy the infectious principle outside of the animal organism, carbolic acid is effective, and, therefore, a good disinfectant; but where the contagious or infectious principle has already entered the animal organism its value is doubtful. Still, wherever there is cause to suspect that the food or the water for drinking may have become contaminated with the contagion of swine-plague, it will be advisable to give every morning and evening some carbolic acid, say about ten drops for each animal weighing from one hundred and twenty to one hundred and fifty pounds, in the water for drinking; and wherever there is reason to suspect that the infectious principle may be floating in the air, it will be advisable to treat every wound or scratch a hog or pig may happen to have immediately with diluted carbolic acid. During a time, or in a neighborhood in which swine-plague is prevailing, care should be taken neither to ring nor to castrate any hog or pig, because every wound, no matter how small, is apt to become a port of entry for the infectious principle, and the very smallest amount of the latter is sufficient to produce the disease.

Still, all these minor measures and precautions will avail but little unless a dissemination of the infectious principle, or disease-germs, is made impossible. 1. Any transportation of dead, sick, or infected swine, and even of hogs or pigs that have been the least exposed to the contagion, or may possibly constitute the bearers of the same. must be effectively prohibited. 2. Every one who loses a hog or pig by swine-plague must be compelled by law to bury the same immediately, or as soon as it is dead, at least four feet deep, or else to cremate the carcass at once, so that the contagious or infectious principle may be thoroughly destroyed, and not be carried by dogs, wolves, rats, crows, &c., to other places.

Another thing may yet be mentioned. which. if properly executed, will at least aid very materially in preventing the disease; that is. to give all food either in clean troughs, or if corn in the ear is fed. to throw it on a wooden platform which can be swept clean before each feeding.

9. TREATMENT.

If the cause and the nature of the morbid process and the character and the importance of the morbid changes are taken into proper consideration, it cannot be expected that a therapeutic treatment will be of much avail in a fully developed case of swine-plague. "Specific" remedies, such as are advertised in column advertisements in certain newspapers. and warranted to be infallible. or to cure every case, can do no good whatever. They are a downright fraud. and serve only to draw the money out of the pockets of the despairing farmer, who is ready to catch at any straw. No cure has ever been found for glanders, anthrax, and cattle-plague. diseases that have been known for more than two thousand years, and that have been investigated again and again by the most learned veterinarians and the best practitioners of Europe. and yet there is to-day not even a prospect that a treatment will ever be discovered to which those diseases, once fully developed, will yield. Neither is there any prospect or probability that fully developed swine-plague will ever yield to treatment. It is true that the *bacilli suis* and their germs can be killed or destroyed if outside of the animal organism, or within reach on the surface of the animal's body. Almost any known disinfectant—carbolic acid, thymic acid, chloride of lime, creosote. and a great many others—will destroy them. But the *bacilli* and their germs are not on the surface of the body, except in such parts of the skin and accessible mucous membranes (conjunctiva and gums) that may happen to have become affected by the morbid process. They are inside of the organism, and not only in every part and tissue morbidly affected, in every morbid product, and in every lymphatic gland. They are also in every drop of blood and in every particle of a drop of blood circulating in the whole organism. Who. I would like to ask, will have the audacity to assert that he is able to destroy those *bacilli* and their germs without disturbing the economy of the animal organism to such an extent as to cause the immediate death of the animal? But even if means should be found by which these *bacilli* and their germs can be destroyed without serious injury to the animal. a destruction of the same will not be sufficient to effect a cure. Important morbid changes must be repaired; extensive embolism is existing in some very vital organs: a rapid, proliferous growth of morbid cells has set in: some of the intestines (cæcum and colon) may have become perforated: exudations have been deposited in the lungs, in the thoracic cavity. in the pericardium, and in the abdominal cavity; the heart itself may have been morbidly changed, and every lymphatic gland in the whole organism become diseased. How, I would like to know. will those quacks who advertise their "Sure Cure" and their high-sounding "Specifics" to swindle the farmer out of his hard-earned dollars and cents—how, I ask, will those quacks restore, repair, stop, and reduce all those morbid changes?

Still, I do not wish to say that a rational treatment can do no good: on the contrary, it may in many cases avert the worst and most fatal morbid changes, and may thereby aid nature considerably in effecting a recovery in all cases in which the disease presents itself in a mild form, and in which very dangerous or irreparable morbid changes have not yet taken place. A good dietetical treatment, however, including a strict observation of sanitary principles, is of much more importance than the use of medicines. In the first place, the sick animals, if possible, should be kept one by one in separate pens. The latter, if movable—movable ones, perhaps six to eight feet square and without a floor, are preferable—ought to be moved once a day, at noon, or after the dew has disappeared from the grass; if the pens are not movable, they must be kept scrupulously clean, because a pig affected with swine-plague has a vitiated appetite, and eats its own excrements and those of others, and, as those excrements contain innumerable *bacilli* and their germs, will add thereby fuel to the flame; in other words, will increase the extent and the malignancy of the morbid process by introducing into the organism more and more of the infectious principle. The food given ought to be clean, of the very best quality and easy of digestion, and the water for drinking must be clean and fresh, be supplied three times a day in a clean trough, and be drawn each time, if possible, from a deep well. Water from ponds and water that has been standing in open vessels, and that may possibly have become contaminated with the infectious principle, should not be used. If the diseased animal has any wounds or lesions, they must be washed or dressed from one to three times a day with diluted carbolic acid or other equally effective disinfectants.

Concerning a therapeutic treatment, I have made several experiments, the principal ones of which I will relate, not because they are illustrative of success, as they are not, but because some interesting features of the disease will be brought to light. A therapeutic treatment—that is, as far as my experiments are able to show—has not been very successful, but the facts will speak for themselves.

1. EXPERIMENTS AT MY EXPERIMENTAL STATION, THE VETERINARY HOSPITAL OF THE ILLINOIS INDUSTRIAL UNIVERSITY.

October 8.—At 5.30 o'clock, p. m., received from Mr. J. A. Hossack eight diseased swine of various size and age for experimental treatment. They were put in pen No. 3, which had been thoroughly cleaned, and were fed three times a day with corn in the ear, and provided with clean water for drinking. I had engaged and had comfortable room for only three or four, but Mr. Hossack thought best to bring me every sick animal he had at that time on his place. So it happened that five of the pigs were in an almost dying condition when they arrived. I numbered them I, II, III, IV, V, VI, VII, and VIII. The therapeutic treatment consisted in giving three times a day about ten drops of carbolic acid in the water for drinking for each hundred pounds of live weight. In deciding upon that amount, it was taken into consideration that some of the water would remain unconsumed. The troughs were emptied and cleaned each time before fresh water was put in.

October 9.—Pig I, a small animal, dead. *Post-mortem* examination was made by Dr. Prentice, and revealed the usual morbid changes—hepati-

4 sw

zation, pleuritis, serum in pericardium, and morbid growths in cæcum and colon.

October 10.—Pig II, a large shoat from eight to ten months old, dead. *Post-mortem* examination by Dr. Prentice. Nearly the same results.

October 11.—Pig III, a small animal, dead. It had probably died on the evening of the 10th; at least it was very much decomposed in the morning, and as pig B had died and had to be examined, no *post-mortem* examination was made.

October 12.—Pig IV, dead; had died during the night. No. V, an old sow, and Nos. VI, VII, and VIII yet alive. No. VIII is the only one that has any appetite. Pig VI is very low, and will soon die. *Post-mortem* examination of No. IV. Externally: skin on lower surface of the body and between the legs purple. Internally: lymphatic glands enlarged; bronchial tubes filled with mucus; both lobes of the lungs, but the left one more than the right, hepatized—red, brown, and gray hepatization; two ounces of straw-colored serum in pericardium, and plastic exudations on the surface of the heart. In abdominal cavity about one pint of serum; spleen enlarged; kidneys normal; mesenteric glands enlarged; intestines free from any morbid growths, and without any lesions whatever; interior of stomach slightly covered with bile.

October 13.—Old sow No. V, and young sow No. VIII (eight months old) have a little appetite. No. VI is very weak, and No. VII is dull; seems to have considerable pressure upon the brain. In the evening No. VI is in a dying condition, and lies motionless in a corner. Sows Nos. V and VIII have some appetite; No. VII breathes with a throbbing motion of the flanks; seems to have headache, is very dull, and holds its nose persistently to the floor.

October 14.—Sow VIII considerably improved; sow V some appetite; VII very low; and VI dead. For *post-mortem* examination of No. VI, see account given in the chapter on Morbid Changes.

October 15.—Old sow No. V and sow No. VIII coughing a good deal; VIII has a good appetite; V has not. No. VII, a sow pig about eight months old, dead in the pen. *Post-mortem* examination of No. VII at 8.30 o'clock, a. m. Externally: Skin on nose, neck, and lower surface of body purple in spots and patches; carcass not very much emaciated. Internally: some adhesion between posterior part of right lobe of lungs and diaphragm; costal pleura and pericardium affected; surface of the lungs exhibit numerous small red specks: both lobes are partially hepatized, and contain considerable exudation yet in a fluid condition. (See photographs, Plates I and II.) External coat of posterior vena cava morbidly changed, inflamed, and coalesced with pulmonal pleura. In abdominal cavity: numerous light-colored nodules or tubercles on the surface of the spleen, some of the size of a millet seed, and others as large as a small pea; mesentric glands very much enlarged: numerous small ulcerous tumors or morbid growths on mucous membrane of cæcum and colon; the whole interior surface of jejunum, for several feet in length one interrupted layer of a morbid growth and subsequent decay of epithelium cells, easily removed with the back of the scalpel, and leaving behind, if thus removed, an uneven villous surface.

October 16.—Old sow No. V and sow No. VIII fair appetite; both cough a great deal. Old sow V discharged yesterday and to-day large quantities of a glassy mucus exuding from the nose. Discovered two ulcerating sores, one in the left middle teat and one in the right forward teat. Her pigs had been weaned a short time before she contracted the disease.

October 17.—Sows V and VIII improving, that is, are less indifferent to surroundings and have better appetite, but still cough a great deal.

October 18.—Sows V and VIII improving; but especially VIII, which has good appetite. In afternoon sow V had some diarrhea, probably caused by feeding on new corn—old corn had been fed before.

October 19.—Old sow V has diarrhea; feces green and semi-fluid. Sow VIII seems to be improving, at least eats a good deal. Sow V is perfectly blind.

October 20.—Sows V and VIII still coughing considerably, but are otherwise improving.

October 21.—Sows V and VIII improving; VIII is already in a little better condition.

October 22.—Sows V and VIII improving.

October 23.—Sow V is still very slow in her movements, but her appetite is much better. Sow VIII still shows difficulty of breathing, but may otherwise be considered as recovered. The diarrhea of sow V has disappeared.

October 24.—Sows V and VIII improving; have good appetite, and are not near so thirsty as formerly; both cough some. Recovery may be considered certain.

October 25.—Sow V very much improved; ulcer in forward teat is healing rapidly (the ulcers have been treated with diluted carbolic acid). Sow VIII shows no morbid symptoms, except some coughing and some difficulty of breathing. She has very good appetite and is very lively.

October 26.—Sow V eats tolerably well, but is still weak. Sow VIII eats and drinks well, and might be looked upon as perfectly healthy if it were not for the yet existing difficulty of breathing. The excrements have gradually lost their peculiar offensive smell.

October 27.—Sow V fair, and sow VIII very good appetite. The latter is getting lively.

October 28.—No perceptible change.

October 29.—Sow V more active, but still partially blind. Sow VIII is gaining in flesh.

October 30.—Both sows have good appetite and are visibly improving.

October 31.—Both improving steadily.

November 1.—Sows V and VIII keep on improving. The ulcers of V have healed, and her sight has been partially restored. The carbolic-acid treatment has been continued to this day (November 1), but is now discontinued.

November 6.—Both sows have been returned to their owners. Sow VIII is like a perfectly healthy pig, but coughs some and also shows a slight difficulty of breathing. Sow V has almost entirely recovered her eyesight; is not in as good condition as sow VIII, and coughs some, but breathes perfectly easy.

October 26.—Received of Mr. D. Burwash, at 6 o'clock, a. m., a Berkshire pig, about five months old, for experimental purposes; it had been sick two or three days. It proved to be very severely affected, but was in a good condition as to flesh. Treatment: about eight or nine drops of carbolic acid in the water for drinking every morning, and about two drams of bisulphite of soda and one dram of carbonate of soda every evening. The pig was designated as No. IX, and put in pen No. 2.

October 27.—Pig No. IX worse; has plain symptoms of pneumonia; died in the afternoon. *Post-mortem* examination three hours after death; four ounces of serum in chest, and also a like quantity in pericardium; trachea filled with mucus; both lobes of lungs congested and gorged with exudation; capillary vessels of the auricles of the heart gorged.

with blood; spleen enlarged, and large numbers of tubercle-like excrescences on its lower surface; cæcum and colon full of hardened feces: a few ulcerous tumors in cæcum, and two large decaying morbid growths in colon; mesenteric glands enlarged; other organs healthy.

Numerous other experiments have been made, and quite a variety of medicines have been tested at different places and in different herds. Some of those experiments have been carried out under my personal superintendence, and some by the owners of the diseased animals in accordance with my instructions. But as the results obtained with any one of them are far from satisfactory, it will be sufficient to mention only a few. The principal medicines tried were carbolic acid, bisulphite of soda, thymol, salicylic acid, white hellebore or *veratrum album*, as an emetic, alcohol, and sulphate of iron, and it has been found that neither of them possesses any special curative value. In a few cases in which most of the lesions were external, applications of very much diluted thymol or thymic acid produced apparently good results; the animals recovered, but might have recovered at any rate. Diluted carbolic acid has been used for the same purpose and with the same results. An emetic of white hellebore or *veratrum album* was given to some shoats (about eight or nine months old, and property of Dr. Hall, at Savoy), in the first stage of the disease, and seemed to have arrested the morbid process immediately, at least the shoats recovered. In other more developed cases it did no good whatever. Bisulphite of soda, salicylic acid, and carbolic acid were used quite extensively, but no good results plainly due to the influence of those drugs have been observed in any case in which the disease had fully developed, neither by myself nor by others. Sulphate of iron has proved to be decidedly injurious. Mr. Bassett used it quite persistently for forty-five nice shoats. Forty-three of them died, one recovered from a slight attack—it had external lesions, which were treated with carbolic acid—and one remained exempted. To bleed sick hogs, in some places a customary practice among farmers against all ailments of swine, has had invariably the very worst consequences, and accelerated a fatal termination. A great many farmers in the neighborhood of Champaign have used several kinds of " specifics " and " sure cure " nostrums, but none of them are inclined to talk about the results obtained, and so it must be supposed that the latter have remained invisible. One case, which should have been related in the chapter on " Prevention," deserves to be mentioned. Mr. Crews had forty-odd hogs, of which he had lost ten or twelve, and was losing at the rate of two to four a day. I advised him to separate those apparently yet healthy, or but slightly affected, from the very sick ones; to put the former in a separate yard, not accessible to the others; to feed them clean food; to water them three times a day from a well, and to give to each animal, two or three times a day, about ten drops of carbolic acid in their drinking water. He did so, and saved every one he separated (fourteen in number), while all others, with the exception of two animals which died later, died within a short time.

Respectfully submitted.

H. J. DETMERS, V. S.

CHICAGO, ILL., *November* 15, 1878.

1. Bacilli and bacillus germs in the different ulceration on the mucous membrane of the jejunum. from pig Nº 7. 6.A.M. 13.10.78. examined 9.P.M.15.10.78. ×850.

...broken down epithelium cells.

In very much diluted. Acetic acid. × 800.

2. Same blood treated with diluted. acetic acid. × 850. Bacilli slightly moving

1. Blood from axillaris of Mr.Bassetts pig. × 850. 22.8.78.

2. The same object 20 minutes later.

2. From an ulcer of the colon. 18.8.78.

IX.

Blood of Mr. J.T.Moore's pig. 8.P.M. 31.10.78. × 850.

X.

1. Blood of Mr.Harris's pig. × 850. 30.8.78.

3. Serum from the lungs of ♀ (Nº IV. × 850. 22.8.78. tined two hours after death.

. of Mr. Bassetts pig. 19.8.78.

2. Bacilli of the intestinal ulcers of Mr.Harris's pig. × 850. 30.8.78.

XI.

1. Bacilli and bacillus germs found in the filtrated exudation of the lungs of Mr. Hossocks pig.

)f Balls farm. × 850.

× 850.

9. P.M.14.10.78.

2. Same serum. after having been exposed for 3 hours to blood-heat.

× 850.

8.P.M.22.10.78.

XII.

1.

2. Process of division.

3.

Froth scraped from the bronchial tubes.

)usey's pig × 850. 29.8.78.
. 2. bacilli moving.
hours after death.

× 850. 18.9.78.

A.Hoen & Co.Lithocaustic Baltimore.

SUPPLEMENTAL REPORT.

Sir: Since the 15th of November, the day on which I forwarded to you my full report. I have devoted my time principally to a solution of some of those questions which had not been fully answered, and have succeeded in ascertaining some additional facts of practical importance. In addition to this the correctness of my conclusions as to the nature of the infectious principle, and the manner in which swine-plague is communicated, has been confirmed by further observations. The vitality of the infectious principle has been tested by experiment; several herds of diseased swine and places where the disease had been prevailing, and where healthy pigs had been introduced a few weeks after the occurrence of the last case of swine-plague, have been visited, and a few more *post-mortem* examinations have been made. In the following, which may be considered as a supplement to my report of the 15th of November, I have the honor of submitting to you, very respectfully, the results of my investigation.

1. THE BACILLI SUIS.

These are found invariably, either in one form or another, in all fluids—such as blood, urine, mucus, fluid exudations, &c.—in all morbidly affected tissues, and in the excrements of the diseased animals, and constitute, beyond a doubt, the infectious principle, or produce the morbid process if transmitted, directly or indirectly, from a diseased animal to a healthy one. These *bacilli* undergo several changes, and require a certain length of time for further propagation; consequently, if introduced into an animal organism, some time—a period of incubation, or a stage of colonization—must pass before morbid symptoms can make their appearance. Three stages of development (a germ or micrococcus stage, a bacillus or rod-bacterium stage, and a germ-producing stage) can be discerned.

The micrococci, globular bacteria, or bacillus-germs, as I prefer to call them, are found in immense numbers in the fluids, but especially in the blood and in the exudations of the diseased animals. If the temperature is not too low, and if sufficient oxygen is present, they soon develop or grow lengthwise, by a kind of budding process—a globular bacterium, or bacillus-germ, constantly observed under the microscope, budded. and grew to double its length in exactly two hours in a temperature of 70° F. (see drawing)—and change gradually to rod-bacteria, or *bacilli*. Some of the latter, finally, after a day or two, if circumstances are favorable, commence to grow again in length, until they appear, magnified 850 diameters, to be from one to six inches long. At the same time, however, they become very brittle, and break into two or more pieces. Where a break or separation is to take place, at first a knee or angle is formed. and then a complete break or separation is effected by a swinging motion of both ends, which move to and fro, and alternately open and close, or stretch and bend the knee or angle. After the division has become perfect, which takes only a minute or two, both ends, thus separated, move apart in different directions. These long bacteria, it seems, are pregnant with new germs; their external envelop disappears or is dissolved, and then the very numerous bacillus-germs become free. In this way a propagation is effected.

Some of the *bacilli* or rod-bacteria move very rapidly, while others are apparently motionless. The causes of this motion I have not been able to ascertain with certainty, but have observed repeatedly that no motion takes place if the temperature of the fluid or substance which contains the bacteria is a low one, and that under the microscope the motion increases and becomes more lively if the rays of light, thrown upon the side by the mirror, are sufficiently concentrated to increase the temperature of the object. So it seems that a certain degree of warmth is required; at any rate I never saw any *bacilli* moving in a fluid or substance immediately after it had been standing in a cold room.

There is, however, also another change taking place, caused probably by certain conditions which I have not been able to ascertain. It is as follows: The globular bacteria or bacillus germs commence to bud or grow in length, but on a sudden their development, it seems, ceases, and partially-developed *bacilli* and simple and budding germs congregate to colonies, agglutinate to each other, and form larger or smaller irregularly-shaped and (apparently) viscous clusters. Such clusters are found very often in the blood and in other fluids, and invariably in the exudations in the lungs: and in the lymphatic gland in pulmonal exudation, and in blood serum, this formation can be observed under the microscope if the object remains unchanged for some time. say for an hour or two. In the ulcerous tumors on the intestinal mucous membrane the clusters are comparatively few, but the fully-developed *bacilli*, many of which move very lively, are always exceedingly numerous. The tumors or morbid growths in the intestines seem to afford the most favorable conditions for the growth and development of the *bacilli* and their germs. That this must be the case is also suggested by the presence of such immense numbers of *bacilli* and bacillus-germs in the excrements, that the latter, beyond a doubt, constitute the principal disseminator of the infective principle. Whether the colonies or viscous clusters of bacillus-germs and partially developed *bacilli* are instrumental in bringing about the extensive embolism in the lungs and in other tissues, by merely closing the capillary vessels in a mechanical way, or whether the presence, growth, development, and propagation of the *bacilli* and their germs produce peculiar chemical changes in the composition of the blood, which disqualify the latter to pass with facility through the capillaries, or which cause a clotting or retention of the same in the capillary system, is a question which I am not prepared to decide. According to my own observations, it appears that the colonies or viscous clusters of bacillus-germs and partially developed *bacilli* get stuck in the capillaries so as to obstruct the passage, and constitute in that way the principal, if not the sole, cause of the embolism. Dr. Orth is of a different opinion. He says: "The principal effect of the 'Schizomycetes' (*bacteria, bacilli*, &c.) is an indirect one, viz., by producing a poison (virus)." (*Archio. fuer wissenschaptliche und praktische Thierheilkunde*, 1877, *page* 1.) It is possible that the circulation of the blood in the capillary system is interfered with by both mechanical obstruction and chemical changes. Still, it seems to me that the observations of Dr. Orth and others apply more to the fully developed *bacilli* in the blood and in the lymph. The vitality of the bacillus-germs, and especially of the *bacilli*, is not a very great one, except where the germs are contained in a substance or a fluid not easily subject to decomposition; for instance, in water which contains a slight admixture of organic substances. If such a fluid is kept in a vial with a glass stop, the germs remain for a long time (over six weeks) in nearly the same condition, or develop very slowly, according to amount of oxygen and degree of temperature. In an open vessel the development is a more rapid one. If oxygen is excluded, or the amount available exhausted, no further change seems to be taking place. In the water of streamlets, brooks, ditches, ponds, &c., the bacillus-germs are not destroyed very soon. How long they retain their vitality I have not been able to ascertain. In fluids and substances subject to putrefaction, the *bacilli* and their germs lose their vitality and are destroyed in a comparatively short time; at least they disappear as soon as those fluids (blood, for instance) and substances undergo decomposition. In the blood they disappear as soon as the blood-corpuscles commence to decompose. That such is the case has been ascertained not only by microscopical observation, but also by clinical experience. The *bacilli* and their germs are also destroyed if brought in contact with, or if acted upon by, alcohol, carbolic acid, thymol, iodine, &c.

2. CLINICAL OBSERVATIONS.

The experimental pigs, Nos. 1 and A, put in pen No. 2, on November 13th (together with experimental pig C), in which pen pig No. IX had died of swine-plague on the 28th of October, remained perfectly healthy, notwithstanding pen No. 2, which was thoroughly infected, had received only an ordinary cleaning, but had not been disinfected. Consequently, it must be supposed that the infectious principle (the *bacilli* and their germs) contained in particles of excrement and in the urine clinging to the floor and lodged in the cracks between the boards must have been destroyed, because I observed repeatedly that the pigs, probably in search of saline substances, licked those parts of the floor which had become saturated with urine.

Mr. Bassett, who had lost nearly his whole herd of swine—of one lot containing originally forty-five animals only two survived—bought, about eighteen days after the occurrence of the last death, two young, healthy pigs, and allowed them to run at large in his orchard, a pasture, and one of his swine-yards, the same premises on which the lot of forty-five animals just mentioned had been kept. The few surviving hogs of his old herd are kept in another yard farther north. Seeing that those two pigs remained healthy, he thought he might risk it and buy some more, and about two weeks later he bought sixty-nine (not ninety-five, as I believe I have stated in my report) healthy Berkshire shoats, from five to six months old, at the auction of the Hon. James Scott, president of the Illinois State Board of Agriculture, and turned them out on the same premises (hog-lot, orchard, and pasture). After these sixty-nine shoats had been there two days they discovered the burial places of the forty-three dead shoats, hogs, and pigs, which, by the way, had been buried only from two to three feet deep. These they commenced to immediately exhume, and soon consumed all the decomposed carcasses. Mr. Bassett would have prevented this had he discovered them before they had accomplished it. Every shoat has remained healthy up to date (November 29th), and as the period of incubation (from five to fifteen days, or on an average seven days) expired some time ago, it must be supposed that the infectious principle, the *bacilli* and their germs, had been thoroughly destroyed by putrefaction. It must be mentioned that there are no straw-stacks, &c., on the swine-range, and that the shoats have no access to any streamlet, ditch, or pool of water.

Mr. Locke's herd of swine has been kept perfectly isolated in a pasture near the city limits of Champaign, and has remained exempt from swine-plague till lately. The hog-pasture is close to the Illinois Central Railroad track. Whether the infectious principle has been introduced into Mr. Locke's pasture by the car-loads of diseased swine which pass by every evening, and which sometimes remain standing on the tracks, at a distance of not much over forty rods from the hog-pasture, for half an hour or longer; whether the vicinity of the rendering establishment has been instrumental in bringing about an infection; or whether the infectious principle has been communicated by other means, I have not been able to ascertain.

The herds of Mr. Clelland (or McClelland), nine or ten miles northwest of Champaign, and of Mr. Allen, six or seven miles northeast of Urbana, have remained exempt for a long time, probably because neither of them has any close neighbors, but finally the disease, spreading from farm to farm, has reached their herds.

Mr. Clay West, three and a half miles northwest from Champaign, living also somewhat isolated, expected that his swine (forty-seven head) would remain exempted. Most of them (forty-two or forty-three) obtained their water for drinking from a running streamlet which, three-fourths of a mile above, passes through the hog-pasture of another farm. On the latter swine-plague made its appearance, and three weeks later Mr. West's swine commenced to die. So it must be supposed that the infection had been brought about by the water in the streamlet. Mr. West, as soon as he found that his hogs commenced to die, sold twenty-seven head to be shipped to Chicago.

3. MORBID CHANGES AFTER DEATH.

Since November 15th I have made some more *post-mortem* examinations, mostly for the purpose of obtaining material for microscopical investigation; but have found nothing not found before, or of any special importance, except in one case, of which, therefore, a full account may not be superfluous. It was a pig of Mr. Clellan's (or McClelland's), who had lost four head out of seventeen within a few days, or after brief sickness. The pig in question, which was a little over four months old, had been sick only two or three days. The *post-mortem* examination was made on November 22d, about sixteen hours after the animal had died.

Externally.—Considerable capillary redness of a purple hue in the skin on the lower surface of the body, between the legs, and behind the ears. *Internally.*—Lower and anterior parts of both lobes of the lungs hepatized (red hepatization); the rest of both lobes gorged with blood-serum or fluid exudation; pericardium coated with plastic exudation; auricles of the heart congested, the capillary vessels tinged with dark-colored blood; lymphatic glands, but especially those of the mesenterium, very much swelled; liver, sclerotic; serous membrane of some of the intestines (cæcum and colon) coated with exudation; ecchymoses and capillary redness in pyloric portion of the stomach; and a few worms (*Trichocephalus crenatus*) in cæcum, but no morbid growths or ulcerous tumors whatever in any part of the digestive canal. This case is worth mentioning, because no morbid growths or ulcerous tumors were found in the cæcum and colon, or in other parts of the intestinal canal; it consequently shows once more that embolism and subsequent exudation in the lungs and in other tissues are more constant and more characteristic of the morbid process of swine-plague than the peculiar morbid growths or ulcerous tumors in the cæcum and colon.

Whether those ulcerous tumors on the intestinal mucous membrane occur only in cases in which the infectious principle has been introduced partly or wholly through the digestive canal, and are absent in those cases in which the *bacilli* and their germs have entered exclusively through wounds or lesions, or whether, finally, this presence or absence depends upon other influences and conditions, is a question which I am not fully prepared to answer. It has decidedly the appearance that the seat and the character of the morbid changes depend, to a certain extent at least, upon the means and parts by and through which the *bacilli* and their germs have entered the animal organism.

My opinion, expressed in my report of the 15th ultimo, that an infection is brought about either through the digestive canal or through wounds or lesions, and probably not through the respiratory mucous membrane and through the skin, if no wounds or lesions are existing, has been corroborated by an observation made at Mr. West's place. I was there on November 20th. The disease had made its appearance on November 10th. Mr. West had lost five animals, had sold twenty-seven more or less diseased, and still had fourteen or fifteen, including four or five older hogs kept in a separate pen, about 12 by 16, which had a wooden floor, and was separated from the hog-lot or hog-pasture only by a board fence. These older animals receive and have received their water for drinking from a well, while all those kept in the hog-lot or hog-pasture, originally forty-two in number, had access to the streamlet before mentioned. None of the older animals, although breathing the same atmosphere as the rest, showed any symptoms of disease, and are still healthy (November 29th), as far as I have been able to learn.

In conclusion, I may say that swine-plague does not seem to be communicable to any other domesticated animals, and must be considered as a disease *sui generis* peculiar to swine.

I intended to make further experiments, by inoculating healthy animals with blood-serum or pulmonal exudations, freed from *bacilli* and bacillus-germs by repeated filtrations and with cultivated *bacilli*, but the time left me (sixteen days) was not sufficient to obtain reliable results. Besides, it appeared to be desirable to use the pigs I had on hand for the purpose of testing the vitality of the infectious principle in such a way as would give the test a direct practical value.

I am, very respectfully, your obedient servant,

H. J. DETMERS, *V. S.*

CHICAGO, ILL., *December* 1, 1878.

REPORT OF DR. JAMES LAW.

HON. WM. G. LE DUC,
 Commissioner of Agriculture:

SIR: I have the honor to submit the following report of experiments and observations on the prevailing fever in hogs.

As you are already aware, my attention has been directed mainly to the pathology of the disease, the nature and vitality of the virus, and its behavior when treated by different disinfectants. Distant as Ithaca was from all infected districts, and seeing it was impossible here to experiment on large herds of diseased and exposed swine, it seemed preferable to leave to others all essays of treatment and prevention of the illness by the use of disinfectants and other sanitary measures. This isolated and noninfected locality offered special advantages for conducting that class of observations which I aimed at, as there was no danger of accidental infection from other sources than the experimental pens. At the same time the number of animals subjected to experiment was limited by the necessity for the most perfect isolation of the healthy and diseased, for the employment of separate attendants for each, and for the disinfection of instruments used for scientific observations, and of the persons and clothes of those who conducted these.

The experimental pens were constructed in a high open field, with nothing to impede the free circulation of air: they were large and roomy, with abundant ventilation from back and front, with perfectly close walls, floors, and roofs, and in cases where two or more existed in the same building the intervening walls were constructed of a double thickness of matched boards with building pasteboard between, so that no communication could possibly take place excepting through the open air of the field. When it seemed needful disinfectants were placed at the ventilating orifices. On the pigs showing the first signs of illness, infected pigs were promptly turned over to the care of attendants delegated for these alone, and the food utensils, &c., for the healthy and diseased were kept most carefully apart. When passing from one to the other for scientific observations, the healthy were first attended, and afterward the diseased, as far as possible in the order of severity. Then disinfection was resorted to, and no visit was paid to the healthy pigs until after the lapse of six or eight hours, with free exposure in the interval. In the pens the most scrupulous cleanliness was maintained and deodorizing agents used so as to keep them perfectly sweet.

I may be allowed to add that I have received most valuable assistance from two of my students, Messrs. A. M. Farrington and A. G. Boyer, in conducting the daily observations, as well as in making *post mortem* examinations, and in the examination of diseased products.

Our experiments have shown this to vary greatly, though in the great majority of cases it terminated in from three to seven days after inoculation. As shown in the table appended, one sickened on the first day, three on the third, two on the fourth, one on the fifth, two on the sixth, four on the seventh, and one each on the eighth and thirteenth days respectively. A comparison of these results with those obtained elsewhere seems to show that we have reached the two extremes. Dr. Sutton, observing the result of contact alone in autumn, sets the period at from thirteen to fourteen days; my own observations in Scotland, in summer, indicated seven to fourteen days; Professor Axe, in London, in summer, concluded on five to eight days; Dr. Budd, in summer, four to five days; and Professor Osler, in autumn, four to six days.

SYMPTOMS.

The cases observed were of all degrees of severity, from a slight access of fever, with some loss of appetite, irregularity of the bowels, and alternations of heat and cold on the surface, to violent attacks, terminating fatally after eleven days' illness.

Early symptoms.—In an average case, one of the earliest signs of illness was an elevated temperature of the body, amounting to one or two degrees above the former indications furnished by the same animal. This qualification appears requisite, as the temperatures of healthy pigs were found to vary widely under different conditions of life. After active exercise or excitement 104° F. is not unfrequent, while in a close pen where they are quiet and still, 100° to 102° F. is quite as common. On more than one occasion, when a pig got accidentally fixed in a narrow space where he had barely room to stand, the temperature was reduced to 99° and even 98° F. The body heat was raised by a hearty meal and lowered by abstinence. Generally a sudden rise of temperature and saturation of the atmosphere with moisture led to an elevation of the body heat, in other cases a reduction of the temperature of the air led to the same phenomenon. (See table of Meteorological Observations and Temperatures.) In connection with the rise of temperature there was generally a diffuse redness of the skin, with increased warmth, alternating with cold, especially in the ears, nose, tail, and limbs. The pulse usually rose perceptibly, sometimes reaching 120 per minute, while the breathing was little if at all affected. The snout was often drawn back, giving a wrinkled or pinched appearance to the face: the movements were less active, sometimes decidedly stiff and slow; there was perceptible falling off in appetite, and the bowels were usually costive.

Disease at its height.—The temperature rose in most cases to 105° F., and exceptionally only to 107° or 198° F. (Dr. Osler records 110° F.), to be followed after a variable length of time (three to twenty days) by a a descent to the natural standard, or even lower. The pulse also rose to 120–130, and the flushes of heat on the skin were much more frequent and extreme. At the same time certain changes appeared in the skin, varying greatly in degree in different cases, but which may be described as follows:

First. A pink or scarlet rash in spots averaging about one-tenth inch in diameter, but often becoming confluent so as to form an extended blush. Many such spots disappeared momentarily under pressure, showing that the minute blood-vessels were not yet completely blocked, but only dilated. Many, however, could not be even temporarily obliterated by pressure, showing already existing embolism if not even rupture and the escape of the blood-elements into the tissue.

Second. In some, though by no means in all, there appeared black spots on which pressure had no effect. The cuticle of such spots dried up and shrunk, and if the pig survived long enough was finally detached.

Third. In nearly all there were slight pointed elevations, mostly around the roots of the bristles, which over the whole body had become more erect, rough, and harsh.

Fourth. Scattered more or less abundantly over the surface were black concretions, hardening in most cases into a scab, but in others, and particularly on the inner side of the thighs, accumulating as a soft, greasy inunction. Where this was not diffused as a uniform black incrustation, it showed as small black particles mostly at the roots of the bristles, and was evidently a product of the diseased sebaceous glands.

Fifth. The skin showed at many points, and above all on the pendent margins of the ears, on the hocks and knees, on the rump and abdomen, an unbroken blue or violet tint, which could not be effaced by pressure. In bad cases this was associated with considerable swelling of the ears, and in one with rupture of the integument and loss of blood.

Finally. A great accumulation of scurf took place along the back, and with the tough, rigid state of the skin contributed much to the unthrifty look of the subject.

The arching of the back, the drawing up of the flank, the advance of the hind toward the fore feet, and the stiff movements of the hind limbs sufficiently attested abdominal suffering, while the contractions of the rectum resisting the introduction of the thermometer testified in most cases to the irritability of the bowels, if not to the thickening and corrugation of their mucous membrane. The gait was stiff and uncertain, and the patient inclined to lie in its litter, by preference stretched on its belly. The bowels at this stage were mostly irritable. In the milder cases they were mostly costive, or if the dung was of natural consistency it smelt strongly. In the worst cases, and in several of the milder ones, they became relaxed with a semi-solid fetid discharge, or a yellowish white or slaty yellow watery flow, alternating with more confined or costive conditions. Vomiting was noticed once or twice, but was altogether exceptional. One patient ground its teeth, but one only. Several had a cough, occurring in paroxysms, but the majority had none, and this is the more remarkable that several of those that appeared to show this immunity harbored numerous lung-worms. In most cases the inguinal glands could be felt to be enlarged.

Stage of sinking.—When patients were approaching death, the temperature, after reaching its highest point, suddenly descended to below the natural, the pulse increased to 130 or even 160 per minute, extreme weakness supervened so that the animal could barely rise or drag itself around; in some cases the nervous powers were so dulled that the pig lay in a stupor, hardly disturbed when pricked to obtain a drop of blood for examination, and in others there seemed to be active delirium, with sudden starting and screaming. Nervous disorder was further shown by general tremors and muscular jerking of the limbs or body. If formerly purging, the anus became relaxed, and the liquid feces escaping involuntarily smeared the thighs and bed. In two this state of things lasted for two days before death supervened. At this stage moving bacteria were repeatedly detected in the blood.

Subsidence of fever.—In cases which seemed to promise recovery, including a majority of the whole, the temperature declined gradually toward the natural standard, the bowels became more regular, the appetite improved, the skin cleared up, and all the bad symptoms steadily diminished. As it was not our object to preserve them they were either sacrificed or again inoculated, so that the too frequently tardy and imperfect or uncertain convalescence was not verified in our pens.

POST-MORTEM LESIONS.

In considering the morbid anatomy of the disease, the lesions of the skin referred to above under the head of symptoms need not be again recorded.

The characteristic lesions were found especially in the digestive organs, the lymphatic glands, and the lungs, though the serous membranes and other tissues were by no means always exempt.

Digestive organs.—In four cases the tongue was the seat of spots of a deep-blue color, ineffaceable by pressure, and in three cases it bore distinct ulcers, similar to those to be described later as existing in the large intestine. Similar ulcers appeared on the soft palate, in two cases, and on the tonsils in one. In four cases the pharynx bore indelible blue spots of extravasation, but no distinct ulceration. In one instance a white concretion in four minute lobes, like pins' heads, was found on the mucous membrane on the back of one arytenoid cartilage, consisting of rounded nucleated cells and granular matter. In one case only did the gullet show patches of congestion. The stomach always contained a fair amount of food, usually smelt intensely acid, the exhalation fuming with ammonia, and presented on the mucous membrane of its great curvature a mottled, dark-brown discoloration, as is often seen in pigs that have been starved for some time prior to slaughtering. In four cases this membrane bore patches of thickening from $\frac{1}{4}$ to 1 inch in diameter, of a deep-red color, from blood extravasation into and beneath the mucosa. In two cases it bore a dirty yellowish-white pellicle of diphtheritic-looking false membrane, the microscopic characters of which will be noted hereafter. In one case slight erosion of the membrane had ensued, but without the formation of any slough.

The small intestines constantly presented spots of congestion, and sometimes extended tracks of the same, with softening of the mucous membrane and excessive production of mucus. The spots were easily overlooked unless when the entire length of the gut was slit open and carefully examined, but when closely examined they presented not only the branching redness resulting from coagulation of blood in the capillary blood-vessels, but also microscopic extravasations of the blood out of thin natural currents. Another point which served to characterize these limited congestions was a greater or less hæmorrhagic reddening of the mesenteric glands immediately adjacent to the congested spots. In three cases only were distinct erosions found on the small intestines, and in one, ulceration with the dirty-white central slough so common in the large intestines. The edge of the ileo-cæcal valve was twice the seat of a sloughing ulcer, and in four subjects the glandular follicles of Peyer's patch were enlarged at this point, a condition which is, however, not uncommon in pigs killed in health.

In the large intestines the lesions were at once more constant and more advanced. The cæcum was the seat of dark-red patches from congestion and extravasation in six cases, the colon in six, and the rectum in five. Ulcers appeared on the cæcum in eight cases, on the colon in seven, and on the rectum in three. In two cases the whole length of the large intestine was the seat of great thickening of the mucous membrane, which was of a deep, dark-red color, and thrown into prominent transverse folds, that considerably diminished its internal caliber. The large intestine was more entirely free from slight congestion of the mucous membrane, and in two cases only were no ulcers found on this part.

The variety of these ulcers deserves a passing notice. In a certain number of cases the mucous membrane, though comparatively free from congestion, showed a number of small conical swellings, with yellowish depressed centers, and about the diameter of one-half a line. To the naked eye these appear like enlarged solitary glands, but have been shown by Dr. Klein, of London, to be enlarged and diseased mucous crypts (follicles of Lieberkühn.) Next, erosions of larger size were not uncommon. In these, the surface layer of the mucous membrane was destroyed, leaving a depressed, red, congested base, and swollen, slightly congested, and reddened edges. Then there are the older ulcers in which, with a more or less reddened base and margin, there is a central dirty-white product, arranged in concentric layers, and usually projecting above the line of the adjacent mucous membrane, and even overlapping it. This appears like a slough, and though sometimes stained with blood contains no pervious vessels. In one instance this slough, in place of occurring in rounded isolated forms, extended transversely to the direction of the intestine, occupying the limits of its morbid transverse folds for half the circumference of the canal, or even more. These bands were abundant in the cæcum and colon, and at intervals two adjacent ones would merge into each other at their widest parts. Finally, in one case, a great part of the surface of the cæcum and colon was covered by a yellowish-white dipthheritic-looking pellicle, in patches of several inches in length, and projecting above the surface of the mucous membrane at its free border.

In one case only was there a blood-colored liquid effusion into the peritoneum. In another, a transparent exudation between the folds of the mesentery contained a microscopic embryo worm; but the most careful search could detect no others at this point, nor in the coats of the intestines. In one case, whitish concretions were found on the mesentery, projecting from the surface and composed of granular cells like those of the concretion on the larynx.

Liver.—Slight ecchymosis on the surface of the liver was common, but extensive congestion, and above all softening, were virtually absent. When congestion existed the acini were most deeply colored in the center, showing the implication of the hepatic veins and intralobular flexus rather than the portal system. In two cases this organ contained slight caseous deposits, in one an *acephalocyst,* and several times *hydatids.*

The *pancreas* appeared to be uniformly healthy.

The *spleen* appeared unduly black and gorged with blood on two occasions only, and in the worst of these the blood was alive with actively-moving bacteria.

The *lymphatic glands* of the mesentery and of the abdomen generally may be said to have been uniformly altered. Those in the vicinity of congested or ulcerated patches of intestines were usually of a dark blood-red, confined to the surface of the gland, or in the worst cases extending through its entire substance. In cases where the disease had passed the crisis, and the subject was advancing towards recovery, there was often simply a grayish discoloration of the surface of the gland, where such hæmorrhagic discoloration would have been found in the earlier stages. In all cases the glands appeared to be materially enlarged.

These remarks would equally apply to the lymphatic glands in the chest, throat, or other parts where congestion and ecchymosis existed.

SWINE FEVER.

Report Commissioner of Agriculture for 1878.

Fig 1.
Microscopic section through skin and slough.

Fig 2
Microscopic section of skin in purple spot

Respiratory organs.—Congestions and ecchymosis were common on the larynx, windpipe, and pleuræ. Though the lungs never entirely escaped, in one case only was an entire lung hepatized. Exudation and consolidation of the lung-tissue were in a few instances confined to the anterior lobes, but as a rule a few of the posterior lobulettes only were affected. In some cases exudation was confined to the interlobular spaces, which accordingly appeared as broad lines circumscribing the lighter-colored lobes, with which they contrasted strongly in color because of their dark blood-stained exudate. Even when the lobules were also the seat of exudation, they were mostly lighter than the interlobular spaces, in this differing from the ordinary inflammation of the lungs, in which the latter appear as yellow lines. The bronchia of the affected lobules were invariably filled with a frothy mucus, while in eight subjects they contained numerous lung-worms (*Strongylus elongatus*). It is worthy of notice that in nearly all cases in which lung-worms were found, the lobules into which the exudate had taken place were invariably connected with the infested bronchia. In one case the windpipe presented along its whole length a yellowish-white false membrane similar to that described as existing on the large intestine. In another instance a blocked bronchium presented a small circular slough not unlike the commencing slough of the intestinal mucous membrane. In no case did I meet with the caseous blocking of the bronchia recorded by Klein.

In one case only was there extensive liquid effusion into the pleuræ. This was of a dark blood color, and, besides, the blood-globules contained myriads of actively-moving bacteria. False membranes of recent formation also connected the pulmonic to the phrenic pleuræ in this case. The right lung was hepatized throughout. In the same subject the pericardium was the seat of a similar exudate, and fibrinous coagula connected the cardiac to the mediastinal layer. In three cases the lining membrane of the heart was the seat of spots of ecchymosis, by preference on the papillary muscles. The right heart usually contained a clot of blood which showed a buffy coat in three cases only. In two cases there was a clear translucent exudation around the auricule ventricular furrow, which, under the microscope, showed fat cells and granules, and a network of capillary vessels in which the blood-globules moved freely, and showed no tendency to adhere.

Brain.—In one case there were four hæmorrhagic spots on the dura-mater, averaging about one line in breadth.

MICROSCOPIC OBSERVATIONS.

Skin.—Microscopic sections through the affected portions of skin showed the various grades of congestion; congestion with blocking of the capillaries, and excess of lymphoid and large granular cells and granules staining deeply with coloring agents; and congestion, with extravasation and the formation of necrotic spots. (See Plate IX, Fig. 1.) With the earlier congestion there is more or less dropsy of the skin and consequent separation of its intimate textures, while in the later or more severe conditions a fibrinous exudation takes place, and this may even exude from the surface and concrete there in dark scabs. In no instance did I meet with the formation of pus in the skin, and notwithstanding the numerous minute extravasations into the true skin and cuticle, in one case only was there sufficient destruction of a superficial vessel to lead to a temporary hæmorrhage. One feature which I have not seen mentioned by other observers is the implication of the bristle follicles. It has been already stated that the pink papular eruption is mostly observed around the roots of the bristles, and it may be added that the bristles always stand erect and harsh. Moreover, in addition to the general unthriftiness and scurfiness of the skin. it tends early to become coated with greasy exudation, resulting usually in the black concretion already mentioned and soluble in ether. This is manifestly a product of the hair follicles and their sebaceous glands. and accordingly a section through one of these shows the deep congestion of the capillary plexus. (See Plate IX, Fig. 2.)

Intestine.—Sections through those portions of the mucous membrane which are merely congested and reddened, but without ulceration. shows stagnation and blocking of the capillary vessels in the mucosa and submucosa, with thickening and softening of the textures. and especially of the epithelial layer. This last contains a great excess of granules and aggregations of granules into cell forms (giant cells of Klein), while the epithelial cells themselves are reduced in size and contain enlarged nuclei. As formerly pointed out by Klein, the degeneration is often greatest around the openings of the crypts of Lieberkühn. and in their interior, while their cavities are not unfrequently filled with extravasated blood. Besides the above are found lymphoid and wandering blood cells, crystals of hæmatine and closely aggregated masses of granules staining deep purple blue in hæmatoxylon and insoluble in caustic potass —the micrococci of Klein. These last are especially abundant on the surface, but extend into the deeper fibrous layers as well. In severe cases the epithelial layer may be raised from the mucosa by a considerable dark-red clot. though the escape of blood in large amount is more frequent under the mucous membrane, so as to separate it from the muscular coat.

The ulcers with a central slough present at their base the same characters as the congested mucous membrane, as regards cellular and granular proliferation. blocking of vessels, exudation. and microscopic extravasation. The slough may be shown to be made up mainly of small nucleated cells and granules. but it retains under the microscope its close laminated appearance, caused by the gradual extension in depth and breadth by the death of successive layers of the mucous membrane. It contains numerous groups of the granular bacteria already referred to. and extending down to its deepest strata.

Lymphatic glands.—As regards the lymphatic glands. I need only repeat the statement of Klein, that the blocking of vessels and extravasation of blood is most commonly into the outer or cortical portion alone : in the more severe forms in which the medullary part is also implicated, the blood effusion is often confined to the lymph-channels and the connective tissue-partitions. while the glandular cylinders escape. It is in cases of longer standing that the cell changes are the most marked. Then there may be found in the lymph-channels the giant cells already mentioned. and the groups of granular-looking micrococci. similar to those found in the intestinal ulcers. as well as lymph-cells of an abnormally dark granular aspect.

Organs of respiration.—The characteristic lesion of the lungs is lobular pneumonia. the exudation taking place most abundantly into the connective tissue between the lobules, and there assuming a dark color by reason of the abundant escape of blood-globules. On making a microscopic section across the smaller air tubes and air sacks, we find in the connective tissues generally. and in the walls of the alveoli and around the bronchia an exudation containing an excess of small round lymphoid cells and granules, and in the air cells themselves accumulations of similar rounded cells (Klein's giant cells), granular matter, and clumps of granular bacteria.

In one instance the wind-pipe from larynx to lung had its superior wall covered by a yellowish-white diptheritic-looking layer similar to that which I found on another occasion throughout nearly the whole large intestine. A section of this under the microscope showed mainly small rounded granular cells, Klein's large granular uniocular cells, and clusters of the granular masses of bacteria, staining deeply with hæmatoxylon. The liver sometimes showed congestion and blocking of its intralobular capillaries and an escape of small rounded granular cells (lymph) into the interlobular spaces, the latter affording a marked contrast to the redness in the center of the acini.

Kidneys.—These were, with one exception, pale in their cortical portion, and a cloudy swelling existed in the walls of the tubules. Spots of blood-staining were common on the papillæ, and at those points the capillaries were blocked by coagula to a greater or less extent.

Blood.—In most cases no alteration of the blood was detected. In one pig, however, on the second day before death, the blood swarmed with bacteria, showing very active movements. In the subjoined drawings (Plate XIII, Fig. 3) may be seen the various forms presented by one bacterium in a few minutes only. The blood of another pig, which had been inoculated from this one showed the same living germs in equal quantity. They were further found in the blood of a rabbit and sheep inoculated from the first-mentioned pig. In an abscess of a puppy which had also been inoculated the germs were abundant. The blood was not examined. In the blood of healthy pigs no such organisms were found. It may be added that the greatest precautions were taken to avoid the introduction of extraneous germs. The caustic potass employed was first fused, then placed with reboiled distilled water in a stoppered bottle that had been heated to a red heat. The glass slides and cover glasses were cleaned and burned, the skin of the animal cleaned and incised with a knife that had just been heated in the flame of a lamp, the caustic solution and the distilled water for the immersion lens were reboiled on each occasion before using, and finally the glass rods employed to lift the latter were superheated before being dipped in them. On different occasions when the animal was being killed I even received the blood from the flowing vessels beneath the skin into a capillary tube which had just been purified by burning in the flame of a lamp. With these precautions it might have been possible for one or two bacteria to get in from the atmosphere, but not for the swarms I found as soon as the blood was placed under the microscope.

PARASITIC WORMS.

In view of the fact that the swine-fever has been repeatedly ascribed to the ravages of worms, it may be well to notice specially those that were found in the pigs subjected to experiment.

Strongylus elongatus (Dry.), *Paradoxus* (Mehlis), *Lung-worm.*—The first eight pigs were purchased of a butcher, and had been fed on offal from his slaughter-house. The lungs of all these contained these worms in numbers varying from ten to forty full-grown specimens, and one pig died, apparently from this cause, on the seventh day. The worms were mostly found in the terminal part of the main bronchium in the posterior lobe of one or both lungs. Others of the air-tubes were, however, occasionally infested. The infested tubes were filled with a glairy mucus, rendering them totally impervious to air, and containing the white thread-like worms and myriads of microscopic eggs. In every case the lobules to which such obstructed air-tubes led were red, congested, and solid, or, as in one or two instances, dropsical, and of a slightly translucent, grayish color. Sections of the diseased portion showed the air-cells partially filled with an exudate in which small rounded cell-forms predominated. The walls of the air-cells were the seat of congested and blocked capillaries and granular cells, while in most cases there were superadded the more specific characters of the fever—the presence of the worms and their irritation having evidently determined the lesions of the specific fever to the infested lobules.

The worms may be thus shortly described: Head slightly conical: mouth terminal, small, circular, with three papillæ; body like a stout thread, white or brownish, skin nonstriated; œsophagus short, 0.63 millimeters, enlarged posteriorly, club-shaped (Plate XIII, Fig. 4); intestine slightly sinuous, and longer than the body; anus opening on a papilla a little in front of the tail. *Male*, 8 to 9 lines in length; tail curved, furnished with a bilobed membranous pouch supported by five rays, two of them double, and two long delicate spiculæ with transverse markings (see Plate XIII, Fig. 5). *Female*, 1 to 1½ inches long; tail turned to one side, narrowing suddenly to be prolonged as a short, curved, conical point; genital orifice in the anterior half of the body, yet close to the middle; oviducts very much convoluted. The *ova* are slightly ovoid $\frac{1}{500}$ inch in diameter, and appear as if they filled the entire body of the adult female (see Plate XIV, Figs. 6, 7, and 8).

Habits.—Like other *strongyli*, these worms attain sexual maturity in the body of their host, and they lay their eggs in the bronchia, to be carried out in all probability and hatched in pools of water and moist earth. It is worthy of note that though I found in the bronchia and air cells eggs in all stages of segmentation, and those containing fully-formed embryos, I did not find a single free embryo worm. The presumption is that, like other closely related worms, they are only hatched out of the body, and that the microscopic embryos live for a variable length of time in water or moist earth, and on vegetables, to be taken in with these in feeding and drinking.

That these worms are injurious there can be no doubt. Pigs infested by them thrive badly, and many die, as did the poorest of my first experimental lot. Like all parasites, they multiply rapidly wherever their propagation is favored by the presence of large herds of swine, and especially if these are kept on the same range and water season after season. In such circumstances they will produce a veritable plague, proving especially destructive to the younger pigs. There is little doubt that many outbreaks of alleged hog-cholera, in which the lungs alone are affected, are but instances of the ravages of these lung-worms, but that they are the cause of the specific fever which we are investigating is negatived by the complete absence of these worms in all of my second experimental lot.

Tricocephalus Dispai (Creplin) Whip-Worm of Swine.—This I found in large numbers in the cæcum and colon of the experimental pigs, and especially of the first lot—those that had been fed on raw offal. This worm is characterized by a long, delicate, filiform anterior part of the body, and a short, thick, posterior portion. The narrow portion is 0.02 millimeters broad and exceedingly retractile; the posterior portion may be almost 1 millimeter thick. The tegument is very finely striated across, and has a longitudinal papillated band. The œsophagus is very wide and slightly tortuous. The *male* is about 1½ inches long but the thick portion does not much exceed ½ inch, and is curved in a spiral. The

Forms assumed in rapid succession by bacterium; also head and tail of lung worm.

Fig. 3 Forms assumed in rapid succession by a bacterium from the blood of a sick pig. x1000.

Fig. 4. Head of Lung Worm Strongylus Elongatus.

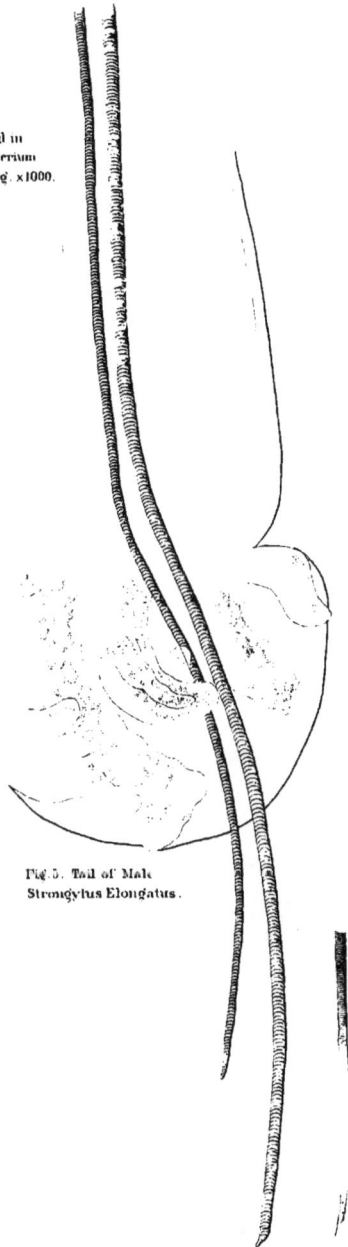

Fig. 5. Tail of Male Strongylus Elongatus.

A. Hoen & Co. Lithocaustic Baltimore.

spiculum measures about 1 line, and is furnished with a funnel-shaped membranous sheath. The *female* is 1½ to 2 inches in length, the thick portion varying from ½ to ¾ of an inch. The posterior portion is brownish, filled with eggs, and ends in a blunt point. The *ova* are 0.052 millimeters in diameter, with a transparent button-like prolongation at each pole.

Like as with other round worms, the ova are laid in the body of the host, but passing out are hatched in water, &c., the young spending their early life in pools, streams, &c., and gain access to the body in food and drink. The worm we are at present considering is especially injurious because of its infesting the human being as well as the pig. Living in the large intestine, it bores its head and much of its anterior filiform body deeply (¼ inch) into the mucous membrane and sucks the blood. When present in large numbers it determines active inflammation of the large intestines, with costiveness or diarrhea, and a rapidly-advancing bloodlessness. Inasmuch as the seat of its ravages, the cæcum and colon, is specially obnoxious to the lesions of the true hog-fever, epizootics caused by the undue prevalence of these worms are very liable to be confounded with the latter disease. The worms are so small that they are easily overlooked among the solid contents of the viscera, unless special care is exercised in the search.

Sclerostomum dentatum (Diesing).—This is another small worm of the cæcum and colon of pigs, found on one occasion only in my experimental animals. It varies from ⅛ to ½ inch in length and is about ¼ line in thickness, hence perhaps more easily overlooked than is the whip-worm, but no less injurious. The body is of a dark gray, brown, or black, according to its contents; the tegument covered with very fine transverse striæ; head broad, mouth terminal, round, and furnished with six very sharp horny teeth, with which to penetrate the mucous membrane. The gullet is broad and club-shaped, and furnished with two salivary glands, opening by delicate canals into the mouth. Intestine wide and sinuous. *Male*, ⅓ inch long, ₆₀ inch in thickness; tail furnished with a bell-shaped membranous expansion, supported by three rays, but open on one side. Testicle single and extended in a sinuous manner from near the gullet to the tail. Two delicate spiculæ. *Female*, 4 to 5 lines in length, tail slowly narrowed and terminated abruptly with a sharp projecting point. Ovaries very tortuous, extend from near the gullet to the tail, where they end in a globular enlargement, beneath which, and close to the point of the tail, is the vulva. The ovoid eggs are laid in the intestines, and carried out with the dung, in which they will hatch, and give exit to the embryo worms on the third day. Like all this family of round-mouthed worms, this fixes itself to the mucous membrane by its mouth, penetrates the tissues with its sharp teeth, and lives upon the blood. If present in large numbers it may establish such a drain that the host becomes pale and bloodless, rapidly loses condition, and perishes from anæmia. It will also, like the whip-worm, irritate the bowels and bring on fatal inflammation, with constipation or diarrhea. In both cases alike the lesions are in the cæcum and colon, the common seat of ulceration, &c., in the specific fever; hence the epizootic is liable to be set down as hog-cholera. It should be added that some members of the family of *Sclerostomata*, and notably the *Sclerostomum equinum* (*Sclerostomum* of the horse), pass a portion of their early life encysted in the mucous membrane and even in other internal organs, and there is some reason to suppose that the *Sclerostomum* of the pig has similar habits, which add materially to the irritation caused by its presence in large numbers. The pigs in Virginia reputed as dying from hog-cholera, caused by microscopic worms in the walls of the bowels, were, in all probability, the victims of an epizootic of *Sclerostomata*.

That the genuine hog-fever is not caused by either of these worms is best illustrated by the fact that in my second lot I found very few whip-worms and no *Sclerostomata*, though both were diligently sought for.

Cysticercus Zemicollis.—This *hydatid* I found in considerable numbers in the abdominal cavity (in the omentum, peritoneum, liver, kidneys, &c.), in the pelvis, perineum, and pleuræ of my first lot of pigs.· It consists of an ovoid bag of liquid $\frac{3}{4}$ to 1 inch in length, with an opening at one end, through which the head is drawn back into the sack. The head is supported on a very attenuated thread-like neck, whence the name. The membrane of the sack is marked by fine transverse striæ, and if placed in tepid water will often undergo active contractions, during which the head can be seen to rise and fall in the interior. The head and neck contain an abundance of dark calcareous particles, soluble with effervescence in a strong acid.

Seventeen of these *hydatids* were fed to a Newfoundland puppy, fresh from its mother, ten having been kept for some time in a solution of common salt, while seven were fresh from a newly-killed pig. After twenty-five days the puppy was sacrificed, and seven tapeworms (*Tænia Marginata*) were found attached by their hooked snouts to the mucous membrane of the jejunum. Exposure to a strong solution of common salt for less than a week in some cases had been sufficient to destroy the first ten, while all the seven cysticerci, grown fresh, developed into tape-worms. These had the globular head with four sucking disks and re-tractile proboscis, surrounded by a double row of 36 hooklets, having the characteristic long posterior process as shown in the accompanying lithograph (Plate XIV, Figs. 9 and 10): also the calcareous markings in the head and neck already referred to.

It is well known that when several ripe segments of this tapeworm are given to a sheep or goat, the myriads of resulting embryo worms that bore their way into the liver and other organs will give rise to such destructive changes in them that death may ensue in ten days. But here again we have the counter evidence in the entire absence of these para-sites in my later lot of pigs, showing that they were in no way responsi-ble for the specific hog-fever.

Other parasitic worms of swine.—It is needless to open up the question of the causation of this disease by the other worms of swine. Many years ago Dr. Fletcher called attention to the destructive effects of the *lard worm—Stephanurus Dentatus—*(misnamed *Sclerostoma Pinguicula*) on the liver and other internal organs, and even attributed the hog-cholera to its ravages. Doubtless he was dealing with an epizootic of this worm, but in many instances since, as in my own recent cases, this worm has been sought for in vain.

So with the *Trichina Spiralis*, the *Hook-headed Worm* (*Echinorhynchus Gigas*), the common measle *hydatid* (*Cysticercus Cellulosa*), and the liver flukes (*Fasciola Hepatica*, and *Distomum Lanciolatum*); however de-structive they may be to pigs in infested localities, their entire absence in my experimental pigs sufficiently excludes them from the causation of the specific hog-fever.

EXPERIMENTS ON THE PROPAGATION OF THE DISEASE BY INOCULA-TION AND OTHERWISE.

Virulence of dried virus.—In experimenting on the hogs it was sought, first, to ascertain the tenacity of life of the dried virus. This was indi-

SWINE FEVER.

Ova, hooks, and head and tail of lung worms.

Fig 8. Head of Female Strongylus Elongatus.

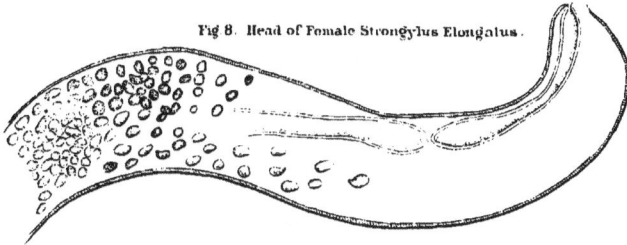

Fig. 6. Tail of Female Strongylus Elongatus.

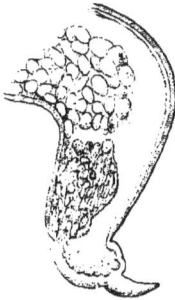

Fig 7. Ova of Strongylus Elongatus.

Fig.10. Long and Short hooks of Taenia marginata × 240.

Fig. 9. Head of Taenia marginata × 50.

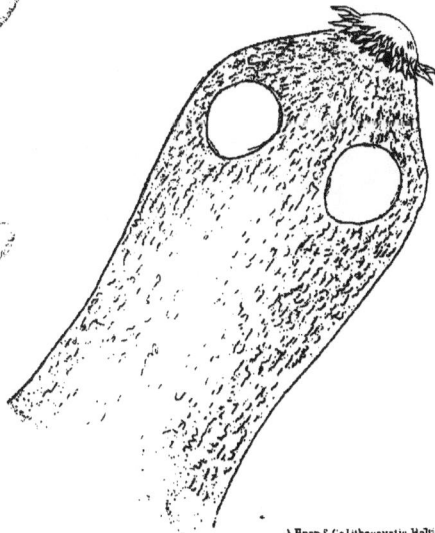

A.Hoen & Co.Lithocaustic Baltimore

cated three years ago by Professor Axe, who successfully inoculated a pig with virus that had remained dried upon ivory points for twenty-six days. It seemed important to test this by further experiment, as upon this question depends the weighty one of arresting or putting an end to the plague by the extinction of its poison.

Three pigs were inoculated with virulent products that had been dried on quills for ONE DAY, one with virus dried on the quill for FOUR DAYS, one for FIVE DAYS, and one for SIX DAYS. The quills had been sent from New Jersey and North Carolina, wrapped in a simple paper covering, and therefore not in any way specially protected against the action of the air. Of the six inoculations, four took effect, and in the two exceptional cases the quills had been treated with disinfectants before inoculation, so that the failure was to be expected.

Virulence of the dried intestine.—In the case of the quills, the virus was dried quickly on account of the tenuity of the layer, and no time was allowed for decomposition. With the diseased intestine the drying in the free air and sun was necessarily slower, and more time was allowed for septic changes. Three pigs were inoculated with diseased intestine which had been dried for THREE and FOUR DAYS respectively. In one case the diseased product was from North Carolina. In all three cases the inoculation proved successful. The morbid product, therefore, even in comparatively thick layers, may dry spontaneously, so as to be the means of transmitting the disease to the most distant States.

Virulence of the moist morbid product if secluded from the air.—A pig was inoculated with a portion of diseased intestine sent from Illinois in a closely corked bottle. The inoculating material had been THREE DAYS from the pig and smelt slightly putrid. The disease developed on the sixth day.

A second pig was inoculated with blood from a diseased pig that had been kept for eleven days at 100° Fahrenheit in an isolation apparatus, the outlets of which were plugged with cotton wool. Illness supervened in twenty-four hours.

The exclusion of air, or more probably the prevention or retardation of putrefaction, therefore, probably favors the longer preservation of the poison.

Probable non-virulence of morbid products that have undergone putrefaction.—Two pigs were inoculated in one day with the elements of an ulcer from a portion of intestine sent from New Jersey in a box. The product was TWO DAYS from the pig and distinctly putrid. Neither seemed to suffer at any time.

A third pig was placed in a pen with a portion of the same diseased intestine, and some manure sent with it. The intestine disappeared after the second day, and was probably eaten, but the pig showed no evil effects.

It should be stated that each of these pigs had been formerly inoculated, and two appeared to pass through a mild form of the disease, while the third had showed an elevated temperature on three alternate days only. It may therefore be questioned whether they had not attained to a certain degree of insusceptibility which insured the negative results. In other cases, however, I have found a second inoculation to take though the first had been successful, and Dr. Osler records cases of the same kind. The results obtained in the three above-mentioned pigs would demand further investigation in this direction, as they suggest a probable explanation of any varying virulence of the disease in wet and dry seasons, in sheds and in the fields.

If we can accept Dr. Klein's theory of the baccillar origin of the disease, the harmless nature of thoroughly putrid products may be explained on the known principle that in preserved or cultivated products the propagation of the septic bacteria leads to the disappearance of the infecting ones.

Virulence of the blood.—A solitary experiment of Dr. Klein's having appeared to support the idea that the blood was non-virulent, I tested the matter by inoculating two pigs with the blood of one that had been sick for nine days. They sickened on the seventh and eighth days respectively, and from one of these the disease was still further propagated by inoculating the blood on three other animals as recorded below. It may, however, still be questioned whether the blood is virulent at all stages, as in the animals infected in the above experiments it was found to contain numerous actively moving bacteria, which had not been found in certain of the milder cases. This subject demands further inquiry.

Infection through the air.—Only one experiment was instituted on this subject. A healthy pig placed in a pen between two infected ones, and with the ventilating orifices within a foot of each other front and back, had an elevated temperature on the ninth, tenth, and eleventh days, with lameness in the right shoulder, evidently rheumatic. On the twenty-fourth day the temperature rose 2°, and remained 104° F. and upward for six days, when it slowly declined to the natural standard.

Infection of sheep, rabbit, and dog.—A merino wether, a tame rabbit, and a Newfoundland puppy were inoculated with blood and pleural fluid, containing numerous actively moving bacteria, taken from the right ventricle and pleura of a pig that had died the same morning. Next day the temperature of all three was elevated. In the puppy it became normal on the third day, but on the eighth day a large abscess formed in the seat of inoculation and burst. The rabbit had elevated temperature for eight days, lost appetite, became weak, and purged, and its blood contained myriads of the characteristic moving bacteria. The wether had his temperature raised for an equal length of time, and had bacteria in his blood, though not so abundantly. He did not seem to suffer materially in appetite or general health. The sheep and rabbit had been each unsuccessfully inoculated on two former occasions, with the blood of sick pigs, in which no moving bacteria had been detected. It remains to be seen whether the virus can be conveyed back to the pig and with what effect. Should further experiment show that other domestic animals than swine are subject to a mild form of the disease, and capable of thus conveying it and transmitting it with fatal effect to pigs at a distance, it will be a matter for the gravest consideration in all attempts to limit the spread of the malady or to secure its extinction. (Since the above was written, I have noticed that Dr. Klein has succeeded in transmitting the disease to rabbits, guinea-pigs, and mice.)

Results of disinfection and inoculation of diseased products.—Under this head eight experiments were conducted with as many different disinfectants, the morbid products being in every case such as had proved successful by direct inoculation on other swine. The object being to test first the most available and least expensive of the disinfectants, the virulent matters were treated with $\frac{1}{5}$ per cent. solution of each of the following agents : Bisulphite of soda, carbolic acid, sulphate of iron, chloride of zinc, and chloride of lime. The materials to be inoculated were in the thinnest layers, in four cases upon quills and in two in thin sections to be inserted under the skin. They were kept in contact with the disinfectants for five minutes, so that the virulent material was thoroughly moistened, softened, and partially dissolved in the five cases in which a solution was used. In the sixth case the thin slice was only kept in the fumes of the burning sulphur for five minutes. In all cases a portion of the disinfectant was necessarily introduced into the wound along with the virulent agent. In four out of the six pigs the disease developed and ran its course as shown in the table, the disinfectants thus proving ineffectual being carbolic acid, sulphate of iron, sulphurous acid, and chloride of lime.

The pig inoculated with virus, treated with bisulphite of soda, died on the seventh day, evidently from lung-worms, and without any distinct symptoms of the plague. There remains the possibility that had it lived longer these would have appeared.

One agent only out of the six can be set down as having proved an efficient disinfectant as used, namely, the chloride of zinc. The virus, treated with this agent, produced no appreciable illness; and though the pig's temperature was raised on the fourth, sixth, and ninth days, this was probably accidental, as it showed no tendency to become permanent. Finally, two pigs were subjected to a hypodermic injection of a few drops of the blood of a diseased subject, mixed in a dram of a solution of permanganate of potassa for the one, and of bromide of ammonium for the other. Both inoculations took effect, and one of the pigs thus infected furnished the blood which conveyed disease to the sheep, rabbit, and dog, as recorded above.

NATURE OF THE HOG FEVER.

Though long confounded with *typhoid fever, anthrax (malignant pustule), erysipelas, measles, scarlatina,* &c., this malady is distinct from all of them. In my report for 1875 I pointed out my reasons for declining to recognize in it either of the above maladies, and claiming it to be "a disease *sui generis*"; and this position has been fully indorsed by the recent researches of Klein, Osler, and others, as well as by my own experiments. This affection may be defined as a specific, contagious fever of swine, characterized by a high but variable temperature, by congestion, exudation, ecchymosis, and ulceration of the intestinal mucous membrane, especially that of the cæcum and colon, and, to a less extent, of the stomach; by congestions and exudations in the lungs in the form of lobular pneumonia; by general heat and redness of the skin, the latter effaceable by pressure; by darker red and black spots unaffected by pressure; by a papular eruption and abundant dark sebaceous exudation; by ecchymosis on the mucous and serous membranes generally; by swelling and ecchymosis of the lymphatic glands; by irregularity of the bowels, costiveness alternating with a fetid diarrhea; and perhaps most important of all, by the presence of colonies of minute globular micrococci in the various seats of morbid change.

An experiment of Dr. Klein, in 1877, in which he cultivated the micrococcus for seven successive generations in the aqueous humor taken from the eyes of rabbits, using only a speck on the point of a needle to inoculate every new portion of the humor, and finally inoculated the product of the fifth and seventh generations successfully on two pigs, seems to establish that these microphytes are the ultimate cause of the disease. My own experiment, in which the disease was conveyed by blood that had been kept for eleven days in an incubator at the temperature of the body, goes to support the same conclusion; but I hope still to subject this question to a more crucial test. If we accept this hypothesis of the pathogenic action of the bacteria, it would almost of necessity follow that the blood, the channel through which these must be carried to the various organs in which they are found, must prove virulent. One of Dr. Klein's experiments appears to negative this conclusion, whereas three of mine go to support it. From what we know of the generation of microphytes, it seems not improbable that at certain stages of its development this specimen may fail to be injurious, or more probably the germs may be filtered from the blood, being arrested in the capillaries, where they determine the morbid changes, and thus many specimens of blood may be obtained which are destitute of the morbid element, until that is again produced in abundance by proliferation in the tissues. By reference to my experiments, it will be seen that the blood with which the successful inoculations were made was taken from pigs in the last stage of the disease, or just after death. That the blood is virulent at certain stages is unquestionable, and in the nature of things this can scarcely fail to be the case, even if we were to set aside experiments and reach our decision from the lesions alone.

CAUSES.

It has been no part of my purpose to investigate the causes of this disease apart from the one specific cause of contagion. It was indeed impossible to pursue such a line of inquiry at a distance from any district where hogs are largely raised, where the disease prevails extensively, and where, presumably, new generations of the poison are taking place. One instance, however, of probable generation *de novo* has been brought under my notice, and the attendant circumstances were such that I think it important to publish the principal facts. In the end of April, 1871, Colonel Hoffmann, of Horseheads, purchased a large herd of swine to consume the buttermilk of his creamery. The swine were supplied with sheds, the open range of an orchard, with plenty of shade under the trees, on a gravelly soil, rising abruptly 10 to 15 feet above the general level of the valley, and were fed fresh buttermilk and corn meal. All went well until late in June or early in July, when the hogs began to sicken and died in large numbers, with the general symptoms of the hog fever. I have mentioned this mainly to negative the widespread belief that the source of the trouble is in the exclusive feeding upon corn. Here we had a laxative and otherwise model diet, supplemented only to a slight extent by corn. It may be well to state that in other years, when he has purchased Western hogs, the disease has always appeared within ten days or a fortnight after their arrival. When New York State hogs only have been bought the pestilence has not broken out.

In view of the strong assertions that pigs will not contract the disease when fed in part on green food or on succulent vegetables—turnips, beets, potatoes, apples, &c.—I had some subjects of experiment freely supplied with potatoes and apples, but whenever the poison was introduced by inoculation I could detect no difference in the period of incubation or the severity of the attack.

It may be added that all unwholesome conditions of feeding and management will favor the development of this as of other specific fevers, by deranging the nutrition, disturbing the balance of waste and repair, loading the blood and tissues with effete and abnormal products, raising the body temperature, and on the whole bringing about a state of the system extremely favorable to the propagation and growth of disease germs. But while the importance of all these may be recognized as accessories, we must not allow them to withdraw our attention from the one condition essential to the development and propagation of the malady—the presence of the specific poison. To quote from my report of 1875, "The important point is this: We know this is a contagious affection, to the propagation of which all possible insalubrious conditions contribute. So soon as we concentrate our attention on this point we have the key to its prevention, if not to its entire extinction."

IS THE TREATMENT OF HOG FEVER GOOD POLICY?

In taking what I know to be an unpopular position on this subject, I am led by the strongest convictions of duty. I well know how popular would be an investigation into the curative powers of different systems, and even nostrums, in this disease, and how many breeders and dealers in swine will readily spend more than the value of the sick hog in the purchase of boasted specifics, to say nothing of the cost of attendance, and how they will rejoice over the wretched unthrifty animal whose life is at times preserved. It is not that recovery is impossible. A certain proportion, 20, 50, or even 80 per cent., will often survive. In my experimental cases only 21 per cent. died and over 28 per cent. recovered from the first attack, so that they were used for further experiment, and this without any attempt at medication or treatment further than wholesome food, cleanliness, and disinfection of the pens. I am convinced that a still better showing could be made in the majority of cases if the sick animals were submitted to careful and intelligent medical treatment.

Were the question of the preservation of the infected pig the only one or the main one to be considered, I would strongly advocate medicinal treatment. But the question is rather one of comparison between this one sick hog or herd and all the healthy swine in the same town, county, State, or nation. This is not a question of morality, but a problem in political economy, and when dealt with by a government must be decided on the ground of what is best for the whole nation. If, then, the preservation and treatment of a single sick hog means the incessant and incalculable increase in its body and secretions of a poison which is in the last degree deadly to other hogs; if this poison can be dried and preserved for a length of time, and carried meanwhile to a distance of a thousand miles, and if not hogs alone but sheep, guinea-pigs, and even wild animals like rabbits and mice, can contract the disease and convey the poison to any distance in their bodies, then the best interests of the nation demand that the sick animal shall not be preserved, but promptly sacrificed to the good of the community.

This point is so important that I may be permitted to dwell on it a little further. Some of my experimental pigs were successfully inoculated with quills that had been dipped in the morbid exudations of sick pigs in New Jersey and North Carolina, and had been dried and preserved for from one to six days in this condition. Here we had the thinnest possible film, such as might have adhered to the clothing of man, the hair of an animal, the feet or bill of a bird, the legs or prehensile organs of an insect, to a dried leaf, or even to a floating thistledown, and might have been thus carried in a great many different ways to infect distant herds. What was actually conveyed some hundred miles on a dried quill, and preserved its virulence for six days in this condition, can be as certainly preserved on any other dry object, and if brought by accident in contact with a raw surface, will produce disease as surely as did the quills in my inoculations. My own observations in this respect have been more than corroborated by one of Professor Axe, of the Royal Veterinary College, London. He produced the disease by inoculating from ivory points on which the cutaneous exudation had been dried up for the long period of twenty-six days.

That the poison can be preserved even in the liquid state when the germs of putrefaction are excluded, may be inferred from my successful inoculations with blood that had been kept in an isolation apparatus, at the ordinary body temperature, for the period of eleven days. As directly to the point is the cultivation of the poison in aqueous humor for seven days, by Klein, and its subsequent successful inoculation. This experiment of Klein is, however, possessed of vastly greater importance, inasmuch as by it it was first shown that the poison can be cultivated and indefinitely increased out of the animal body as well as in it. On seven successive days he inoculated seven successive portions of aqueous humor with as much of the inoculated liquid of the previous day as would adhere to the point of a needle, the first having been similarly inoculated from the sick pig. From the cultivations of the fifth and seventh days, respectively, a drop was taken and two pigs were successfully inoculated therewith. In the cultivation of each day were found myriads of *bacillus*, but no other organization, and thus Klein was the first to show that the *bacillus* is the probable cause of the disease. Had there been no reproduction and increase of the poison, it must have been rendered inconceivably dilute, an approximate ratio of the poison added to the first day's cultivation, and that added to the last, being about as 1 is to 1,000,000,000,000,000,000,000. That such a dilution could be operative seems utterly incredible, and as modern research shows that virulence resides not in simple liquids, but in the solid particles contained in them, and as the only definite organisms in the cultivation liquids were the *bacilli*, it seems inevitable that these are the active cause of the disease. But if so, they cannot only be preserved, but increased in suitable fluids outside the animal body. It is true they disappear when the active organisms of ordinary putrefaction (*bacterium termo*) become numerous, but they are not necessarily destroyed. From what we know of the life of these mycrophytes it is to be feared that so far as the *bacillus* has advanced to the production of spores, it will be preserved in a dormant state, like so many dried seeds, until conditions favorable to its growth shall transpire. On the other hand it may be recollected that my attempts to propagate the disease from a putrefying bowel failed, so that further observation is wanted before we can say that the *bacillus* or its spores are preserved in a septic liquid. However that may be, the possibility of its increase in a non-septic normal fluid is an additional argument for the total destruction of all diseased pigs and morbid products.

In the case of high-priced pigs, where expense is no object, and where the patients can be kept in thoroughly disinfected pens, under the most rigid seclusion, treatment may sometimes be commendable; but in the case of common herds, and as viewed from the standpoint of the greatest good to the greatest number, there can be no question at all that the treatment of the sick is the most ruinous policy, while the most stringent measures for the extinction of the poison is the only economical one. The universal experience of veterinarians supports this conclusion, and nearly every European government has now reached the same conviction, and absolutely prevent the preservation and treatment of the victims of those fatal contagious diseases which most threaten their flocks and herds.

MEASURES TO ARREST AND EXTIRPATE THE DISEASE.

To put a stop to the ravages of the fever concerted measures are essential. One farmer may easily eradicate it from his own herds; but so long as his neighbors continue to harbor it his stock is daily subjected to the danger of renewed infection. His personal sacrifice is all in vain, so long as he is liable to have his herds infected by a chance visitor, a wandering animal or bird, or even a favorable wind. What is true of the individual farmer is equally true of the township, county, and State. One may crush out the disease at a cost of immense effort and outlay only to find it reappearing the next day, as the result of carelessness on the part of an adjoining or even distant State or district. In our Eastern States this plague is almost invariably the result of importation, and though from the lack of pigs it never gains a wide prevalence, it sufficiently illustrates how the disease is propagated in the West, where its more extended ravages are liable to blind the eyes to the fact. To secure a complete or even partial immunity active measures must be taken over the entire land, and while this cannot be done by States, districts, counties, or even towns, separately, it will be rendered the more effectual in the precise ratio that it is inaugurated as a uniform system over the entire country, and under one central controlling authority.

Without entering at this time into all the details of the necessary restrictive measures, the following may be especially mentioned: 1st. The appointment of a local authority and inspector to carry out the measures for the suppression of the disease. 2d. The injunction on all having the ownership or care of hogs, and upon all who may be called upon to advise concerning the same, or to treat them, to make known to such local authority all cases of real or suspected hog fever, under a penalty for every neglect of such injunction. 3d. The obligation of the local authority, under advice of a competent veterinary inspector, to see to the destruction of all pigs suffering from the plague, their deep burial in a secluded place, and the thorough disinfection of the premises, utensils, and persons. 4th. The thorough seclusion of all domestic animals that have been in contact with the sick pigs, and in the case of sheep and rabbits the destruction of the sick when this shall appear necessary. 5th. Unless, where all the pigs in the infected herd have been destroyed, the remainder should be placed on a register and examined daily by the inspector, so that the sick may be taken out and slaughtered on the appearance of the first signs of illness. 6th. Sheep and rabbits that have been in contact with the sick herd should also be registered, and any removal of such should be prohibited until one month after the last sick animal shall have been disposed of. 7th. All animals and birds, wild and tame, and all persons except those employed in the work, should be most carefully excluded from infected premises until these have been disinfected and can be considered safe. 8th. The losses sustained by the necessary slaughter of hogs should be made good to the owner to the extent of not more than two-thirds of the real value as assessed by competent and disinterested parties. 9th. Such reimbursement should be forfeited when an owner fails to notify the proper authorities of the existence of the disease, or to assist in carrying out the measures necessary for its suppression. 10th. A register should be drawn up of all pigs present on farms within a given area around the infected herd—say, one mile—and no removal of such animals should be allowed until the disease has been definitely suppressed, unless such removal is made by special license granted by the local authority after they have assured themselves by the examination of an expert that the animals to be moved are sound and out of a healthy herd. 11th. Railroad and shipping agents at adjoining stations should be forbidden to ship pigs, excepting under license of the local authority, until the plague has been suppressed in the district. 12th. When infected pigs have been sent by rail, boat, or other mode of conveyance, measures should be taken to insure the thorough disinfection of such cars or conveyances, as well as the banks, docks, yards, and other places in or on which the diseased animals may have been turned.

Other measures would be essential in particular localities. Thus in the many places where the hogs are turned out as street scavengers and meet from all different localities, such liberty should be put a stop to whenever the disease appears in the district, and all hogs found at large should be rendered liable to summary seizure and destruction.

The great difficulty of putting in practice the means necessary to the extirpation of the disease will be found to consist in the lack of veterinary experts. No one but the accomplished veterinarian can be relied on to distinguish between the different communicable and destructive diseases of swine, and to adopt the measures necessary to their suppression in the different cases. In illustration I need only recall the numerous reports in which what is supposed to be hog cholera has been found to depend on *lung worms*, on any one of the four different kinds of *intestinal round worms*, on the *lard-worm*, on *embryo tape-worms*, on *malignant anthrax*, on *pneumonia*, or on *erysipelas*. To class all these as one and apply to all the same suppressive measures would be a simple waste of the public money, but to distinguish them and apply the proper antidote to each over a wide extent of territory would demand a number of experts whom it would be no easy matter to find. This state of things is the natural result of a persistent neglect of veterinary sanitary science and medicine as a factor in the national well-being, and must for a time prove a heavy incubus on all concerted efforts to restrict and stamp out our animal plagues. It will retard success under the best devised system, and will sometimes lead to losses that might have been saved, yet if an earnest and prolonged effort is made the obstacle should not be an insuperable one, and the United States should be purged not of this plague only, but of all those animal pestilences which at present threaten our future well-being.

Respectfully submitted.

JAMES LAW.

ITHACA, N. Y., *January* 2, 1879.

SWINE FEVER.

Microscopic section showing exudation in the cæcal mucous membrane beneath an ulcer

Microscopic section through skin, showing hair follicle containing effused blood
The bristle was detached in mounting.

Microscopic section of lung with exudate filling the air cells
and thickening the alveolar walls.

Microscopic section of congested gut, showing villi with excess of granular matter,
stained in hæmatoxylon. Detached round cells.

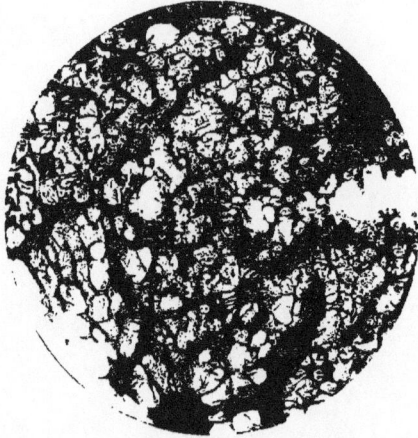

Microscopic section of lung, showing thickened walls of air-cells; blocked vessels;
exudate into cell-walls, and a few of the cells.

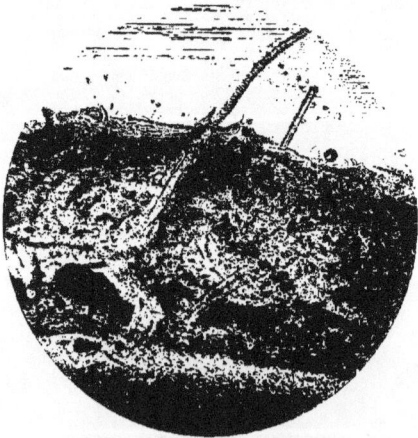

Microscopic section from ear, showing cartilage and skin with broken surface,
and crust-entangling bristles.

APPENDIX.

Record of Experiments.—No. 1.

Male white pig, eight months old; no special breed. Formerly fed offal from a slaughter-house.

Date.	Hour.	Temperature of body.	Remarks.
Sept. 30	3 p. m.....	104.75° F.	Had escaped and was caught after a good chase.
Oct. 1	9 a. m.....	103.25	
1	6 p. m....	103.5	
2	9.30 a. m..	102.5	
3	9.30 a. m..	102	
5	4 p. m.....	102.75	Inoculated from quill charged with dried liquid from infected lung; matter from North Carolina, and five days old; quill dipped five minutes in solution of bisulphite of soda—: 1 :: 500.
6	5 p. m.....	103.25	
7	11 a. m....	100	
8	12 noon ...	101.5	
9	11 a. m....	103.5	
10	5 p. m.....	101.25	
11	10 a. m....	102	
12	4 p. m.....	99	

Was found sprawling upon its belly unable to stand; breathing slow, deep, panting, and labored; snout hot, dry, and of a leaden color; ears and feet warm, bluish, but without any rash, eruption, blotches, or extravasations. Blood appears at the arms. An hour later this pig died.

Post-mortem examination thirty-six hours after death.—Body in excellent preservation; condition low; skin scurfy along the back; snout livid blue, but without petechiæ.

Digestive organs: Tongue has papillæ, at its base reddened; a similar blush appears on the fauces and pharynx.

Stomach and bowels normal.

Liver firm and sound. Kidneys and bladder sound.

Urethra (intrapelvic) deeply congested, almost black, but without any obstruction.

Parasites in abdomen: A few *tricocephali* (*whip-worms*) in the large intestines; a *hydatid* in the pelvic fascia.

Chest: Pleura normal; *pericardium* healthy, with a small quantity of serum.

Right heart: Auricle and ventricle filled with dark clotted blood.

Left heart: Auricle contains a small clot of black blood; ventricle empty.

Lungs: A great part of these is in a condition of carnification or infarction. This is confined to definite lobules or groups of lobules, the collapsed, red, fleshy aspect of which is in marked contrast with the full form and pale pinkish-white color of the remainder.

The air passages (bronchi and bronchia) contain small portions of the contents of the stomach which have been vomited up and drawn into the lungs in the last violent efforts to breathe. The air-passages leading to the collapsed lobules contain large quantities of a watery mucus and pellets of worms (*strongylus elongatus*) which completely block them. The obstructed terminal bronchia are dilated, and have their mucous membrane variously reddened and congested. Around these bronchia the connective tissue is strongly congested and filled with extravasated lymph, by which the vessels passing to and from the lobuletts are compressed and obstructed. In view of this state of things, the explanation of the process of infarction in the lobules is easy; the irritation and congestion caused by the worms in the infested air-tubes extended to the surrounding connective tissue and the sheaths of the accompanying blood-vessels; the exudation of lymph compressed and obstructed the vessels, inducing stagnation, congestion, and exudation in the whole substance of the lobule or lobuletts to which these led. Hence the invariable connection of the infarcted lobule, and the blocked, congested, and worm-infested tube that led to it.

EXPERIMENT No. 2.

White male pig, eight weeks old, smallest of litter. Formerly fed offal at a slaughter-house.

Date.	Hour.	Temperature of body.	Remarks.
Sept. 30	3 p. m........	104. ° F.	Has just come one mile in a wagon.
Oct. 1	9 a. m........	103. 24	
1	6 p. m........	102. 5	
2	9.30 a. m	102.	
3	....do	101.	
5	4 p. m........	102.	
6	5 p. m........	101.	
7	11 a. m.......	100. 75	Bowels quite loose; rain.
8	12 noon	102.	Inoculated from quill dipped in liquids of diseased lungs forty-eight hours ago in New Jersey; quill treated with chloride of zinc before inoculating.
9	11 a. m.......	101.	
10	5 p. m........	103. 25	
11	10 a. m.......	101. 5	
12	4 p. m........	105. 25	
13	12 noon	102. 75	
14	4 p. m........	104.	
15	10 a. m.......	102. 5	
16	....do	102. 5	
17	....do	104.	
18	....do	102. 5	
19	....do	103. 3	
20	....do	103.	Scouring; placed in pen with semi-putrid ulcerated intestine and manure of diseased pig.
21	....do	102.	
22	....do	102. 5	
23	....do	102. 75	
24	....do	103.	
25	....do	101.	
26	....do	102. 75	
27	....do	101.	
28	....do	102. 25	
29	9.30 a. m	103.	
30	2 p. m........	100. 5	
31	9 a. m........	102. 5	
Nov. 1	10 a. m.......	101. 75	
3	9 a. m........	101. 25	Inoculated with quill charged with liquid from lungs of pigs having no bowel lesions; sent from Indiana.
4	....do	102.	
5	9.30 a. m	101.	
6	10 a. m.......	100. 5	
7	....do	103.	
8	....do	100. 9	
9	....do	100. 5	
10	....do	103. 5	Pining; gets lighter daily.
11	....do	102. 9	
12	....do	103.	Wasting, but lively.
13	....do	102. 5	
14	....do	102. 2	
15	....do	102. 8	
16	....do	102. 5	
17	....do	102.	
18	....do	100. 5	
19	....do	103.	
20	....do	102.	
21	....do	102.	
22	....do	101. 75	
23	....do	100. 5	
24	....do	100. 5	
25	....do	97. 5	
26	....do	98.	Very weak and exhausted; surface cold; breathing slow and rattling; left its bed, but was unable to get back without assistance. An hour later breathing seemed to have ceased, but when removed for dissection it returned in a gasping manner; killed by bleeding.

Post-mortem examination.—*Skin:* Pale, bloodless, withered, and inelastic, covered almost universally with black concretions or unhealthy-looking and thick, dirty, white scurf. Snout beneath the nostrils blue, but not ecchymosed.

Digestive organs: Tongue healthy ; beneath the right tonsil is a considerable collection of dirty, grayish-yellow, cheesy matter, consisting of pus-cells and much granular matter.

Stomach: Moderately full, contents fetid and slightly acid, firmly adherent to the mucous membrane, and bringing off part of the epithelium when detached. The mucous membrane on the great curvature is congested, and bears several patches of deep, blood-red extravasation.

Small intestines: Red and congested throughout. The contents are small in quantity and dry, being collected in dry masses at considerable intervals, and partly frothy. The duodenum and first half of the jejunum contains twenty-two ascarides (*A. Suilla*), one extending to 11 inches in length. At different points the bowel is completely blocked by the rolls of these worms.

Large intestine: Ilio-cæcal valve normal. Cæcum and colon, like the small intestine, congested throughout nearly its whole extent, with patches of extravasation and erosion at intervals, but none of the characteristic sloughs nor ulcers, with thick indurated base. The cæcum and upper portion of the colon contains thirteen whip-worms (*tricocephalus crenatus*), their heads firmly imbedded in the mucous membrane, and requiring considerable force to withdraw them.

Liver: Small and of healthy aspect. Gall-bladder full of a dark-green, tenacious bile. Spleen small, black, and somewhat soft. Pancreas normal. Mesenteric glands apparently little altered. Some were slightly congested.

Kidneys: Normal. In the prepuse is a slight, fetid, concretion-like false membrane. On the *omentum* are two *hydatids*.

Respiratory organs: The whole interior of the larynx is of a dull brownish-red, excepting where covered by an extensive false membrane. Along the upper wall of the windpipe, where the ends of the cartilages overlap, is a false membrane about a third of an inch in breadth, and extending from the larynx as far as the lungs. This has a firm consistency, and a dirty yellowish-white color, tinged with green, and stands out prominently from the adjacent mucous membrane by an abrupt margin on each side. Under the microscope it is seen to consist of large quantities of granular matter, granule cells, epithelial and pus corpuscles, blood globules, and numerous crystals. It also contains eggs of the lung-worm beneath this morbid product.

Lungs: Whole anterior lobe of the right lung carnified, of a deep-red color, and sinks in water. The special bronchus for this lobe, and its divisions, are filled with a tenacious mucus, but contain no worms. Several lobulettes in the anterior lobe of the left lung are in a similar condition. On the posterior border of each lung several lobulettes are consolidated, being of a dirty-gray color and semi-transparent. They present, in short, the appearance of pulmonary œdema. The bronchia leading to these lobulettes are completely filled with a thick mucus and numerous worms (*strongylus elongatus*) and their eggs.

The *bronchial lymphatic glands* appear normal.

Blood: The blood is very black, coagulates slowly but firmly, and without buffy coat, and has its globules full-sized and rounded. The right side of the heart beat, when touched, for nearly five hours after the death of the animal, and of its removal from the body.

EXPERIMENT NO. 3.

White pig, eight weeks old; no special breed. Has been fed on raw offal at a slaughter-house

Date.	Hour.	Temperature of body.	Remarks.
Sept. 30	3 p. m........	103.5° F.	Has just come a mile in a box-wagon.
Oct. 1	9 a. m........	103	
1	6 p. m........	102.5	
2	9 a. m........	101.5	
3	....do	101	
4	....	....	No observations.
5	4 p. m........	102.3	Blood taken from saphena vein for cultivation experiment; then inoculated with quill-point charged with liquid from diseased lung, five days old, from North Carolina.
6	5 p. m........	103	Slightly costive.
7	11 a. m......	100.75	Bowels natural.
8	12 noon	102.5	
9	11 a. m......	102.5	
10	5 p. m........	103	
11	10 a. m......	103	
12	4 p. m.......	104	
13	12 noon	103	Dung very fetid.
14	4 p. m.......	104.25	
15	10 a. m......	102.25	
16	....do	101.5	
17	....do	103.25	
18	....do	103	
19	....do	102.75	
20	....do	103	Inoculated with putrid intestinal ulcer from diseased pig in New Jersey. Fed a portion of same.
21	....do	100	
22	....do	101.5	
23	....do	102.25	

EXPERIMENT No. 3—Continued.

Date.	Hour.	Temperature of body.	Remarks.
Oct. 24	10 a. m	101° F.	Appears to suffer from introduction of thermometer.
25	... do	102. 5	
26	9 a. m.......	100. 5	
27	....do	101	
28	10 a. m......	101. 5	
29	9.30 a. m	100. 75	
30	2 p. m.......	102. 25	
31	9 a. m ......|	102. 5	
Nov. 1	10 a. m......	101. 5	
2	....	....	No observations.
3	9 a. m	100. 25	Inoculated with dried diseased intestine sent from North Carolina. Dried in sun and air.
4	....do	101	
5	9.30 a. m	101. 75	
6	10 a. m......	101. 75	
7	....do	103	
8	....do	102	
9	....do	100. 5	
10	....do	104. 5	
11	....do	102. 5	
12	....do	102. 5	
13	....do	103	
14	....do	103. 6	
15	....do	103	
16	....do	103. 5	Limited pink papular eruption on skin.
17	....do	103	
18	....do	103	
19	....do	103	
20	....do	102. 5	
21	....do	101	
22	....do	102. 5	
23	....do	102	
24	....do	101. 5	
25	....do	102	
26	....do	103	
27	....do	104	
28	....do	104	
29	....do	101. 75	
30	....	....	Killed by bleeding.

Post-mortem examination.—Skin: The seat of some papular eruption and black incrustations, but without any patches of purple.

Digestive organs: Mouth and throat sound.

Stomach: Is mottled, of a dark brown along the great curvature, but without any extravasations or erosions.

Small intestines: Has several limited patches of slight congestion, but no erosions. It consains twenty *ascarides*.

Large intestines: Shows some slight congestions, but no slough, erosion, or ulcer. A dozen whip-worms are present in the cæcum and colon.

Mesenteric lymphatic glands: Generally healthy, but a few were unusually red and congested near to the congested patches of the small intestines.

Hydatids: The abdomen contains eight of these.

Liver: Firm and of nearly a natural appearance.

Spleen and *pancreas:* Sound.

Kidneys: Have cortical substance blanched, but are firm and apparently sound.

Lungs: Have some lobulettes solidified red, impervious to air, and sinking in water In the main terminal bronchia towards the posterior part of the lungs are numerous worms (*strongylus elongatus*), though not always in the air-tubes leading to the consolidated lobulettes.

Heart: Sound.

Brain: Sound.

EXPERIMENT No. 4.

White female pig, eight weeks old; no special breed. Formerly fed on raw offal at a slaughter-house.

Date.	Hour.	Temperature of body.	Remarks.
Sept. 30	3 p. m......	103.75 ° F.	Just come one mile in a wagon.
Oct. 1	9 a. m......	102.75	
1	6 p. m......	104	
2	9.30 a. m	102	
3	... do	100.5	
5	4 p. m.......	102.	
6	5 p. m.......	101	
7	11 a. m......	103	Bowels quite loose.
8	12 noon	101.5	Inoculated with quill charged with lung-fluids of a pig that had died suddenly in New Jersey. Virus one day on quill.
9	11 a. m......	102.5	
10	5 p. m	104	
11	10 a. m......	102.75	
12	4 p. m......	104.5	
13	12 noon	103	
14	4 p. m	105.75	
15	10 a. m	105	
16	...do	104.5	
17	... do	107	
18	...do	106	
19	....do	104.25	Scouring. Cold north gale, rain and frost.
20	.. do	105	Do.
21	...do	105.25	
22	....do	104	Skin covered with purple and black spots with red areola. The cuticle or black spots is dead and easily separated.
23	...do	105.25	
24	...do	105.75	Extensive purple blotches on ears, flanks, and abdomen, and a pink rash one to two lines in diameter; appetite poor.
24	5 p. m......	105	
25	10 a. m	106	Killed to-day by bleeding.

Post-mortem examination.—Has been purging; feces fetid and bright yellow.

Skin: Nearly covered with black spots of from one to two lines in diameter, and evidently formed by sloughs or small necrotic patches of cuticle, infiltrated with blood and dried up. The median line of the belly between the rows of teats is almost devoid of these spots.

A purple rash in spots averaging one line across exists in different parts of the body, but is most abundant on snout, ears, buttocks, root of tail, and limbs, especially on the lower parts and inner sides. At certain points, as on the pendant half of the ears, on the hocks, in the region of the arms, and on part of the snout, there is a uniform leaden discoloration. The inner sides of the arms have similar but more circumscribed patches.

Digestive organs: A deep purple blush extends along the line of papillæ on the right border of the tongue. Similar spots exist in the posterior nares. Salivary glands are pale and normal. The guttural lymphatic glands have spots of congestion on their surface, but not extending into their interior.

Abdomen: No effusion. Three *hydatids* are found attached respectively to the posterior surface of the stomach, to the back of the liver, and to the mesocolon.

Stomach: Full of undigested food, yellow at pylorus. No marked congestion nor softening. No parasites.

Small intestine: Duodenum without extra vascularity; its epithelium gray, pigmented, and easily detached. Jejunum and ilium had circumscribed spots of congestion one-half inch in diameter on an average, and in one case slightly eroded.

Large intestine: Cæcum presents three ulcers, each one-fourth inch in diameter, having a circular elevated mass of dirty-white deposit, apparently non-vascular, and a very slightly reddened base. The matter on the surface of the ulcer consisted of cells, round, angular, and of other forms, much granular matter and myriads of round and linear moving bacteria. None of these ulcers appear to be situated on the solitary glands. The same remark applies to the congestions and erosions of the small intestines. Colon and rectum natural.

Parasites: The small intestines contain three *ascarides* (*A. Suilla*). The colon contains a young whip-worm (*tricocephalus crenatus*). The coats of the intestines at the points of congestion and elsewhere were carefully examined for parasites, but without result. The muscular tissue of the diaphragm was also examined in vain.

Liver: Two small cysts, each one-half line in length, exist on the middle lobe. They had thick fibrous walls and liquid contents in which the microscope detected cell forms.

The general substance of the liver is firm and natural, a few acini only isolated and in groups, being congested. The color predominates in the center of the acinus. The liver cells are granular.

Gall-bladder: Is full, but not to excess, with bright yellow bile. The bile-ducts in the liver are also full.

Pancreas: Normal, pink. Pancreatic lymphatic gland blotched; deep red on the surface.

Kidneys: Normal, unless it be in extra pallor of the cortical substance.

Chest: Heart, right auricle and ventricle contain clots showing a buffy coat. Left auricle and ventricle empty. A few petechia exist on the septum ventriculorum.

Lungs: Petechia exist on the pleura. A number of lobulettes are solidified or infarcted, and of a deep red flesh color. The bronchia leading to such lobulettes are blocked by numerous worms (*strongylus elongatus*) and their eggs, embedded in an abundant tenacious transparent mucus. In some cases the bronchia appear dilated, the mucous membrane congested, and the epithelium degenerating, round and ovid granular cells predominating in its structure. There is no visible stasis (coagulation) of blood in the capillaries of the bronchia. The worms are confined to the smaller bronchia, and are only exceptionally found in the otherwise sound portions of the lungs.

Blood: That from the gluteal vein contains no bacteria nor free hæmatine so far as can be detected. Red globules are crenated and shrunken.

EXPERIMENT No. 5.

Female white pig, eight weeks old, no special breed.   Formerly kept on raw offal at a slaughter-house.

Date.	Hour.	Temperature of body.	Remarks.
Sept. 30	3 p. m.	103.75° F.	Just brought one mile in a wagon.
Oct. 1	9 a. m.	103.75	
1	6 p. m.	103	
2	9 a. m.	102	
3	9 a. m.	101.5	
5	4 p. m.	102.25	
6	5 p. m.	102	Inoculated with a quill dipped in liquids of diseased lung (five days old). Before the inoculation. quill was dipped ten seconds in solution of carbolic acid : 1 :: 500.
7	11 a. m.	103	
8	12 noon.	103.5	
9	11 a. m.	103.75	
10	5 p. m.	104	
11	10 a. m.	105	Bowels natural. Lively.
11	5 p. m.	105	Lively. Hungry.
12	4 p. m.	103.75	
13	12 noon.	104.3	
14	4 p. m.	102.25	
15	10 a. m.	104	
16	...do	105	
17	...do	104	Coughs.
18	...do	104.25	Bowels loose. Feces fetid.
19	...do	103.75	Scouring.
20	...do	103	Inoculated with substance of a firm intestinal ulcer, sent from New Jersey, and slightly putrid.
21	...do	102.75	
22	...do	103.25	
23	...do	103.75	
24	...do	103	
25	...do	102.25	
26	9 a. m.	101	
27	...do	102	
28	10 a. m.	103	
29	9.30 a. m.	102.75	
30	2 p. m.	103	
31	9 a. m.	103.75	A slight pink rash on skin.
Nov. 1	10 a. m.	101.5	
3	9 a. m.	102	Inoculated with intestinal mucus and ulcer from Illinois, very slightly putrid.
4	...do	100.75	
5	9.30 a. m.	101.5	
6	10 a. m.	101	
7	...do	103.25	
8	...do	102.5	
9	...do	101	
10	...do	104.75	
11	...do	103.8	
12	...do	102.75	
13	...do	104	

EXPERIMENT No. 5—Continued.

Date.	Hour.	Temperature of body.	Remarks.
Nov. 14	10 a. m.	104 ° F.	Shedding black scales, leaving red conical papules.
15	....do	103.75	Abundant pink papular eruption, excessive between the thighs.
16	...do	103.8	
17	... do	104	
18	....do	104	
19	....do	104	
20	....do	103.75	
21	....do	103.2	
22	....do	103	
23	....do	102.75	
24	....do	103.2	
25	....do	103.8	
26	....do	104	
27	....do	103.5	
28	....do	104	
29	....do	102	
30	....do	103.2	
Dec. 1	....do	102	
2	....do	103.2	
3	....do	102.5	Killed by bleeding.

Post-mortem examination.—*Skin:* Presents many papules or slightly pink conical elevations, just raised enough to be felt by the finger; also black concretions like pinheads and up to twice or thrice that size. It is, however, much cleaner than it was a week ago.

Digestive organs: Mouth normal, likewise the pharynx, larynx, and adjacent lymphatic glands.

Stomach: Has its mucous membrane dark brown along the great curvature, but without any extravasation, ulcer, or recent lesion.

Small intestines: Have a few spots of congestion, but these are very circumscribed. They contain twelve *ascarides.*

Large intestine: With few and slight patches of congestion. No enlargement of Peyer's patches, nor solitary glands; no erosions. The cæcum contains six whipworms.

Lymphatic glands of the mesentery are mostly gray on the outside from pigmentary deposit, but normal in their interior. The pigmentation is evidently the result of a former blood extravasation, as is so constantly seen in the earlier stages of the disease. The blood coloring matter is being transformed into black pigment, as a concomitant of convalescence.

Liver: Presents several hard yellow concretions as large as peas, also spots and patches of purple. Similar rounded yellow concretions are found in the mesocolon. They are covered by a reticulated membrane, and are probably the remnant of some parasite. *Gall-bladder* very full (the pig had been killed fasting), bile green, glairy.

Spleen and *pancreas:* Normal.

Kidneys: One contains two *hydatids;* excepting marked pallor of the cortical substance they are otherwise normal.

Hydatids: Nine of these are found in different parts of the peritoneum.

Heart: Right side normal; contains a small clot.

Left ventricle: Has numerous patches of extravasation, of a deep claret color, situated mostly on the *carneæ columnæ* and *musculi papilares.* These have their seat in and beneath the serous lining, and barely extended into the muscular substance. The margin of the bicuspid valve is slightly thickened.

Lungs: Have a very few red consolidated lobulettes; of the remainder many are only partially dilated, though they have nearly their normal color.

Parasites: The terminal main bronchium of the right lung contains from thirty to forty worms (*Strongylus elongatus*). The lobules corresponding to this bronchium were slightly collapsed, but not consolidated nor congested.

Lymphatic glands of chest almost unchanged.

Brain: Healthy.

A microscopic section from a petechia on the heart showed, in addition to the blocked capillaries and blood extravasations, a fine example of the curious ovoid parasites long known as Rainey's cysts.

6 SW

EXPERIMENT No. 6.

Male white pig, eight weeks old; no special breed; has been hitherto fed raw offal at a slaughter-house.

Date.	Hour.	Temperature of body.	Remarks.
Sept. 30	3 p. m.....	103° F.	Has just come one mile in a wagon.
Oct. 1	9 a. m.....	103.25	
1	6 p. m.....	103.5	
2	9.30 a. m ..	101.75	
3	....do	101.5	
5	4 p. m.....	102.25	
6	5 p. m.....	100	
7	11 a. m....	103.25	
8	12 noon ...	102.25	Inoculated with quill dipped in pulmonary exudation of a pig that had been sick for a week or two. Infected quill sent from New Jersey.
9	11 a. m....	101.5	Rectum very red, and bleeds easily.
10	5 p. m.....	103.75	
11	10 a. m....	102	
12	4 p. m	102.5	
13	12 noon ...	102	
14	4 p. m.....	104	
15	10 a. m....	103	
16	....do	103	
17	....do	101.75	
18	....do	103	
19	....do	102	
20	....do	102.5	
21	....do	103.5	
22	....do	105	
23	....do	103.5	
24	....do	104	Shows extensive blue patches on ears, flanks, and belly; also a pink rash, spots one to two lines in diameter. Appetite impaired.
24	5 p. m.....	105	
25	10 a. m ...	105.25	
25	6 p. m.....	105.75	
26	9 a. m	105	Off feed, but active; ears partly purple; feces dark but moderately firm; struggles when the thermometer is used.
26	p. m6.....	104.75	
27	9 a. m	105	Ears cold, livid in their outer half; pulse 120 per minute; breathing natural; is bright and feeds when up, but is inclined to lie, and shows much weakness; has always resented handling, but to-day, when caught, threw itself on its side and lay to have its temperature taken.
28	10 a. m	103.5	Costive; dung in firm round balls, but of good color, and not specially offensive; runs around readily, but is weak; discoloration mainly on ears.
29	9.30 a. m ..	104.3	Still costive; ears cold and very blue.
30	2 p. m......	106	Weak on limbs; ears very dark purple; legs, tail, and rump badly blotched; bowels costive; dung in yellow balls.
31	9 a. m	103.75	Skin extensively blotched with dark purple; bowels costive; weak on limbs, especially the hind.
Nov. 1	10 a. m	103.75	Very weak; disinclined to move; sways on its hind limbs when up; bowels quite soft.
3	9 a. m	99.75	Very dull; weak; evidently sinking; pulse 132 per minute; grits its teeth continually when up; breathing slow; nervous tremors and jerking constant.
3	6 p. m.....	99.5	Evidently delirious; screams when its door is opened, or when approached or touched; stands with difficulty, having its hind feet drawn forward to the level of the fore, or in front of them; muscular jerking constant, and prevents us from taking the pulse; no grinding of teeth; has not eaten since morning.
4	....	Found dead.	

Post-mortem examination, November 4.—Skin : Almost universally scarlet, passing to dark purple on ears, belly, and hocks. Inner sides of the fore-arms and thighs have the skin white, but blotched with indelible purple spots one-half to one line in breadth. Many of these spots have a dark red or purple areola, with a firm black central scab or slough, evidently resulting from extravasation into the cuticle and superficial layers of the true skin. A section made perpendicularly to the surface shows much redness from blocked branching blood-vessels, especially around the hair follicles, and numerous minute spots of blood extravasations.

The *snout* is of uniform dark red, but with deeper purple spots ineffaceable by pressure.

Margin of the arms deep purple, almost black.

Digestive organs: Tongue, left border has an extensive slough near the tip. Right border has a number of firm elevated points, with purple areola and yellow centers.

Soft palate: Lower or buccal surface has its follicles deeply stained with blood and surrounded with purple areola ; some follicles are filled with a yellowish material.

Right tonsil: Is swollen and has its ducts distended with a thick, tenacious, transparent mucus, containing great numbers of rounded granular cells.

Throat: Epiglottis bears spots of congestion ineffaceable by pressure.

Gullet: Healthy.

Stomach: Moderately full; acid. The mucous membrane on the great curvature presents patches of extravasation and erosion, the latter varying from one to three lines in diameter. Contains a worm (*ascaris Suilla*).

Small intestine: Contains twelve *ascarides*, one as much as ten and one-eighth inches in length. The mucous membrane presents along its whole course patches of redness, congestion, and softening, which are especially numerous and extensive towards its lower portion.

Ilio-cæcal valve: Bears a sloughing ulcer completely encircling it.

Cæcum: Contains a number of ulcers with white sloughs, many of them confluent, and forming bands or belts tending to encircle the gut, being situated on the summits of the transverse folds.

Colon: The anterior portion is much ulcerated, some of the ulcers being confluent and tending to form transverse bands as in the cæcum, while others are mere circular masses, two or three lines in diameter, with white necrotic center, and very little vascularity around the margin.

Rectum: Has patches of congestion and extravasation one line and upwards in breadth; in the case of one, advanced to the formation of a firm white slough and ulcer as in the cæcum. Close to the anus the entire mucous membrane is very deeply congested and thickened by exudation and extravasation.

Parasite: The cæcum contained one whipworm (*Tricocephalus crenatus*).

Parasites in the peritoneum: In the cavity of the abdomen were found twelve *hydatids* in connection with the liver, stomach, omentum, mesentery, meso-colon, and pelvic fascia. Three others were lodged in the perineum near the urethra.

Kidneys: Softened slightly and of an unusual pallor in their cortical portion.

Bladder sound. Intrapelvic urethra deep red, almost black, from petechial extravasation.

Urine about two ounces, turbid, strongly acid, albuminous; density, 1020; urea, 2 per cent.

Chest: Heart has a gelatinoid material filling the auricula-ventricular groove similar to that seen in No. —.

Right heart has a considerable buffy clot in both auricle and ventricle. Left auricle contains a small clot, almost the entire substance of which is pale or buffy. It further contains some very dark fluid blood.

Lungs: A few lobulettes only are infarcted or consolidated. In all cases the bronchia leading to the consolidated lobulettes are blocked by worms (*S. elongatus*). The other bronchia are clear of worms excepting in the immediate vicinity of the infarcted lobulettes. The great bulk of the lung is healthy, and of a soft white color, slightly tinged with pink.

Parasites: Attached to the pleura were two *hydatids*.

EXPERIMENT NO. 7.

Female pig, eight weeks old, no special breed.   Formerly fed raw offal at a slaughter-house.

Date.	Hour.	Temperature of body.	Remarks.
Sept. 30	3 p. m.....	103.75° F.	Has just come one mile in a wagon.
Oct. 1	9 a. m.....	103.3	
1	6 p. m	103	
2	9.30 a. m ..	102	
3	....do	100.75	
5	4 p. m.....	102.2	
6	3 p. m.....	103	Inoculated with quill charged with matter from diseased lung from New Jersey, six days old; quill treated with solution of copperas : 1 : : 500.
7	11 a. m....	104	
8	12 noon ...	103.25	
9	11 a. m....	104.20	
10	5 p. m	103.25	
11	10 a. m....	105.75	
11	5 p. m.....	105.75	
12	4 p. m.....	104	
13	12 noon ...	104	
14	4 p. m.....	103.75	

84   DISEASES OF SWINE AND OTHER ANIMALS.

EXPERIMENT No. 7—Continued.

Date.	Hour.	Temperature of body.	Remarks.
Oct. 15	11 a. m....	107° F.	Lively; good appetite.
16	10 a. m....	105. 75	
17	....do	102 25	
18	....do	104. 25	
19	....do	103	
20	....do	103. 75	Scouring.
21	... do	104. 75	
22	....do	104. 25	
23	....do	105. 50	
24	....do	105	Shows blue patches on the rump and flank, and a red rash on belly.
24	5 p. m....	105. 5	
25	10 a. m....	106. 5	Pulse 108 per minute. Will scarcely move from bed.
25	6 p. m....	104. 75	Very dull; skin hot.
26	9 a. m......	103	Dull; lies much; does not struggle when handled; ears deep purple; bowels loose; dung fetid; skin cool.
26	6 p. m....	105. 3	Dull, very hot skin.
27	9 a. m....	107	Skin very hot, hips stained with feces. Defecations semi-fluid, dark greenish, with clayey aspect, and fetid. Pulse 160 per minute. Breathing 28 per minute; deep, rather labored; wheezing inspiration, terminated by a snore. Can scarcely be roused, and crouches in the litter at once when released.
28	10 a. m....	104. 5	Scouring. Feces offensive. Lies constantly on belly. When lifted hangs helpless with no attempt at struggling. Discoloration is very marked on ears, snout, belly, and thighs.
29	9.30 a. m ..	102. 75	Ran from bed to avoid being caught, but hangs helpless in hands when lifted. Feces very soft; fetid. Skin more deeply colored than before, but cool.
30	2 p. m....	99. 75	Very sick; stupid; stands constantly with fore limbs drawn back and hind advanced, so that all four feet meet. Flanks hollow. Skin on discolorations very deep purple, almost black on rump. Bowels loose. Fetid.
31	9 a. m....	94. 5	Lies in stupor, with limbs and body jerking every instant. Breathing slow, sighing, rattling. Feces and urine discharged involuntarily, and have soaked the left (lower) thigh, which, in consequence, shows a much brighter red than the other parts of the body. The general surface, excepting some white patches inside the arms and thighs, was of a dark purple, almost black on the ears, snout, median line of the abdomen, rump, and hocks. Killed by bleeding.

Post-mortem examination.—Blood: Scanty; that from axillary vein is neutra or slightly alkaline. Red globules deeply crenated and shrunken very disproportionately to the white globules, which are large and rounded, but appear deficient in numbers : 1 : : 80.

Skin: Section of the blue skin of the ear shows cutis, cuticle, and bristle follicles deeply congested, most of the capillaries being blocked by coagulated blood, and microscopic extravasations appearing at short intervals. The red globules in this part are full, rounded, and of the usual size.

Digestive organs: Tongue has a series of white sloughs along its tip and right margin, resembling those of the intestines, being yellowish-white, laminated, non-vascular, and with very slight congestion and redness around them. Microscopically these sloughs are composed of epithelial cells with much granular matter. In one a central red spot presents stagnation and coagula in the capillaries and microscopic extravasations. It is manifest these form in the same manner with the sloughs in the intestines. Circumscribed spots of the mucous membrane become the seat of congestion, resulting in coagulation of the blood in the capillaries and exudation and extravasation alike into the epithelial and sub-epithelial layers, leading to thickening and induration of the deeper strata, and death of the more superficial ones.

Soft palate: The buccal or lower surface bears a similar slough, while many of its follicles are red, swollen, and filled with a yellowish-white (cheesy?) matter.

Throat: The laryngeal surface of the epiglottis is congested, the redness being ineffaceable by pressure. The mucous membrane on the back of the right arytenoid cartilage bears a four-lobed warty looking excrescence like a small pin's head, which, under the microscope, discloses only round granular cells and free granules.

Abdomen, Stomach: This contains a few ounces of half-digested food. This, together with the lower portion of the gullet, is of a deep yellow hue, apparently from regurgitated bile. No marked congestion of the mucous membrane.

Small intestine: Shows circumscribed spots and patches of congestion and small petechia, but no erosions.

Large intestine: One sloughing ulcer on the ilio-cæcal valve, three on the cæcum, and a considerable number in the colon. The colon and rectum also bore numerous patches of extravasation one to two lines in diameter. The last inch of the rectum is of an uniformly deep dark red. The mucosa and sub-mucosa are alike gorged with blood, and at one point a bleeding pile projects into the passage.

Liver, pancreas, and spleen are firm and seemingly healthy.

Kidneys: Firm and apparently sound; cortical part rather pale.

Bladder: Sound; moderately full.

Urine: Strongly acid; density, 1026; albuminous; urea, $3\frac{24}{100}$ per cent.

Parasites in abdomen: Attached to the peritoneum of stomach, liver, and spleen are seven *hydatids*.

Chest: Right heart contains clots; left heart empty. Auriculo-ventricular furrow filled with a gelatinoid material, which, under the microscope, appears as a loose fibrous stroma, its open meshes filled with a nearly homogeneous material, together with a few fat cells, granule cells, and abundant capillary net-work filled with uncoagulated blood. The white corpuscles are more abundant in these than in the axillary vein. No parasites nor ova could be found in this gelatinoid material.

Lungs: Mostly healthy. Isolated lobules and at certain points a few adjacent ones are infarcted and solid, and all such have their bronchia filled with worms (*Strongylus elongatus*) and a thick mucous. The plugged bronchia are mostly dilated, and on the mucous membrane of one such is a white patch about a line in diameter, resembling the sloughs on the intestines, but not so thick.

EXPERIMENT NO. 8.

White pig, eight weeks old; common breed. Formerly fed raw offal.

Date.	Hour.	Temperature of body.	Remarks.
Sept. 30	3 p. m	104° F.	Just come a mile in a wagon.
Oct. 1	9 a. m........	103	
1	6 p. m.........	103	
2	9.30 a. m	101. 5	
3	9 a. m	101	
4	(*)	(*)	
5	4 p. m........	98. 75	Pigs in next two pens inoculated. Was found between door and bars, where it could not move.
6	5 p. m	99	Again between door and bars.
7	11 a. m	99	Costive.
8	12 noon	101	
9	11 a. m......	104. 5	Still very costive.
10	5 p. m	102 75	Bowels natural.
11	10 a. m......	102. 5	
12	4 p. m........	103. 25	
13	12 noon	103	Feces fetid.
14	4 p. m	104	
15	10 a. m.......	105	Lame in right fore limb.
16	....do	104. 25	
17	....do	103. 5	
18	....do	101. 5	Scours. Feces fetid.
19	....do	102. 5	
20	....do	103. 75	Pigs in adjacent pens reinoculated.
21	....do	104. 25	
22	....do	103. 5	
23	....do	103. 75	
24	....do	103	
25	....do	103	Placed in new pen, with infected pen on each side.
26	....do	103	
27	9 a. m........	103	
28	10 a. m.......	103	
29	9.30 a. m	105. 3	Slight cutaneous rash
30	2 p. m........	104	Lively.
31	9 a. m........	104. 2	No skin eruption.
Nov. 1	10 a. m.......	105. 75	Still looks well.
3	9 a. m........	104. 8	Stiff in hind limbs.
3	6 p. m........	104	
4	9 a. m........	103	Placed in pen just vacated by dead pig.
5	9.30 a. m	101	
6	10 a. m.......	103. 5	
7	....do	102. 73	
8	....do	102. 6	Dull; no appetite; skin covered with black spots one-third to one line in diameter. Right ear has purple spots. Killed by bleeding.

* No observation.

Post-mortem examination.—Skin: Nearly covered with black spots from one-third to one line in diameter, consisting of minute sloughs of epidermis, infiltrated and discolored with blood. In a number of these the subjacent layers of true skin are congested, and even the seat of microscopic extravasations of blood, while in some cases the black necrotic cuticle is covered by a dried crust of exuded lymph of a dark brown color.

The right ear is of a deep purple color, and purple patches of various sizes are found inside forearms and thighs, on the hocks, and beneath the chest. In these purple patches the true skin is the seat of extensive congestion with stagnation and coagulation of the blood in many of the capillaries, and numerous microscopic clots of extravasated blood, while all the tissues are stained with hæmatine.

Blood: That from the jugular is very dark and forms slowly a soft diffluent clot: red globules round and large. That from the carotid is crimson, and clots quickly and firmly; red globules crenate, small and shrunken. Blood from both vessels is slightly alkaline.

Tongue: On the posterior third of the right border is a purple spot one-half line in diameter, which cannot be effaced by pressure. Under the microscope this shows the same congestion and microscopic extravasations with the spots on the skin. The conical papillæ on the upper surface of the organ near its base have their tips of a very deep purplish red.

Larynx: There is purple punctiform discoloration on the posterior surface of the epiglottis, which cannot be removed by pressure.

Lymphatic glands: Those around the throat are deeply stained with blood, some only superficially and some throughout. This is true also of the glands of the chest, groin, and abdomen, but especially of the mesentery. In several cases the glands appear to be enlarged. Microscopically, they present congested capillaries filled with coagulated blood, minute extravasations, and a profusion of granules and granular cells.

Abdomen—parasites in peritoneum: Two *hydatids* were found respectively in the omentum and mesentery.

Stomach: Well filled; great curvature of a deep dark red; contents strongly acid.

Small intestine: Congested in some parts, but with no observed extravasation nor deep discoloration; contents not abundant, but at intervals stained of a deep biliary yellow, and with excess of mucus throughout.

Ilio-cæcal valve: With Peyer's follicles dilated, and contents in some slightly yellowish.

Cæcum: Close to the ilio-cæcal valve a considerable erosion, with raised center and margin, but no excess of vascularity.

Colon: Six inches from the cæcum is a sloughing ulcer, one and one-half lines in diameter, raised above the adjacent membrane, the superficial layers being of a dirty white color in the center, and non-vascular, while around the margin of the ulcer is no marked redness.

Liver, colon, and rectum: Several extravasation patches averaging one line in diameter, bright red, and evidently quite recent.

No intestinal parasites.

Liver: Firm; solid; considerable portions are of a deep purple hue, the deep coloration being mostly confined to the center of the acini.

Kidneys: Cortical portion soft and of a very light brown, almost parboiled, appearance. Papillæ and medullary parts of a very deep red.

Muscles: Contained no parasites.

Brain: Normal.

EXPERIMENT No. 9.

Female pig, eight weeks old; breed, Chester White.

Date.	Hour.	Temperature of body.	Remarks.
Nov. 5	9.30 a. m	103. 75? F.	
6	10 a. m	103. 75	
7	....do	103. 75	Inoculated with part of small intestine of pig that died November 4, the virulent product having first been brought for five minutes in contact with a solution of chloride of lime (: 1 :: 500).
8	....do	100. 75	
9	....do	101	
10	....do	104	
11	....do	105	
12	....do	105	Costive.
13	....do	104	Bowels loose.
14	....do	103. 8	
15	....do	104. 6	
16	....do	104. 75	
17	....do	104	Scours.
18	....do	105	
19	....do	105	
20	....do	105	Skin hot.
21	....do	106	Killed by bleeding.

Post-mortem examination, November 21, 11 *a. m.*—Body in good condition.

Skin: Almost devoid of eruption. The ears alone present increased vascularity, with a moderate blush and excess of scurf.

Digestive organs: Natural above the stomach. Guttural lymphatic glands in part congested and the seat of microscopic blood extravasations. Stomach mottled of a deep brown for a span of two and one-half inches by three inches along the mucous membrane, covering its greater curvature. Contents abundant, intensely acid, and fumes with ammonia.

Duodenum: Bears a small erosion near the pylorus.

Jejunum and ilium: Have patches of congestion and microscopic extravasation at intervals.

Ilio-cæcal valve: Has its edges thickened and of a dark bluish gray. Many follicles in Peyer's patch covering the valve are distended with a yellowish-white product, but there is no extra vascularity nor erosion.

Cæcum, colon, and rectum: Bear at intervals patches of congestion and microscopic extravasation in the mucous and submucous layers, over which the epithelial layer is softened and easily detached. No ulcers are found.

Liver: Discolored in parts by blue punctiform spots involving individual acini or several adjacent ones. Toward the lower margin of the gland the deep redness is mostly confined to the center of the acini.

Spleen: Seems large, but not unduly gorged with blood nor softened.

Pancreas: Healthy.

Kidneys: Pale in their cortical part, present punctiform petechiæ on the medullary portion and papillæ.

Bladder: Empty and normal. Ovaries and womb sound.

The mesenteric, sublumbar, and inguinal lymphatic glands appeared enlarged and more or less stained, of a deep blood-red color.

Parasites in the abdomen: Two ascarides in the small intestine; one *tricocephalus* in the cæcum.

Lungs: Present numerous congested lobules varying in color from brownish pink to a dark purple (almost black). The bronchia leading to these lobules are pervious and without parasites. The congested lobules seem less solid than when worms have been present.

Heart and pericardium: Normal.

Brain: Sound. Dura mater bears four patches of extravasation on the right side near the vertex. The average breadth of these is one line.

Spinal cord: Sound; subarachnoid fluid, about two drachms.

EXPERIMENT No. 10.

White male pig, eight weeks old; breed, Chester White; condition, fine.

Date.	Hour.	Temperature of body.	Remarks.
Nov. 4	12 noon	....	⎧ Inoculated with mucus and congested and softened mucous‑
5	9.30 a. m	104.75° F.	⎨ membrane of the small intestines of No. —, found dead this
6	10 a. m......	103.75	⎩ morning.
7	....do	103.8	
8	....do	103.75	
9	....do	102.5	
10	....do	104.5	
11	....do	103.5	
12	....do	104	Ears red.
13	....do	104.5	
14	....do	105	
15	....do	105.1	Losing condit...n. The skin shows the customary black necrotic spots of epidermis. Ears blue at edges.
15	3 p. m.......	103.5	Respiration 36. Killed by bleeding.

Post-mortem examination.—*Skin:* Slight eruption on the ears and blueness on the margins.

Digestive organs: No lesions in the mouth or pharynx.

Pharyngeal lymphatic glands: Stained of a deep blood-red color.

Stomach: Well filled with food. Contents strongly acid. On the great curvature a space of two and one-half inches square has a brownish mottled discoloration, and numerous deeper brownish markings, as if from altered hæmatine.

Small intestine: Epithelium is thick, soft, and easily detached. Contents liquid, with a great excess of mucus. The bowel is reddened and congested around its entire periphery, and for a considerable distance at intervals, the congested portions being mostly empty and contracted.

Ilio-cæcal valve: Peyer's patch, which passes over the valve, has many of its follicles filled up with a yellowish-white matter. The whole patch is swollen, but not very vascular to the naked eye.

Cæcum and colon bear petechiæ: Many solitary glands in the colon are unusually large; some excessively dilated, filled with yellowish matter, and apparently commencing to form ulcers. Spots of congestion scattered over the mucous membrane show minute extravasations when placed under the microscope.

Mesenteric glands: Some unchanged; some stained of a deep blood color. Inguinal glands large.

Kidneys: Normal.

Liver: Is firm and solid. Bears numerous punctiform petechiæ on the posterior surface of its right lobe, and a large dark-purple patch on the posterior aspect of its middle lobe.

Gall bladder: Moderately filled with a straw-colored, glutinous bile. Membranes of the bladder unchanged.

Pancreas and spleen: Normal.

Chest—heart: Left ventricle contains petechiæ. Right auricle just above the auriculo-ventricular valve presents a brownish-red spot which, under the microscope, is seen to contain much granular matter in the sub-serous connective tissue.

Lungs: The right has two dark, blood-colored spots on its posterior part. The left shows similar colorations, mostly in lines along the inter-lobular spaces. The bronchia leading to such points contained no parasites nor exudation.

Bronchial lymphatic glands: Normal.

Brain: Normal.

EXPERIMENT No. 11.

White male pig, eight weeks old; breed, Chester White.

Date.	Hour.	Temperature of body.	Remarks.
Nov. 5	9.30 a. m	102.75° F.	
6	10 a. m.......	103	
7	....do	102	Inoculated with small intestine of pig that died November 4 the gut having been fumigated five minutes with sulphurous acid.
8	....do ....|	100.5	
9	....do	100.75	
10	... do	101.75	
11	....do	104.5	
12	....do	102.5	
13	....do ....|	103.5	
14	... do ....|	103.5	
15	....do ....|	103.25	
16	....do	104.75	
17	...do	102.75	Scouring.
18	...do ....|	104.5	Fetid scouring.
19	... do	104.5	
20	....do ....|	105	Feces still soft; unusually fetid; skin hot
21	....do ....|	105	
22	....do ....|	103	
23	....do ....|	103.75	
24	....do	103.3	
25	...do	104	
26	....do ....|	104.25	
27	....do	103	
28	....do	104	
29	....do	103.5	
30	....do	103	
Dec. 1	....do	102.5	Red ears; dull; thriftless.
2	....do	103.2	
3	....do ....|	102.25	
4	...do	100.75	Scours.
4	3 p. m.....	102	
5	9.30 a. m	102.25	
6	....do ....|	102.5	Killed by bleeding.

Post-mortem examination.—Skin: In great part covered by the usual black concretion. Has patches of purple on ears and legs.

Digestive organs: Some deposit exists on the lower surface of the tongue, to the left of the frenum, composed of granular matter and cells having more than one nucleus; evidently the remnant of a small abscess. On the fauces, to the right side, is a purple patch not removed by pressure, extending to an inch in length and a quarter of an inch in breadth.

Pharynx and larynx: Normal.

Stomach: Full; contents moderately acid. Shows the usual brownish discoloration of the mucous membrane covering the great curvature.

Small intestines: Show only a few patches of congestion. The follicles of Peyer's patch just above the ilio-cæcal valve are considerably enlarged.

Large intestines: Show a great many enlarged solitary glands, yet but little congestion. The rectum is much congested and presents two ulcers: one with raised edges and raw, depressed center; the other, with a firm, dirty-white slough in the center.

Mesenteric lymphatic glands: Enlarged and thickly streaked with gray. Those near the ilio-cæcal valve, and those above the rectum, are congested and deeply reddened.

Inguinal glands: Are also greatly enlarged and streaked dark-gray with pigment.

Liver: Of normal consistency and color, excepting some few patches of deep purple. Gall-bladder moderately filled with a yellowish-green, viscid bile.

Pancreas: Healthy.

Spleen: A portion very dark colored (nearly black) extending its whole length and about half its breadth; is evidently gorged with blood; but is not raised above the level of the remaining part.

Kidneys: One contains an acephalocyst in its pelvis. The cortical substance of both is pallid, but no other change is noticeable.

The *lungs, heart,* and *brain* appeared healthy.

EXPERIMENT No. 12.

Male pig, eight weeks old ; breed, Chester White.

Date.	Hour.	Temperature of body.	Remarks.
Nov. 19	10 a. m....	104.5° F.	Costive. Inoculated with blood of sick pig (No. 1) after treating the same with a solution of bromide of ammonia: 1::500.
20	....do	104.75	
21	....do	104.2	
22	....do	104.75	
23	.. do	104.2	
24	....do	103.8	
25	....do	104	
26	....do	104.3	
27	....do	105.75	
28	....do	105.75	
29	....do	105.75	
30	....do	106	
Dec. 1	.. do	106.2	Edges of ears purple. Purple spots on scrotum.
2	....do	106	Right ear a deep purple, bleeding at the point where exudation had formed a black scab.
3	....do	105	
4	....do	105	Ears blue; skin has purple blotches only partially effaceable by pressure. Feces liquid; yellowish white.
4	5 p. m.....	105	
5	10 a. m.....	105	
6	...do	101	Very prostrate; can barely rise.
7	...do	....	Found dead in pen this morning.

Post-mortem examination.—Skin : Of ears, throat, breast, belly, and legs, of a uniform dark purple ; white patches remain inside the forearm and thigh, and along the back, which is covered by a very thick scurf. The discoloration which is due to congestion of capillary vessels, the coagulation of blood within them, and numerous minute extravasations, is confined to the integument. The skin is also abundantly covered with the usual black concretions.

Digestive organs : Tongue blue, but with no abrasions.

Tonsils, fauces, and pharynx: The seat of general congestion and discoloration. Œsophagus has some spots of slight congestion.

Stomach : Distended with solid food; not so strongly acid as in many other cases. Its great curvature has the mucous membrane covered with patches of blood extravasation, such patches standing out in greater part as dark-red clots.

Small intestine: Exceedingly contracted, almost empty, and congested throughout in varying degree, from a simple branching redness, with softening of the mucous membrane and excessive production of mucus, to distinct circumscribed extravasations with decided thickening; in several instances the redness and the thickening is most marked on Peyer's patches. The duodenum contains three *ascarides.* Several small ulcers exist just above the ilio-cæcal valve.

Large intestine: Cæcum remarkably small and contracted. Neither cæcum nor colon contains much ingesta. The mucous membrane along the whole large intestine is inflamed, greatly thickened by exudation, and thrown into prominent circular folds. Its general color is of a dark brownish red, in many points verging upon black. At different points it shows the characteristic ulcers with a firm, dirty, white slough in the center of each, but these have in no case attained a large size, nor any marked thickening nor induration of their base, and without special care in the examination might be easily overlooked. The rectum contains numerous blood extravasations and some considerable ulcers with the central whitish necrosed portions.

Mesenteric glands: Almost universally enlarged and of a deep red, from congestion and extravasation.

Liver: Of a very deep purplish brown, gorged with blood, but not materially softened nor moderately friable. It is especially dark near the margin of the lobes.

Gall-bladder: Moderately full, bile dark green and viscid.

Pancreas: Sound.

Spleen: Enlarged, gorged with blood, and almost black.

Kidneys: Nearly normal as examined externally. Corticle substance of a darker red than in most of the diseased pigs, and the papillæ bear black extravasations, punctiform and up to half a line in breadth. The right kidney contains a small cyst in its pelvis.

Left supra-renal capsule is enlarged to about one-third the size of the kidney, and has a clot of blood and a collection of cheesy matter superposed in its anterior end.

Lungs: Nearly normal; some congestion in the posterior lobes is evidently quite recent, and the cut surface freely exudes a frothy liquid.

Heart: Right ventricle slightly discolored by punctiform petechiæ beneath the endocardium. The great aorta contains a very firm clot, partly buffed.

Blood under a No. 10 Hartnack immersion shows no moving bacteria, but a great excess of granular matter.

EXPERIMENT No. 13.

White female pig, eight weeks old ; breed, Chester White.

Date.	Hour.	Temperature of body.	Remarks.
Nov. 19	10 a. m	105.5° F.	Inoculated with the blood of sick pig No. —, five drops being mixed with a drachm of a watery solution of potassium permanganate (:1 :: 500) and injected.
20	....do ....|	104	
21	....do........	103.25	
22	....do	103	
23	....do	104.75	
24	....do	103.25	
25	....do	104	
26	....do	104.8	
27	....do	104.75	
28	....do	104.5	
29	....do	104.75	
30	....do	105.3	
Dec. 1	....do	105	Deep-red ears; black concretions on skin.
2	....do	105.3	
3	....do	104.25	
4	....do	104.5	Stiff. unsteady gait; humped back; blue ears: costive.
4	5 p. m	103.5	
5	10 a. m	103.5	
6	....do	105	
7	....do	102.5	
8	....do	105	
8	6 p. m	104	
9	9.30 a. m	104	
10	....do	105	
10	4.30 p. m	104.5	
11	9.30 a. m	104	Very dull and quiet.
11	5.30 p. m	103.5	
12	10 a. m	107.75	Very languid and prostrate.
12	5 p. m	107.75	Does not rise when handled ; breathing 28 per minute.
13	11 a. m	107	Feces soft. fetid. yellowish. Pig very prostrate. eats nothing. and scarcely moves when pricked to obtain a drop of blood. Blood contains moving bacteria.
13	5 p. m.......	107	

Pig found dead on the morning of December 14.

Post-mortem examination.—Skin: Blue spots on the belly, legs, rump, perineum, and ears. Free portions of the ears of a dark purple. Pink papillary eruption, and black concretions on the ears.

Digestive organs: Tongue has an ulcer, with slough a little to the left of the tip—size one and a half lines in diameter.

Tonsils and soft palate: The seat of a uniform bluish congestion. Submaxillary lymphatic glands in part reddened and congested.

Gullet: Contains clots of a stringy. fibrinous material.

Stomach: Near the left *cul de sac* is a dirty, yellowish-white false membrane of about one inch square. The great curvature is of a dark-brownish red, with some brighter red spots of more recent blood extravasation.

Small intestines: Nearly empty, though at intervals were round. hard pellets of ingesta. The coats of this bowel were more or less congested, with softening of the membrane at different points.

A large ulcer is forming on the edge of the ilio-cæcal valve, in which the outline of the follicles can still be seen of a yellowish color.

Large intestines: Cæcum and colon congested throughout, but much more at some points than at others. In the upper part of the colon are extensive deposits of false membrane of a dirty yellowish-white color, in places in spots of small size, and in others in extended patches of several inches in length. The cæcum has smaller spots of the same kind. The rectum is very much thickened and of a deep red throughout, the thickening existing mainly in the mucous membrane. It presents, further, nine small ulcers, with the characteristic dirty sloughs in the centers.

Parasites: The cæcum contains one whip-worm.

Liver: In the main firm, but contains bluish patches.

Pancreas: Apparently unchanged.

Spleen: Black, full of blood, but not apparently enlarged.

Mesenteric and sublumbar lymphatic glands: Are almost universally of a dark red, almost black color.

The left kidney: Has a cyst one-half inch in diameter in the anterior part of its pelvis. In common with the right kidney, it also presents numerous black petechia on the medullary portions and papillæ.

Chest and respiratory organs: Larynx shows considerable congestion, especially on the epiglottis and on the arytenoid cartilages.

Pleuræ: Contained an abundant blood-colored liquid exudation, especially in the right sac, where the lung had contracted extensive adhesions by newly-formed false membranes. The liquid effusion contained numerous white and red blood lobules and actively-moving bacteria, which assumed the most varied forms in rapid succession. A loose coagulum forms in the exposed fluid.

Bronchia: Filled with froth having a perceptibly pink tint.

Left lung: Anterior lobes congested and consolidated by recent exudation. Posterior layer lobe sound.

Right lobe: Consolidated throughout; sinks in water; but has not yet become firm, granular, nor friable. The color of this lung varies from a light brick-red to a deep red, approaching black, the darker shades mostly occupying the spaces of connective tissue between the lobules, these spaces being often stretched by the exudation to the breadth of a line or more. On making a section of the lung a considerable pulmonary vein was found to contain a friable granular grayish clot which had evidently existed for some time before death.

Pericardium: Contains a large amount of blood-colored effusion, in which blood-globules and moving bacteria abound. The parietal and visceral layers were connected by loose false membranes. Loose dark clots and some fluid blood existed in the right side of the heart, and spots of extravasation on the walls of the left ventricle.

Lymphatic glands: In the region of the throat are of a very deep red. The same remark applies to the bronchial and subdorsal glands.

Table showing the duration of incubation in different cases.

No.	Inoculated.	Apparently ill.	Duration of incubation, in days.	Percentage on different days.	Remarks.
1	Nov. 19	Nov. 20	1	6. 6″	Inoculated with old blood that had been kept eleven days in an incubator.
2	Oct. 6	Oct. 9	3	....	
3	Oct. 6	Oct. 9	3	20	
4	Nov. 7	Nov. 10	3	....	
5	Oct. 8	Oct. 12	4	} 13. 3	Temperature raised for three days only.
6	Nov. 7	Nov. 11	4		
7	Oct. 5	Oct. 10	5	6. 6	
8	Oct. 8	Oct. 14	6	} 13. 3	
9	Nov. 4	Nov. 10	6		
10	Nov. 3	... do....	7	)	
11	... do......	... do....	7		
12	... do......	... do....	7	} 26. 6	
13	Nov. 19	Nov. 26	7	)	
14	Nov. 19	Nov. 27	8	6. 6	
15	Oct. 8	Oct. 21	13	6. 6	

Table of experiments undertaken to ascertain the relative virulence of the products of Hog Fever after exposure to the air for different periods of time.

No. of experiments.	Nature of inoculated material.	Date of inoculation.	Date of first signs of illness.	Duration of fever.	Died.	Killed.	Recovered.	Remarks.
1	Inoculated from quill charged with diseased lung fluids of a pig which died five days before, in North Carolina.	October 5	October 10	6 days			1	
2	Inoculated from quill charged with pulmonary exudation of a pig that died suddenly in New Jersey, twenty-four hours ago.	October 8	October 14	11 days		1		
3	Inoculated from quill-point dipped in pulmonary exudation of a pig which had been sick for a week or two. Virus from New Jersey.	October 8	October 21	11 days	1			Temperature was abnormally low for two days before death.
4	Inoculated from quill dipped in diseased lung liquids of a pig having the disease of the lungs. Virus from Indiana, and four days from pig.	November 3	November 10	11 days		1		Temperature was abnormally low for two days before death.
5	Pig placed in pen along with diseased intestines (semi-putrid) and manure from diseased pig. Both products from New Jersey, and forty-eight hours from pig.	October 20	No apparent effect.					When inoculated this pig had but barely recovered from the slight effects of a former inoculation.
6	Inoculated with the firm fibrous substance from the lume of an intestinal ulcer. Also fed a portion of the same. Products from New Jersey forty-eight hours from the pig, and slightly putrid.	October 20	No apparent effect.					
7	Inoculated with a firm fibrous substance from an intestinal ulcer. Product sent from New Jersey, forty-eight hours from the pig and slightly putrid.	October 20	Temperature unaffected.			1		A pink rash appeared on the skin October 31.
8	Inoculated with portions of dried intestine from North Carolina, and at least four days from the pig. Product dried in sun and air.	November 3	November 10	18 days		1		Rise of temperature very moderate.
9	Inoculated with intestinal ulcer and mucus from Illinois. Product sent in closely corked bottle, containing liquid; was at least three days from the pig and slightly putrid.	November 3	November 10	18 days		1		Temperature high throughout.
10	Inoculated with mucus and congested soft	November 4	November 10	5 days	1			

	ened mucous membrane from pig No. 14 found dead in its pen the same morning Product taken not more than twelve hours after death.			
11	Inoculated by injecting under the skin one drachm of blood taken from a diseased pig and kept in an incubator for eleven days having had communication with the air only through narrow tubes blocked with cotton-wool. The blood smelt stale, but not putrid.	November 19	November 20	20 days	Maintained a high temperature for twenty days with diarrhea, purple ears, purple blotches on body, pustules of a pink color and black concretions. After this temperature was normal, but diarrhea and skin eruption and discoloration remained for four days, when it was again inoculated with blood swarming with bacteria (moving).

Experiments undertaken to test the effect of disinfectants on virulent products inoculated.

No. of experiments.	Disinfectant.	Nature and treatment of inoculated material.	Date of inoculation.	Date of first signs of illness.	Duration of fever.	Died.	Killed.	Recovered.	Remarks.
12	Bisulphite of soda	Pig inoculated from a quill dipped in the liquids of a diseased lung in North Carolina, five days ago. Before inoculation the quill was placed five minutes in a solution of bisulphite of soda: 1::500.	Oct. 5	No febrile change.		1			Died the seventh day from lung worms.
13	Carbolic acid	Pig inoculated with quill charged like the last, in North Carolina, six days from the pig, and before inoculation dipped for five minutes in a solution of carbolic acid: 1::500.	Oct. 6	Oct. 9	9 days			1	
14	Sulphate of iron	Pig inoculated with quill charged with virus six days ago, in New Jersey, and before inoculation dipped five minutes in a solution of copperas: 1::500.	Oct. 6	Oct. 9	20 days		1		Killed October 31, after two days of abnormally low temperature, and when already in *articulo mortis*.
15	Chloride of zinc	Pig inoculated from a quill charged with virus from the lungs of a New Jersey hog that had been sick for a week or two; virus one day from the pig. Before inoculation was steeped five minutes in a solution of chloride of zinc: 1::500.	Oct. 8	(?)					High temperature on fourth, sixth, and ninth days only.
16	Sulphurous acid	Pig inoculated with diseased intestine three days removed from the sick hog No. 14, and smoked for five minutes with the fumes of burning sulphur.	Nov. 7	Nov. 11	20 days		1		Killed twenty-sixth day.
17	Chloride of lime	Pig inoculated with diseased intestine, three days from the sick hog No. 14, and steeped five minutes in a solution of chloride of lime: 1::500.	Nov. 7	Nov. 10	11 days		1		Fever ran very high.
18	Bromide of ammonium	Pig inoculated with blood from No. 17, a few drops being added to a drachm of a solution of bromide of ammonium (:1::500) and thrown under the skin.	Nov. 19	Nov. 27	8 days	1			
19	Permanganate of potassa	A few drops of blood from No. 17 were mixed with a solution of permanganate of potash and thrown under the skin.	Nov. 19	Nov. 26	18 days	1			Fever ran very high. The day before death the blood contained numerous moving (septic) bacteria.

EXPERIMENT NO. 20.

Experiment undertaken as a test of the propagation of the disease-poison through the air.

October 5.—A pig was placed in a pen between two infected ones, and separated from each only by an impervious double wall of matched boards, with building-paper between. The only means of communication was through the open air by means of ventilators at the front and back of each pen, and the openings of which in adjacent pens were less than a foot apart. On the ninth, tenth, and eleventh days the pig had an elevated temperature and was lame in the right shoulder, the illness being evidently rheumatic.

On October 29th, the twenty-fourth day, the temperature rose 2° and remained at 104° F. and upward for six days (till November 3rd). It then showed a daily diminution, and by November 8th, having attained the natural standard, the pig was destroyed.

Experiments on sheep, rabbit, and dog. Inoculation with fresh virulent pig's blood, containing moving bacteria.

Subject.	Date of inoculation.	First signs of illness.	Duration of incubation.	Duration of illness.	Remarks.
Merino wether..............	Dec. 14	Dec. 15	1	......	
Adult female rabbit..........	Dec. 14	Dec. 15	1	......	
Newfoundland puppy, seven weeks old.	Dec. 14	Dec. 15	1	1	Temperature rose 2.25°, but was normal on the second day.
Female pig, twelve weeks old.	Dec. 14	....	......	......	Purged actively for three days. When inoculated the pig was in advanced non-febrile stage of the fever, and the temperature did not rise above the normal.

Inoculation with fresh virulent blood in which no moving bacteria had been observed.

Subject.	Date of inoculation.	Remarks.
Merino wether..............	Nov. 21	Scouring and rise of temperature 1° on fourth and sixth days only.
Do.....................	Dec. 7	No appreciable effect.
Adult female rabbit..........	Nov. 21	Do.
Do.....................	Dec. 7	Temperature rose 1° on the first day only.

Table showing relation of body-temperature to the weather during October.

Date	Thermometer in open air.				Rain and snow.	Barometer.				Humidity.						Temperature of pigs.					
										7 a.m.		2 p.m.		9 p.m.		Healthy.			Sick.		
	7 a.m.	2 p.m.	9 p.m.	Mean.		7 a.m.	2 p.m.	9 p.m.	Mean.	Dry.	Wet.	Dry.	Wet.	Dry.	Wet.	Rose.	Fell.	Un ch'nged.	Rose.	Fell.	Un ch'nged.
Oct. 1	58.0	71.7	62.7	64.13		29.336	29.262	29.204		580	555	717	635	627	562		8				
2	60.8	67.5	60.1	62.80		29.340	29.134	29.126		618	573	675	630	601	580		8				
3	50.6	71.2	58.5	60.10		29.242	29.144	29.204		506	493	712	573	585	560		8				
4	45.6	63.8	57.5	55.63		29.252	29.192	29.302		456	435	638	528	575	521		1				
5	44.3	55.8	51.6	50.56		29.354	29.254	29.248		443	418	558	495	516	501	7	3			2	
6	50.8	53.8	48.8	49.46		29.208	29.318	29.418		508	475	598	468	438	438	7	1		4	1	
7	56.6	56.3	51.0	47.96	Three-hours snow, in. .11.	29.418	29.278	29.290		366	341	563	477	510	490	2	1		2	2	
8	50.6	62.6	60.9	58.03	One hour's rain, in. .02.	29.302	29.328	29.140		506	484	626	525	609	590	2	2			3	
9	62.6	66.3	56.5	61.80		29.076	28.940	29.156		626	565	663	615	565	550	1	1	1		3	
10	48.2	55.0	40.6	47.93		29.256	29.352	29.302		482	458	550	458	406	392				5	4	
11	37.4	56.8	51.5	48.56	Two hours' rain.	29.250	29.152	29.186		374	365	568	493	515	486				3	2	
12	49.4	55.6	45.6	46.5		29.244	29.226	29.248		494	406	556	465	465	457	1			2	4	
13	40.3	51.8	52.8	51.63		29.334	29.286	29.368		403	390	618	514	528	492	1			1	2	
14	54.7	71.2	63.2	63.03		29.358	29.264	29.272		547	495	712	568	632	615				4	4	
15	60.0	74.2	64.2	61.30		29.244	29.186	29.210		603	534	742	638	672	600				2	2	
16	61.0	72.5	67.2	68.30	Two hours' rain.	29.298	29.196	29.182		630	580	775	625	672	563				1	4	
17	62.4	73.8	60.4	66.20	Two hours' rain.	29.170	29.030	29.092		624	580	758	621	604	590				2	3	
18	47.5	40.2	40.8	42.83	Two hours' rain, in. 1.2.	29.104	29.104	28.940		475	455	402	396	402	402			2	3	3	
19	42.5	40.2	43.5	42.06		28.872	29.026	29.104		425	420	402	385	435	422				1	5	1
20	46.8	55.8	49.9	51.30		29.154	29.146	29.210		488	432	558	448	490	465	1	1	1	5	3	
21	41.7	64.5	58.2	53.20		29.376	29.306	29.292		417	390	645	522	535	510	1	1	1	3	3	
22	42.0	67.2	54.2	36.80		29.680	29.176	29.004		490	448	672	546	542	520				3	2	1
23	48.2	49.8	49.8	48.56	Sixteen hours rain, in. 1.80.	28.880	28.864	28.864		462	475	498	497	498	472	1	1	1	4	2	2
24	43.3	51.8	40.2	45.10		29.244	29.338	29.456		433	412	518	453	402	392		1		3	4	
25	33.8	51.8	49.3	44.86		29.538	29.436	29.448		338	325	513	453	495	477		1		1	4	
26	47.8	64.8	61.5	48.03		29.452	29.362	29.270		478	431	648	556	615	590		1		2	2	
27	56.5	49.8	41.8	49.36	Eight hours, in. .47.	29.208	29.186	29.202		565	527	448	480	418	413		1		2	4	
28	37.3	42.5	48.8	38.23		29.310	29.328	29.348		373	335	425	357	383	370				2	2	
29	28.8	47.8	41.3	39.36		29.394	29.284	29.214		286	278	478	394	413	387				3	3	
30	40.7	49.5	47.2	45.66	Sixteen hours, in. .32.	29.032	28.944	28.910		107	395	495	472	413	451				3		
31	41.6	48.6	52.5	30.23		28.888	28.866	29.068		416	386	436	383	325	321				3	3	
Mean for month				52.67																	

Table showing relation of body-temperature to the weather for November.

Date	Thermometer in open air — 7 a.m.	2 p.m.	9 p.m.	Mean	Rain and snow	Barometer — 7 a.m.	2 p.m.	9 p.m.	Mean	Humidity — 7 a.m. Dry	Wet	2 p.m. Dry	Wet	9 p.m. Dry	Wet	Temperature of pigs — Rose	Fell	Unchanged
Nov. 1	32.8	42.8	39.5	38.36		29.210	29.142	29.186		328	280	428	368	385	370	21	21	1
2	36.5	32.8	42.0	45.70		29.214	29.100	29.260		365	325	528	430	420	395	1		
3	34.3	36.0	35.0	55.30		29.466	29.456	29.390		343	311	366	317	350	335	1	4	
4	34.2	32.2	29.2	31.96		29.322	29.344	29.342		342	313	325	272	292	271	2	3	
5	26.8	33.2	32.5	30.73		29.438	29.358	29.214		265	258	332	287	325	312	1	3	
6	29.5	35.2	34.0	32.96		29.072	28.946	29.050		292	287	355	342	342	328	2		
7	31.5	37.2	37.0	33.63	Three hours	29.008	28.838	28.866		315	303	372	334	325	315	2		2
8	55.8	35.2	29.4	28.96	Four hours	28.944	29.058	29.110		283	278	292	285	294	272	1	6	1
9	29.8	42.3	36.5	31.81	Snow	29.190	29.070	29.084		298	294	352	312	365	357	1	4	1
10	39.1	50.5	43.7	41.70	Snow 2.5 inches	29.000	29.054	29.020		391	362	423	397	437	402	2		
11	45.8	49.5	43.8	45.43	Snow	28.976	28.732	28.722		405	395	500	432	458	438	1	2	2
12	46.4	40.6	44.0	46.70	Snow .02 inch	29.656	28.750	28.786		464	451	495	422	412	430	4	2	1
13	41.4	36.6	34.0	28.33		28.784	28.998	28.500		412	340	408	348	342	306	2	2	
14	21.4	40.8	28.5	22.06	Five hours	29.436	29.500	29.520		314	308	366	325	282	273	1	2	1
15	33.5	43.5	43.5	32.36	Three hours .18 inch	29.642	29.368	29.544		235	226	408	362	325	317	4	3	2
16	36.5	51.8	46.3	49.20	Four hours .07 inch	29.522	29.484	29.438		362	341	455	410	432	432	1	1	1
17	47.5	47.6	46.8	46.73		29.376	29.278	29.156		475	462	518	505	483	476	3	4	
18	45.8	48.3	47.7	47.06	Rain .26 inch	29.214	29.160	29.170		498	450	476	475	468	465	2	4	
19	45.4	49.8	47.0	47.06	Rain .07 inch	28.454	28.892	29.010		454	434	483	466	475	466	1	4	2
20	43.8	37.5	46.7	47.30	Rain .98 inch	29.010	28.972	28.920		462	438	483	456	407	460	2	5	2
21	43.6	37.5	36.3	40.16	Rain .93 inch	28.682	28.364	28.783		435	430	498	455	476	470	1	1	1
22	37.8	43.6	35.6	36.96	Rain .04 inch	28.372	28.456	28.302		468	438	372	264	365	358	3		
23	38.7	43.2	32.0	36.43		28.920	28.636	28.670		378	382	436	405	356	343	2	4	1
24	39.2	34.5	36.7	39.56	Twelve hours .02 inch	29.056	28.946	29.138		397	376	432	435	320	315	1	4	2
25	32.8	35.0	32.8	33.36	Ten hours .46 inch	29.204	28.372	29.170		392	218	345	305	363	305	5	1	1
26	28.2	38.2	37.2	33.46		28.634	28.948	29.388		328	270	350	336	328	305	1		
27	28.2	41.7	57.0	39.13		29.208	28.850	29.004		282	391	382	343	372	367	7	3	2
28	37.6	38.6	50.2	40.16		29.208	29.184	29.670		422	342	437	367	370	338	2	4	1
29	37.6			36.30		29.420	29.472	29.270		376			333	392	380	3	4	1
30	56.4		33.6					29.406		364	335	366	333	336	321	3	1	1
Mean for the month				38.76														

7 SW

SUPPLEMENTAL REPORT.

As an addendum to my former report, I would respectfully submit the following further observations on the fever of swine, commonly known as hog cholera:

EXPERIMENTS IN FEEDING THE VIRULENT MATTER.

A healthy pig was fed the substance of an intestinal ulcer and a little manure from the same bowel. but showed no evil results for fourteen days, when it was put to other uses. It should be added that the ulcer fed to this pig was partially putrid, and was inoculated on two other swine without success.

A second pig was fed a portion of dried intestine and its contents, both of which had remained packed in wheat-bran for a month. Notwithstanding this, the animal retained good health for seventeen days, when it, too, was put to other uses. The material fed to this pig acted with fatal effect on two other pigs on which it was inoculated.

These experiments can only be taken as showing that a small quantity of poison may pass through the intestinal canal with impunity, but they would not warrant the conclusion that similar materials would be equally harmless when taken in larger quantities and with every meal, as invariably happens when swine are fed in the ordinary manner and plunge their filthy feet and noses fresh from the pestiferous manure into the feeding-trough. Dr. Osler has succeeded in developing the disease by feeding the diseased intestine, but as the feeding was accomplished by force there is just the possibility of abrasion and direct inoculation. Abrasions are indeed so common in the mouth from injuries by the teeth and by hard objects masticated and derangements of the epithelial covering of the mucous membrane of the stomach and intestines, are so frequent in connection with slight gastro-intestinal disorders, that it is needless to calculate on an immunity which can only be secured by the entire absence of such lesions. If to secure immunity in feeding we must provide that not even a worm shall bite the mucous membrane of the stomach or intestine, any guarantee rests on an exceedingly slender basis and had best be rejected at once.

SUCCESSFUL INOCULATION WITH FROZEN PRODUCTS OF THE DISEASE.

In two cases I have successfully inoculated virulent products which had been frozen hard for one and two days respectively. In both instances the resulting disease was of a very violent type, and would assuredly have proved fatal if left to run its course. The freezing had certainly failed to impair the virulence: it had rather sealed it up to be opened and given free course on the occurrence of a thaw; for, once it is frozen, it is manifest that no further change could take place until it was again thawed out, and if it was preserved for one night unchanged in its potency, it would be equally unaffected after the lapse of many months, provided its liquids had remained in the same crystalline condition throughout. In this way undoubtedly the virus is often preserved through the winter in pens and yards, as well as in cars and other conveyances, to break forth anew with returning spring. This is precisely what we find to be the case with the other fatal animal plagues, the virus of rinderpest, lung fever, anthrax, and aphthous fever. being often bound up through the winter with frozen manure to reappear with undiminished power on the access of warmer weather. This is a matter of no small moment inasmuch as the long-continued frosts of our Northern States prevent any such destruction of the poison as takes place so readily in summer in connection with the alternate wetting and drying and the resulting putrefaction.

I have had instances brought under my notice in which, after the prevalence of the fever in a herd in early summer, new swine were introduced into the open yard a month or two after all trace of the disease had disappeared and had continued to preserve the most perfect health. This is quite in keeping, too, with my failure in the attempts to convey the disease by feeding and inoculating with a semi-putrid intestine. It serves, moreover, to explain my failure, as the exposure and wet at a moderately high temperature would lead in both cases alike to decomposition and destruction.

The bearing of this upon the prevention of the disease is self-evident. Infected yards and other open and uncovered places may be considered safe after two months' vacation in summer, provided that sufficient rain has fallen in the interval to insure the soaking and putrid decomposition of all organic matter near the surface, and that there are no great accumulations of manure, straw, hay, or other material in which the virus may be preserved dry and infecting. In winter, on the other hand. the yard or other open infected place may prove non-infecting for weeks and months, and yet retain the virus in readiness for a new and deadly career as soon as a thaw sets in. Safety in such circumstances is contingent on a disuse of the premises so long as the frost continues and for at least one month thereafter. Even during the continuance of frost such places are dangerous, as the heat of the animals' bodies or of the rays of the sun at mid-day may suffice to set the virus free. Again, while they are especially dangerous on the accession of warmer weather, yet, when once the temperature has risen permanently above the freezing point, we may count upon the rapid putrefaction that ensues in all organic bodies that have been frozen and on a disinfection almost as speedy, and it may be at times even more speedy than in the extreme heat of summer. The course of safety is to hold all places that have been infected in late autumn or during winter as still infected until one or two months after the frost has gone out of the ground in spring.

This, of course, has little bearing upon the question of covered pens, barns, cars, &c., in which the poison may be preserved dry, active, and accessible in winter and summer alike. On this question of infection through pens in winter I instituted the following experiment:

CONTAGION FROM AN INFECTED PEN.

A healthy pig was placed in a pen from which a sick one had been removed thirteen days before. The pen had been swept out, but subjected to no disinfection other than the free circulation of air; and as the pig was placed in the pen on December 19, all moist objects had been frozen during the time the apartment had stood empty. The pig died on the fifteenth day without having shown any rise of temperature, but with *post mortem* lesions that showed the operation of the poison. This case was an example of the rapidly fatal action of the disease, the poison having fallen with prostrating effect on vital organs—the lungs and brain—and cut life short before there was time for the full development of all the other lesions. It sufficiently demonstrates the preservation of the poison in covered buildings at a temperature below the freezing point.

SUCCESSFUL INOCULATION OF PIGS WITH VIRUS THAT HAD BEEN KEPT FOR A MONTH IN DRY WHEAT-BRAN.

Appended will be found the daily record of two pigs infected by inoculation with bowel ingesta and mucous membrane that had been preserved for a month in dry wheat-bran. In both cases the disease followed the inoculations promptly and ran a severe course, one case proving fatal, while in the other death was anticipated by killing the animal. At the autopsies the usual characteristic lesions were found.

Here, as in the case of the virus preserved on quill-tips, we find the poison preserved without the slightest impairment of its potency. Thus two series of inoculations with dried virus show how careful and thorough must be the disinfection in dry seasons, and indoors in all seasons, and the importance of the destruction by fire, or in other certain manner, of all dry fodder and litter in which the poison may have been secreted.

COHABITATION WITH SICK PIGS IN DIFFERENT STAGES OF THE DISEASE.

A healthy pig was inclosed in a pen with a sick one which had been inoculated with virulent blood on two occasions; the first thirty days and the last five days before. After the first inoculation the pig had suffered from a slight fever and the characteristic phenomena of the disease. Before the second inoculation the temperature had been normal for eight days, and it was not materially affected by the operation. In short, the disease had manifestly spent itself in the system of the pig, though it had left it a most shrunken, emaciated, and wretched spectacle.

The two pigs occupied the same pen, lay on the same bed, and fed from the same trough for sixteen days, during which no unequivocal sign of disease was manifested in the healthy pig. It seemed indeed to have successfully resisted the contagion.

It was now removed to another pen and placed in company with a pig in which the disease had just reached its height. On the twelfth day thereafter its temperature permanently rose, and it passed through a sharp attack from which it is now recovering.

This seems to show that the poison is much less virulent after the febrile stage of the malady has passed, and that the danger from the recuperating animal decreases with advancing convalescence. At the same time it must not be too hastily concluded that a mild form of the disease did not exist in this pig during the occupancy of the first pen. It appears unquestionable that the poison may be present in the system, and yet give rise to so little disorder that the most careful observer would fail to detect anything amiss.

OCCULT FORMS OF THE DISEASE.

On *post-mortem* sections I have found the characteristic lesions of the bowels and lymphatic glands, in cases where no cutaneous rash or discoloration, no rise of temperature, no loathing of food, nor constitutional disorder had betrayed its presence during life. The occurrence of such slight and occult forms of the disease must present a serious obstacle to all attempts to stamp it out. In most of the plagues of animals' and notably in lung fever, in aphthous fever, and in rinderpest out of its native home, the rise of the body temperature precedes all outward manifestations of the disease. In these affections the indications of the thermometer alone enable us to separate the sick and healthy before the disease has attained to a stage of material danger to their fellows. But in the pig fever the earliest symptoms will vary according to the vagaries of the poison and its primary seat of election. Perhaps the most common initial symptom is the enlargement of the inguinal glands, but it may be some derangement of the digestive organs, or it may be the elevation of the body temperature, or it may be the appearance of red spots or blotches on the skin, or finally the poison may be operating in the system in the absence of all external manifestations. It is noticeable that since the access of extremely cold weather the cutaneous discoloration has been much less extensive than during the warmer season. Even when the temperature has been abnormally raised it will rise and fall in such an irregular manner that no single observation will be always successful in detecting the disease. To detect such cases the investigation must be conducted from day to day, and in view of all possible manifestations of the disease, to be successful. Then again the temperature, even in health, varies widely in different swine and under different conditions of life, so that a knowledge of the body heat of the individual in the existing environment is essential to the drawing of sound deductions from thermometric indications.

INFECTION OF OTHER ANIMALS THAN SWINE.

I consider the most important part of my researches to be that which demonstrates the susceptibility of other animals than swine to the fever we are investigating. Dr. Kline of London, England, claimed, nearly a year ago, that he had conveyed the disease "with difficulty" to rabbits, Guinea-pigs, and mice, but he gives no hint as to whether he had subjected the question to the crucial test of reinoculation from these animals back upon the pig. This test it seemed very important to apply, so that the identity or otherwise of the two diseases might be determined. I have accordingly instituted experiments on a rabbit, two sheep, a rat, and a puppy, the three former of which have turned out successfully.

INFECTION OF A RABBIT FROM A SICK PIG.

After two inoculations with questionable results, made with the blood of sick pigs, in which microzymes had been observed, a rabbit was once more inoculated, this time with the pleural effusion of a pig that had died during the previous night, and in which were numerous actively moving bacteria. Next day the rabbit was very feverish and ill, and continued so for twenty-two days, when it was killed and showed lesions in many respects resembling those of the sick pigs. The blood of the sick rabbit contained active microzymes like those of the pig.

SUCCESSFUL INOCULATIONS FROM THE SICK RABBIT.

On the fourth day of sickness the blood of the rabbit containing bacteria was inoculated on a healthy pig, but for fifteen days the pig showed no signs of illness. It was then reinoculated, but this time with the discharge of an open sore which had formed over an engorgement in the groin of a rabbit. Illness set in on the third day and continued for ten days, when the pig was destroyed and found to present the lesions of the fever in a moderate degree.

A second pig, inoculated with the frozen matter which had been taken from the open sore in the rabbit's groin, sickened on the thirteenth day and remained ill for six days, when an imminent death was anticipated by destroying the animal. During life and after death it presented the phenomena of the plague in a very violent form.

It can no longer be doubted, therefore, that the rabbit is itself a victim of this disease, and that the poison can be reproduced and multiplied in the body of this rodent and conveyed back with undiminished virulence to the pig. We may follow Dr. Kline in according a similar sad capacity to the other rodents, mice and Guinea-pigs. The rabbit, and still more the mouse, is a frequent visitant of the hog-pens and yards, where it eats from the same feeding-troughs with the pig, hides under the same litter, and runs constant risk of infection. Once infected they may carry the disease as widely as their wild wanderings may lead them, and communicate it to other herds at a considerable distance. Their weakness and inability to escape, in severe attacks of the disease, will make them an easy prey to the omnivorous hog, and thus sick and dead alike will be devoured by the doomed swine.

PROBABLE SUSCEPTIBILITY OF OTHER RODENTS.

The infection of these rodents creates the strongest presumption that other genera of the same family may also contract the disease, and by virtue of an even closer relation to the pigs may succeed in conveying the malady to distant herds. The rat is at once suggested to the mind as being almost ubiquitous in piggeries, as feeding in common with the swine, as liable to be devoured by the hog when sick or dead, as given to wandering from place to place, and as possessed of a vicious habit of gnawing the feet and other parts of his porcine companion, and 'thus unconsciously inoculating him.

I have up to the present time had the opportunity of inoculating but one rat with the hog-poison. Unfortunately my subject died on the second day thereafter, the body showing some suspicious lesions, namely, congested lungs with considerable interlobular exudation, congested small intestines, dried-up contents of the large intestines, and sanguinous discoloration of the tail from the seat of inoculation to the tip.

INOCULATIONS FROM THE RAT.

With the fresh congested small intestine of the rat I inoculated one pig, and with the frozen intestine one day later I inoculated a second. The first had no appreciable rise of temperature, loss of appetite, nor digestive disorder, but on the sixth day pink and violet eruptions, the size of a pin's head and upward, appeared on teats and belly, and on the tenth day there was a manifest enlargement of the inguinal glands. From what I had seen of the occult forms of the disease I was led to the opinion that this was one of them. Unfortunately, I had at the time no healthy pig available for the crucial test of reinoculation.

In the second pig, inoculated with the frozen intestine, the symptoms were too obscure to be of any real value. As soon as I obtain a supply of rats I propose to subject this question to a further investigation.

SUCCESSFUL INOCULATION OF SHEEP.

Less significant than the infection of rats, yet of immense practical importance, is the susceptibility of sheep to the hog-fever. I have experimented on two sheep of different ages, an adult merino wether and a cross-breed lamb, and in both cases have succeeded in transmitting the disease.

INFECTION OF THE MERINO.

This sheep was inoculated by hypodermic injections of one and a half drachms of blood from a pig just killed. On the fourth day he had elevated temperature, and on the sixth scouring and snuffling breathing, but the symptoms rapidly subsided. On the fourteenth day he had an injection of two drachms more of blood from a sick pig, and on the twenty-first day of one drachm of blood and pleural fluid containing multitudes of bacteria. Next day the temperature was raised and the snuffling breathing reappeared, both symptoms continuing for some time. On the sixth day his blood was found to contain moving bacteria similar to those present in the injected blood. On the twenty-third day from the last inoculation he was reinoculated, this time with the scurf from the ear of a sick pig. This was followed by no rise of temperature, but there existed much irritation of the bowels with redness and swelling of the anus, occasional diarrhea, and the passage of an excess of mucus, sometimes stained with blood. Seventeen days after the last inoculation he had another hypodermic injection of one drachm of blood and pleural fluid from a pig just killed. As before, this led to an extensive rise of temperature while the intestinal catarrh continued.

INFECTION OF THE LAMB.

The lamb was first injected with a saline solution of the scurf and cutaneous exudation from the ear of a sick pig. There followed a slight rise of temperature, a scurfy eruption on the ears and oozing of blood from different points on their surface, so as to form dark red scales.

On the sixth day following it was reinoculated by the hypodermic injection of one drachm of pleural fluid from a pig just killed, the fluid containing an abundance of moving bacteria. Next day there was extreme rise of temperature, some dullness and swelling in the right axilla, but appetite and rumination were not altogether lost nor suspended. On the fifth day there was tenderness and unusual contraction of the rectum with the passage of bloody mucus, and on the eighth day profuse diarrhea with the passage of much mucus.

SUCCESSFUL INOCULATION OF A PIG FROM THE SICK SHEEP.

A healthy pig was inoculated with mucus from the anus of the wether, and showed a slight deviation of temperature for five days, but without any other marked symptom of illness. Eleven days later it was reinoculated with scab from the ear of the lamb, and again three days later with anal mucus from the sheep. The day before this last inoculation it was noted that the inguinal glands were much enlarged, and six days after the temperature was elevated, and purple spots appeared on the belly. This fever temperature has lasted but a few days up to the present time, but, taken along with the violent rash and the enlarged lymphatic glands, it furnishes satisfactory evidence of the disease. We can therefore affirm of the sheep as of the rabbit that not only is it subject to this disease, but that it can multiply the poison in its system and transmit it back to the pig.

Two other pigs have been inoculated from the lamb, but during the few days that have elapsed they have shown no outward symptoms.

UNSUCCESSFUL INOCULATION OF A PUPPY.

A drachm of blood and pleural fluid containing bacteria, from a pig just dead, was injected hypodermically on the side of a Newfoundland puppy. Next day she was very dull and careless of food, while her temperature was abnormally high. The third day the heat of the body was natural, and a fair amount of liveliness had returned. A few days later a large abscess appeared on the seat of inoculation, discharged and healed, and from this time the health seemed to be re-established.

SIGNIFICANCE OF THE INFECTION OF RODENTS AND SHEEP.

Many will, no doubt, be startled at the above developments, and inquire, half incredulously, How is it that the susceptibility of these animals to this affection has never been noticed before? It may even be suspected that we have been mistaken as to the identity of the disease, and that we may be dealing with the *malignant anthrax* (*bloody murrain*) rather than the specific fever of swine. But a slight attention to the phenomena and *post-mortem* lesions of our cases will speedily dispel the doubt. *Malignant anthrax* is more fatal to sheep and rabbits than to the other domestic animals, whereas in my sheep the disease was so mild that its very existence would almost certainly have been overlooked in the ordinary management of a flock, and it was only detected in these cases by the careful thermometric and other observations made day by day on the inoculated animals. In the rabbit the disease was more severe, and would undoubtedly have proved fatal if left to itself, yet even in this animal there was no indication of the rapid course and speedy destruction which characterize the *malignant anthrax*. Again, although in both diseases alike, the lymphatic glands are the seat of morbid enlargement, yet the increase and engorgement of the spleen which are so constant and so characteristic in *malignant anthrax* were altogether absent in my pigs infected from the rabbit. Moreover the disease in the pigs ran the usual comparatively slow course of the pig-fever, rather than the speedily fatal one of the *anthrax* affection. In the inoculated pigs, too, the combined lesions of the skin, lungs, bowels, and lymphatic glands are unquestionably those of the swine-plague, and not those of *malignant anthrax*.

It is not surprising that the disease should have been hitherto unrecognized in the sheep and rabbit. The most obvious symptoms in pigs—the pink, purple, violet, or black spots and patches of the skin—were never observed in these animals, unless we can consider the eruption on the ears of the lamb as of this nature. In the sheep, to which alone much attention would be paid, the constitutional disturbance was so slight as to be easily overlooked, the appetite even, and rumination scarcely suffering for a day.

Again, the failure to recognize the identity of a disease in two different genera of animals is familiar to all who have made a study of comparative pathology. Cow-pox and horse-pox have existed in all historic ages, but it remained for the immortal Jenner to recognize and show their identity in the last century. *Malignant anthrax* has prevailed from the time of Moses, yet in all the older veterinary works we find its different forms described as independent diseases—*blain, quarter evil, putrid sore throat, &c.* Even to the present day many cases of this disease occurring in the human subject (malignant pustule) are mistaken for erysipelas (black erysipelas). Glanders in horses seems to have been known to Aristotle, and was familiar to the ancient Greek Zooiatres and Roman Veterinarii, but its identity with the same disease in man was only shown in 1840 by Waldinger, of Vienna. *Asiatic cholera* has prevailed in the East from time immemorial, but it is only in the present century that its identity with cholera in animals has been shown by Indian and European observers.

It is no wonder, therefore, that the mildness of the hog-fever in the sheep should have masked its true nature, and that the universal disregard of the disease of the small rodents should have led us to ignore it in these as well. Now, however, that the truth is forced upon us, we must recognize it in all further attempts to arrest the course of the disease or to exterminate it. The destruction and burial of infected pigs, and the disinfection of the premises where they have been, can no longer be considered a sufficient safeguard. The extermination of rabbits, wild and tame, of Guinea-pigs, of mice, and probably also of rats, within the infected area, will be equally essential. Sheep must be rigidly excluded from the hog inclosures, and if they have gained admittance they must either be destroyed with the pigs, if few and valueless, or they must be shut up in a secluded place, or sent to a safe distance from all hogs until they can be certified as healthy, when they may be disinfected and released. No danger of a fatal extension among sheep is to be apprehended; the disease appears to be as harmless to the sheep as the fatal glanders is to the dog, yet the infected sheep is evidently dangerous to the hog, and must be carefully secluded in all measures for the suppression of the plague.

RECORD OF EXPERIMENTS—No. 1.

Pig of common race, eight weeks old.

Date.	Hour.	Body temperature.	Remarks.
Nov. 19	10 a. m	104° F.	Costive. Inoculated with blood of pig killed November 8, and kept in inoculator in isolation apparatus, communicating with the air only through plugs of cotton-wool. The blood smells stale, not putrid; its cells have disappeared.
20	... do	104. 5	
21	....do	104. 5	
22	....do	105. 2	
23	....do	104. 75	
24	....do	104	
25	....do	104	
26	....do	104. 75	
27	....do	104. 5	
28	....do	105	
29	....do	104. 75	
30	....do	104. 75	
Dec. 1	....do	103. 5	Quite dull. Purple spots and black concretions on the skin.
2	....do	104. 75	Red and black spots on the skin.
3	....do	104. 25	
4	....do	102. 5	Scours. Ears blue and cold.
4	5 p. m......	104	Do.
5	9.30 a. m	104	Do.
6	....do	105	Do.
7	....do	103. 5	
8	....do	103. 5	
8	6 p. m	104	
9	9.30 a. m	103	Bowels continue loose.
10	....do	103. 5	
10	4.30 p. m	104	
11	9.30 a. m	103	
11	6 p. m	103	
12	10 a. m	102. 25	Feces fluid and of a bright yellow color.
12	5 p. m	102. 75	
13	9.30 a. m	102. 75	Quiet; ears deep red; extensive papular eruption and greasy exudation on the skin; scouring.
13	5 p. m.......	102. 5	
14	9 a. m	100. 5	Hypodermic injection of one dram of blood and pleural fluid from pig just dead. Inoculation liquid contains numerous actively moving bacteria.
15	....do	102. 75	Dull; has not eaten supper of last night.
15	5 p. m	102. 75	Scours.
16	10 a. m	102. 25	Do.
16	5 p. m	102. 5	
17	10 a. m	102	
17	4 p. m.......	103. 25	Slightly costive.
18	10 a. m	101	Sebaceous secretion excessive on the inner sides of thighs and forearms, &c. Has a blackish-brown color, and disagreeable but not putrid odor.
18	4 p. m	103. 2	
19	10 a. m	103. 5	
20	....do	102. 5	Improving; regaining appetite and liveliness.
21	....do	103. 25	
21	5 p. m.......	103	
22	9 a. m	102. 5	
22	4.30 p. m	102	
23	9 a. m	103	
24	....do	103. 25	
25	....do	103. 75	
26	....do	104	
27	....do	102. 5	
28	....do	103	
29	....do	104	
30	....do	102	
31	....do	102. 75	
Jan. 1	....do	102. 5	
2	....do	103	
3	....do	103	
4	....do	102. 5	
5	....do	101. 5	
6	....do	103	
7	....do	102. 75	Killed by bleeding.

Post-mortem examination at once.—Skin: Covered almost universally by a blackish exudation in great part dried into crusts. On the ears are some remnants of the former exudations and extravasations; half an inch of the tip of one ear is necrotic.

Digestive organs: Mouth healthy. *Guttural lymphatic glands* greatly enlarged and gray from pigmentation.

Stomach: Full; contents dry and acid; has reddish discoloration as from blood extravasations and broad lines along its great curvature. The mucous membrane at this point is peeling off.

Small intestine: Contents abundant and liquid. Spots of congestion of about one line in diameter; no ulcers nor erosions; six ascarides.

Large intestine: Presents little abnormal. One or two depressed spots like cicatrices.

Mesenteric glands: Greatly enlarged and mostly grayish from pigmentary deposit. *Inguinal glands* also much enlarged and gray.

Thoracic duct: Is filled with a milky fluid.

Liver: Firm patches of purple. The lower margin very pale; almost transparent.

Spleen: Small, rigid, twisted as if from binding organizing lymph. Its surface is unusually white and fibrous-looking, but there is a deep black line along its anterior border.

Pancreas: Sound.

Heart: Right ventricle marked with bluish discoloration, evidently from former ecchymosis. One flap of the tricuspid valve has a round, blackish nodule beneath the endocardium. Left ventricle with similar bluish surface, and bicuspid valve with a translucent thickening.

Respiratory organs: Larynx and right bronchus have each a dark red ecchymosis. Lungs have black spots of ecchymosis and slight reddening of certain lobules.

Bronchial glands: Enlarged and pigmented.

Subdorsal glands: Enlarged and of a very deep red.

Brain: Generally unchanged.

EXPERIMENT No. 2.

Poland-China pig, nine weeks old.

Date.	Hour.	Temperature of body.	Remarks.
Dec. 19	10 a. m	103. 5 ° F.	Fed infected feces and intestinal mucous membrane preserved for a month in dry bran.
20	....do	104. 25	
20	5 p. m	103. 5	
21	10 a. m	103. 25	
21	5 p. m	104	
22	9 a. m	103. 5	
22	4.30 p. m	102. 5	
23	9 a. m	102. 75	
24	....do	102	
25	....do	101. 75	
26	....do	103. 5	
27	....do	102	
28	....do	100. 75	
29	....do	102	
30	....do	101	
31	....do	101	
Jan. 1	....do	102. 5	
2	....do	102	
3	....do	103	
4	....do	102. 75	
5	....do	103	Inoculated with intestine of pig which died yesterday. The intestine had been frozen over night.
6	....do	103	
7	....do	104. 75	
8	....do	105	
9	....do	104	
10	....do	103	
11	....do	105	
12	....do	104	
13	....do	105. 25	Purple spots on ears and rump; greasy exudation from skin. Enlarged inguinal glands.
14	....do	105	
15	....do	106. 5	Scours: a bright-yellow liquid feces.
16	....do	105	Do.
17	....do	105	Do.
18	....do	105. 5	Scours.
19	....do	105. 5	Do.
20	....do	105. 5	Do.
21	....do	103	Do.
22	....do	107	Great prostration; will not rise for food nor to have temperature taken. Purple blotches are especially abundant on ears and snout, and to a less extent on the head, generally the teats, rump, and hips. When lifted scarcely made a struggle. Killed by bleeding.

Post-mortem examination.—Blood: Dark colored; contained moving bacteria.

Digestive organs: Tongue sound. Tonsils unusually red in their openings.

Submaxillary and guttural lymphatic glands: Of a dark red, merging to a dirty yellow.

Peritoneum: With considerable reddish-brown effusion and bands of recently formed false membrane. The liquid coagulates on exposure. Under the microscope (No. 10 Hartnack) it is seen to contain numerous moving bacteria, also others less active, and two or four segmented chain-like.

Stomach: Full; sour. Great curvature mottled red and brown.

Small intestines: Has considerable tracks of deep congestion. It contains much mucus, and ten ascarides. One ascaris extended into the gall-duct and as far as the center of the right lobe of the liver; a second extended into the middle hepatic lobe. The pressure of these had led to a considerable dilatation of the bile-duct just above its junction with the cystic duct.

Ilio-cœcal valve: Very black, with its follicles enlarged and filled with a yellowish product. The whole length of the large intestine is black from deep pigmentation of its mucous membrane, which is, besides, greatly thickened and puckered. Both conditions imply former active inflammation.

The rectum: Of a dark grayish red; had several caseous deposits under its mucous membrane.

The mesentery: Contains a yellowish caseous deposit as large as a pea.

All the *lymphatic glands of the abdomen* are greatly enlarged, pigmented, and in many cases reddened from recent blood-staining. The *inguinal lymphatic glands* and those of the flank are in a similar condition.

Liver: Has patches of deeper purple discoloration, especially deep in the center of the acini. *Pancreas* sound.

Spleen: Shrunken with puckered edges, and whitish thickening of its proper capsule.

Kidneys: Vascular, congested and softened; corticle part dull brownish yellow. Medullary, more or less purple, with deeper shades in lines radiating from the papillæ.

Respiratory organs: Margin of epiglottis bears a blue patch, surrounded by ramified redness. *Bronchi* and bronchia sound.

Lungs: Of varying shades of light pink in the lobules, excepting one or two, which are of a dark red. The interlobular spaces are of a deep blood-red color, giving a dark marbling over the entire surface. *Right pleura* contains a little effusion with thread-like false membranes, and the same bacteria named as existing in the peritoneum.

The *axillary prepectoral, internal pectoral, bronchial,* and *sub-dorsal lymphatic glands* were enlarged, pigmented, and in some cases blood-stained.

The *heart* bore some purple discolored spots on the internal lining.

EXPERIMENT No. 3.

Poland China pig, nine weeks old.

Date.	Hour.	Temperature of body.	Remarks.
Dec. 19	10 a. m......	102.5 ° F.	Placed in infected pen from which a sick pig had been removed December 6.
20	....do	102.75	
21	....do	103.75	
21	3 p. m.......	103	
22	9 a. m.......	102.8	
22	4.30 p. m	102	
23	9 a. m.......	101	
24	....do	102.75	
25	....do	101.5	
26	....do	102	
27	....do	102.75	
28	....do	101.75	
29	....do	98.8	
30	....do	101	
31	....do	101.5	
Jan. 1	....do	100	
2	....do	101	
3	....do	101	
4	....do	98.5	Eyes very red and prominent. Scarcely able to stand. Screams when touched. (Evident phrenitis.) Died at 2. p. m.

Post-mortem examination the same afternoon.—Skin: Presented little change.

Digestive organs: Mouth sound, fauces and pharynx of a deep blue color, irremovable by pressure.

Stomach: A portion of about an inch square of a deep red, and with an abundant gelatiniform exudation under the mucous membrane.

Small intestines: Empty, much congested, and containing ten ascarides.

Large intestines: Has its mucous membranes congested, reddened, and thickened. At intervals are circumscribed spots of bloody extravasation, covered by a clot of blood on the free surface. These vary from one to two lines in diameter. In a great portion of the colon the contents are very dry and blood-stained. Between the layers of the mesentery, among the convolutions of the large intestines, are translucent gelatinoid exudations.

Liver: Gorged with blood, softened, and somewhat friable.

Spleen and *pancreas:* Normal.

Mesenteric glands: Small, but in some instances partially discolored by blood.

Lungs: Congested throughout, of a brick-red, with circumscribed black spots of extravasation.

Bronchia: Filled with frothy liquid, but without worms.

Heart: The right cavities were gorged with an intensely black clot. The left cavities contained a smaller clot. No ecchymosis was observed.

EXPERIMENT NO. 4.

Poland China pig, nine weeks old.

Date.	Hour.	Temperature of body.	Remarks.
Dec. 19	10 a. m	103. 75° F.	Inoculated with virus preserved one month in wheat bran.
20	....do	104. 2	
20	5 p. m.......	104. 5	
21	10 a. m	104	
21	5 p. m.......	105	
22	9 a. m	104	
22	4. 30 p. m....	104. 75	
23	9 a. m	103. 5	
24	....do	104	
25	....do	102. 25	
26	....do	101. 75	
27	....do	103. 75	Passes bloody mucus from the bowels.
28	....do	102. 75	
29	....do	102	
30	....do	101	
31	....do	105	
Jan. 1	....do	106	
2	....do	103	
3	....do	102	
4	....do	101	
5	....do	101	
6	....do	98. 75	Very low; can scarcely stand. Died during the following night.

Post-mortem examination January 7.—Skin: Extensively covered with purple maculæ and patches. Snout deeply blood-stained, some of the spots extending over the lips into the mouth. The greater part of the skin being black, congestions and extravasations into it are only clearly made out when it is cut into.

Digestive organs: Tongue sound. Pharynx has pellets of food accumulated in front of the epiglottis. Submaxillary and guttural lymphatic glands enlarged and stained of a blood red.

Stomach: Not one-third filled; odor faint, mawkish, not sour. Bears red patches of congestion and ecchymosis on its great curvature.

Small intestines: Congested almost throughout. Peyer's patch just above the iliocæcal valve has some black ecchymosis. On the lower surface of the valve the follicles are enlarged and filled with a yellowish deposit.

Cæcum and, to a still greater extent the *colon* and rectum, are deeply congested, and of a dark red; the mucous membrane is much thickened and thrown into prominent folds and wrinkles.

Two ascarides were found in the small intestine.

Liver: Extensively discolored of a purple hue, the staining being deepest in the center of the acini.

Spleen: Large, gorged with blood. *Pancreas* unchanged.

The lymphatic glands of the liver, stomach, intestines, sublumbar region, pelvis, groin, and flank are much enlarged and of a very deep red, in many cases almost black.

Kidneys: Cortical substance pale; medullary deep red, with spots of ecchymosis. The anterior part of the left kidney contained a cyst as large as a bean. The right contained two cysts, one in the pelvis, the other in the anterior part.

Respiratory organs: The epiglottis bore on its posterior surface some congestion and redness, partly ramified and partly diffuse and ineffaceable by pressure.

The lungs have a few black spots of ecchymosis and blood-colored extravasation in the connective tissue between the lobules. The lobules themselves are only very slightly congested. The left main bronchus present a spot of ecchymosis.

Heart: Empty, presents slight sanguineous discoloration through the lining membrane.

EXPERIMENT No. 5.

Poland China pig, nine weeks old.

Date.	Hour.	Temperature of body.	Remarks.
Dec. 19	10 a. m	104° F.	Inoculated with ingesta from the large intestine; also a portion of the mucous membrane, both having been preserved in dry bran for a month.
20	....do	104. 5	
21	...do	103. 75	
21	5 p. m.....	104	
22	9 a. m	103	
22	4. 30 p. m..	104	
23	9 a. m	101	
24	....do	103. 5	
25	....do	103	
26	....do	102. 5	
27	....do	102. 5	
28	....do	102. 75	
29	....do	101	
30	....do	102	
31	....do	103	
Jan. 1	....do	106. 75	Rump and tips of ears purple.
2	....do	104. 75	
3	....do	102	Scours.
4	....do	102	
5	....do	102 .	
6	....do	103	
7	....do	101. 25	
8	....do	102. 5	Scours; feces fetid.
9	....do	101	Very weak; eats little; fetid diarrhea.
10	....do	100. 9	
11	....do	100	
12	....do	98. 5	
13	....do	102. 5	
14	....do	100. 5	
15	....do	104. 5	Killed by bleeding.

Post-mortem examination at once.—Skin : Ears of a deep purple and thickly covered with concretions. Remainder of the skin has similar concretions, but no ecchymosis is observable. The snout presents scarcely a spot of discoloration.

Digestive organs : Extensive induration and ulcer on the left side of its median part and extending over its border. A similar but smaller ulcer exists on the right margin directly opposite. Small ulcers exist on the dorsum near the hip; also a diphtheritic-looking deposit extending over the margin on to the lower surface. Tonsils, palate, and pharynx sound. Submaxillary and guttural lymphatic glands are enlarged and congested.

Stomach : Has its mucous membrane thick, rugose, and as if water-soaked along its great curvature.

Small intestine : With mucous membrane thickened and puckered throughout; the duodenum deeply congested.

Ilio-cæcal valve : Thickened; its follicles enlarged and filled with a yellowish deposit.

Mucous membrane of *cæcum* and *colon* deeply pigmented and of a dark gray aspect. Some parts of the colon are still red in patches. Rectum pigmented, presents several small ulcers and a caseous deposit beneath the mucous membrane.

Liver : Bears blue patches of various sizes; gall-bladder contains a little bile of a bright yellow color, with greenish flakes.

Spleen : Small and puckered, so that its borders turn inward.

Pancreas sound.

Abdominal lymphatic glands : Hepatic, gastric, splenic, pancreatic, mesenteric, sublumbar, and pelvic, as well as the iliac, are enlarged, pigmented, and partially congested.

Kidneys : Corticle substances pale yellowish, slightly softened; in the case of one, reddened to the depth of one-third line. Medullary portion deeply colored.

Respiratory organs : Larynx and trachea sound; right lung with almost the normal pale pink hue externally, but seems to be congested internally when cut into; left lung nearly normal; *heart* and pericardium normal.

EXPERIMENT No. 6.

Poland China pig, eight weeks old.

Date.	Hour.	Body temperature.	Remarks.
Dec. 19	10 a. m....	104° F.	Placed in pen with pig partially convalescent.
20	....do	103	
21	....do	103.75	
21	5 p. m.....	104.5	
22	9 a. m	103.75	
22	4.30 p. m..	104	
23	9 a. m.....	104.25	
24	....do	102.75	
25	....do	103.75	
26	....do	105	
27	....do	103	
28	....do	104	
29	....do	104	
30	....do	103	
31	....do	102.5	
Jan. 1	....do	102	
2	....do	103	
3	....do	103.25	
4	....do	103	Placed in pen with another pig in height of the disease.
5	....do	103	
6	....do	101	
7	....do	102.75	
8	....do	102.5	
9	....do	103	
10	....do	103	
11	....do	103.25	
12	....do	104	
13	....do	101.25	
14	....do	103.5	
15	....do	106	
16	....do	105	
17	....do	105.5	Feces coated with film of blood.
18	....do	104.8	
19	....do	104.5	Bloody feces.
20	....do	104.25	Do.
21	....do	105	Do.
22	....do	103	Do.
23	....do	103	Do.
24	....do	103	Bloody feces. Inguinal lymphatic glands enlarged.
25	....do	101	
26	....do	104.75	
27	....do	104	
28	....do	103	Appetite improving.
29	....do	102	
30	....do	102	
31	....do	103	

EXPERIMENT No. 7.

Female rabbit.

Date.	Hour.	Body temperature.	Remarks.
Nov. 21	....	....	Inoculated hypodermically with one drachm of the blood of a sick pig just killed.
22	9 a. m	104? F.	
23	....do	104	
24	....do	104.1	
25	....do	104.5	
26	....do	104.5	
27	....do	104	
28	....do	104.5	
29	....do	104	
30	....do	104	
Dec. 1	....do	104	
2	....do	104	
3	....do	104	
5	....do	104	
7	....	....	Hypodermic injection of one drachm of blood of pig which died during last night.
8	9 a. m	105	
9	....do	104.75	
10	....do	103.75	
11	....do	103.75	A firm ovoid nodule in the seat of inoculation.
12	....do	104.5	
13	....do	103	

EXPERIMENT No. 7—Continued.

Date.	Hour.	Body temperature.	Remarks.
Dec. 14	9 a. m......	103. 5° F.	Hypodermic injection of one drachm of blood of pig found dead this morning. Blood swarming with actively-moving bacteria.
15	....do	105. 5	Has not eaten supper.
15	5 p. m......	105. 5	Eats nothing.
16	10 a. m	106. 25	
16	5 p. m......	106. 75	
17	10 a. m	105. 5	
17	4 p. m.......	103	
18	10 a. m	105. 75	
18	4 p. m......	105. 5	
19	10 a. m	104	Blood showed numerous moving bacteria as in the pig. Induration in the right iliac region.
20	....do	104. 75	
21	....do	105. 5	
21	5 p. m......	104. 5	
22	9 a. m	103. 5	
22	4.30 p. m	104. 25	
23	9 a. m.......	103. 5	
24	....do	104	
25	....do	104	
26	....do	104. 75	
27	....do	104. 75	
28	....do	105	Abscess has burst to the right of vulva. A white fibrous extravascular mass exposed.
29	....do	104	
30	....do	105	
31	....do	105	
Jan. 1	....do	104	
2	....do	104	
3	....do	103	
4	....do	103	
5	....do	103	
6	....do	102. 5	Is very low and has eaten little for some days.
7	....do	102	Sore still open. Killed by bleeding.

Post-mortem examination at once.—Connected with the raw sore in the groin was an immense mass of whitish, fibrous material, infiltrated with pus, and extending from the lumbar vertebræ above to the median line below. The mesenteric glands were enlarged and blood-stained. Two had been transformed with yellow, cheesy-looking masses. The stomach and bowels appeared healthy; also the liver and spleen, heart and lungs.

EXPERIMENT No. 8.

Poland China pig, eight weeks old.

Date.	Hour.	Body temperature.	Remarks.
Dec. 18	4 p. m......	102. 75° F.	
19	....	....	Inoculated with blood of sick rabbit hypodermically.
20	....	....	
21	4 p. m......	103. 5	
22	9 a. m	101. 5	
22	4.30 p. m	103. 75	
23	9 a. m	100. 75	
24	....do	101	
25	....do	101	Skin hot. Hides under the litter.
26	....do	101. 5	
27	....do	101	
28	....do	101	Scours.
29	....do	100	
30	....do	100	
31	....do	102	
Jan. 1	....do	102. 75	
2	....do	102	Inoculated with matter from open sore of sick rabbit.
3	....do	101. 5	
4	....do	102	
5	....do	103. 5	Feces fetid.
6	....do	104. 5	
7	....do	104. 75	
8	....do	104. 5	
9	....do	104. 25	Fetid diarrhœa.
10	....do	103	
11	....do	103	
12	....do	102. 5	
13	....do	103	
14	....do	104. 75	
15	....do	105	Killed by bleeding.

Post-mortem examination.—Skin: Naturally black; no purple nor congested spots seen.

Digestive organs: Mouth and throat healthy.

Guttural lymphatic glands: Enlarged and somewhat congested.

Stomach: Moderately full; of a deep brownish red along its great curvature.

Small intestine: Slightly congested in patches; contains twelve ascarides.

Large intestine: Nearly normal.

Mesenteric lymphatic glands: Enlarged and slightly congested. Their surface presents clear, glistening, rounded masses like pins' heads. *Inguinal glands* have the same character.

Lung: Isolated lobulettes are dark red and solid; at some points the interlobular connective tissue is distended by a dark-red infiltration.

In the bronchia of the left lung were twelve strongyli.

EXPERIMENT No. 9.

Common white pig, ten weeks old.

Date.	Hour.	Body temperature.	Remarks.
Jan. 7	10 a. m	104° F.	Inoculated with frozen white product from the groin of the infected rabbit.
8	....do	102. 5	
9	....do	103	
10	....do	103	
11	....do	101. 75	
12	....do	102	
13	....do	103	
14	....do	103	
15	2 p. m......	101	
16	10 a. m.....	102. 25	
17	....do	103. 25	
18	....do	103. 8	
19	.. do	103	
20	....do	103. 25	
21	....do	103. 3	
22	....do	104. 5	Purple spots on rump. Eats little.
23	....do	105	Blue ears.
24	....do	102. 5	Scours, bright-yellow liquid feces. Inappetence.
25	....do	105	Do.
26	....do	98. 75	Do.
27	....do	97	Does not rise when temperature is taken; is stretched on its side with muscular jerking. Killed by bleeding.

Post-mortem examination.—Skin: Margin of snout for one-half line deep of a dark brown, and apparently without vascularity or life. Beneath this is a red congested line.

Ears: Deeply blotched with dark red and purple maculæ, each about one-half inch in diameter, but to a great extent confluent, so as to form extended lines and patches. Stump of tail maculated. Perineum and adjacent parts of hip of a deep purple.

Digestive organs: Tongue with a whitish fur. On the center of its dorsal surface is a dark spot about two lines in diameter, which is found to cover a considerable extravasation and clot on the muscular substance. Glandular follicles on the lower surface of the soft palate filled with a soft yellowish puriform mass.

Submaxilliary lymphatic glands: Greatly enlarged and of a deep purple. *Guttural glands* also blood-stained and moderately enlarged.

Stomach: Full, very fetid, not sour. Great curvature has its mucous membrane much congested with numerous black spots of extravasation projecting beyond the general surface. In the left *cul de sac* the ingesta next the mucous membrane is of a dark baked appearance and firmly adherent to the mucous membrane, the epithelial layer of which comes off with it. It has evidently been adherent for some time.

Small intestines: Have large tracts of congestion, and in the duodenum and commencement of the jejunum are ten ascarides. Seven ascarides have made their way into the gall duct and the different lobes of the liver, but none in the cystic duct nor gallbladder. The biliary duct is greatly distended and coated with a layer of yellowish-green biliary coloring matter.

The ilio-cæcal valve: Has its margin of a deep grayish-black and its follicles enlarged.

The large intestines: Are throughout black from pigmentary deposit, the blackness being especially marked on the agminated gland, extending from the ilio-cæcal valve on the colon. Many round blackish elevations are scattered over the length of the colon, appearing like enlarged solitary glands. On some parts of the colon the dark color is modified by the deep red of a recent congestion. Through the whole length of the large intestine the mucous membrane is considerably thickened and puckered. Near the anus are some caseous deposits beneath the mucous membrane, but communicating with the surface by open orifices.

The liver: Has great patches of a deep purple, deepest in the center of the ascini.

The gall bladder: Is full of dark dreen, thick, very viscid bile.

The inguinal, sublumbar, mesenteric, mesocolic, gastric, and hepatic lymphatic glands: Are greatly enlarged and deeply blood-stained.

The kidneys: Somewhat softened, are of a dull yellowish brown in the cortical portion and of a purple hue, with darker radiating lines in the medullary.

Respiratory organs: Larynx sound. Lungs sound, excepting some slight congestion in particular lobes, and the filling of the bronchia and air-cells with blood evidently drawn in in dying. No pleural effusion.

Heart and pericardium: Sound.

EXPERIMENT NO. 10.

Merino sheep.

Date.	Hour.	Body-temperature.	Remarks.
Nov. 21	2 p. m.......	103° F.	Hypodermic injection of one and a half drachms. Blood from sick pig just killed.
22	10 a. m......	102. 5	
23	....do	103. 75	
24	....do	103	
25	....do	104. 5	
26	....do	103. 25	
27	....do	104. 5	Scouring and snuffling.
28	....do	103. 75	
29	....do	102	
30	....do	102. 5	
Dec. 1	....do	103. 75	
2	....do	102. 5	
3	....do	103. 25	
5	....do	102. 5	
7	....	....	Hypodermic injection of two drachms blood from pig which died during the night previous.
8	....do	103. 75	
9	....do	103. 3	
10	....do	103. 75	
11	....do	100. 25	
12	....do	102	
13	....do	103	
14	....do	103	Hypodermic injection of one drachm blood and pleural fluid of pig which died during the preceding night. Fluids full of actively moving bacteria.
15	....do	105. 5	Snuffling.
15	5 p. m.......	105	
16	10 a. m......	104. 5	
16	5 p. m.......	104. 5	
17	10 a. m......	105. 5	
17	4 p. m.......	103. 5	
18	10 a. m......	103. 75	
18	4 p. m.......	105	
19	10 a. m......	103. 25	
20	....do	105. 2	Blood shows moving bacteria, but less numerous than in the rabbit.
20			
21	10 a. m......	102. 25	
21	4 p. m.......	104	
22	9 a. m.......	104	
22	4.30 p. m....	105. 25	
23	9 a. m	103. 25	
24	....do	102	
25	....do	103	
26	....do	104	
27	....do	103. 75	
28	....do	103. 2	
29	....do	103. 5	
30	....do	102. 75	
31	....do	104	
Jan. 1	....do	103	
2	....do	103. 75	
3	....do	103	
4	....do	102	
5	....do	103	
6	....do	103	Inoculated with scurf from the ear of a sick pig.
7	....do	102	
8	....do	102. 75	Scours.
9	....do	103. 8	Do.

EXPERIMENT No. 10—Continued.

Merino sheep—Continued.

Date.	Hour.	Body-temperature.	Remarks.
10	....do	103° F.	Scours. Anus red and sore. Strongly objects to the thermometer. Has passed bloody mucus.
11	....do	103	
12	....do	102	
13	....do	102.5	
14	....do	103.5	Anus still red and puffy, with abundant mucus.
15	....do	103	
16	....do	103.5	
17	....do	103.5	Scours.
18	....do	103	Do.
19	....do	104	
20	....do	102.75	
21	....do	103	
22	....do	102.5	Anus still red and swollen.
23	10 a. m	102	Same afternoon injected one drachm of blood and pleural fluid from pig just killed. Fluids contained active bacteria.
24	....do	104	Slight subcutaneous swelling in the right axilla. Tenderness of the skin of the abdomen.
25	....do	104.5	
26	....do	104	
26	4.30 p. m ...	105	
27	12 m	105	
28	10 a. m	103	
28	5 p. m	104	
29	10 a. m	105	
30	....do	104	

EXPERIMENT No. 11.

Long wooled (cross-breed) lamb.

Date.	Hour.	Body-temperature.	Remarks.
Jan. 17	10 a. m	104.25° F.	Injected hypodermically in the axilla matter from the ears of two sick pigs, also anal mucus from one of them.
18	....do	104.25	
19	....do	103.8	
20	....do	105.25	
21	....do	103.5	Ears with scurfy eruption.
22	....do	106.5	Bleeding spots on ears.
22	5 p. m	104.75	
23	10 a. m	104.5	Injected hypodermically one drachm pleural fluid containing actively moving bacteria from pig just killed.
24	....do	108	Hard engorgement two inches in diameter in right axilla.
25	....do	107	Axillary swelling more defined; like a hazel-nut.
26	....do	104	
26	4.30 p. m ...	108	
27	12 m	108	
28	10 a. m	105.25	Rectum contracted and tender; thermometer covered with bloody mucus.
28	5 p. m	106	
29	10 a. m	106	
30	....do	104	

REPORT OF DR. D. W. VOYLES.

Hon. WM. G. LE DUC,
Commissioner of Agriculture:

SIR: In conducting an examination of the diseases of swine, as prevailing throughout the State of Indiana during the present season, the following plan was pursued, viz:

A tour of observation and inspection was made through the counties of Floyd, Harrison, Washington, Greene, Morgan, Monroe, Owen, Putnam, and Bartholomew. Some of the most intelligent and leading stock men of each county were sought, and all the information obtained which they had upon the subject of the disease, both in regard to its present manifestation and past history. Speculators in live hogs and large feeders were closely interrogated upon every feature of the disease as coming within the range of their experience and observation. Diseased herds were visited, and in each case the farm minutely inspected in all its bearings upon the health of animals; the methods of breeding, feeding, and general management of swine diligently inquired into; dead animals, where not too far advanced in decomposition, dissected, and living ones, having the disease, were slaughtered for examination, and the pathological indications carefully noted. The month of September was entirely devoted to this branch of the investigation.

The object of this method of inquiry was to ascertain whether the disease, as prevailing throughout these several districts, was uniform in its character, differing only in such modification in type as may be due to local influences; or whether these were to be found separate and distinct diseases in different localities, due to entirely different causes for their production; and if uniformity was found to exist in the character of the disease as now prevailing, to learn from practical and intelligent observers in each district whether, in any essential particular, it differs from the disease that has prevailed in other years.

PREVALENCE OF THE DISEASE.

The several districts visited were all more or less affected by the disease, but to a much less extent than during former years, except, perhaps, in the county of Putnam, where it was prevailing for the first time as a general and wide-spread epidemic, the loss being estimated at from fifty to sixty thousand dollars. In this county the surface is sufficiently undulating to produce good drainage; the soil is red clay on limestone. Springs of pure limestone water are abundant, and woodlawns beautifully swarded with blue grass are seen upon almost every farm. Feeding swine has been an extensive and profitable branch of farm industry in this county, and the herds are, therefore, quite large for a grass-growing section. During the summer months hogs in this county run upon blue grass and clover, and are fed some corn. We found the corn so fed often unfit for use, because of a very reprehensible practice of hauling to the field for convenience in feeding and throwing it in an open rail pen, where, by exposure to heat and moisture, it soon becomes moldy. The mean temperature in this county during the summer was slightly above, and the rain-fall considerably below, the average seasons.

The counties of Floyd, Harrison, and Washington possess much the same kind of soil, and are abundantly supplied with running springs of limestone water; but blue grass and clover are not so extensively or generally grown. In these three counties hog-raising is not a branch of farm industry sufficiently remunerative to induce the farmers to generally engage in it, and the herds are, therefore, usually small and the animals very imperfectly cared for.

The observations made in the counties of Greene, Owen, Monroe, Morgan, and Bartholomew were on a line with the White River Valley. This and the Wabash Valley constitute pre-eminently the hog-growing sections of Indiana. It is in this part of the State that the disease has prevailed to the greatest extent. Hog-raising being the leading business industry, the herds are ordinarily quite large.

No observations were made in the Wabash country. In the White

8 sw

River Valley the disease has prevailed during the present season to much less extent than for several years past. This is due in part to the fact that there are not so many hogs here as formerly—great loss having greatly discouraged hog-raising, a branch of agricultural industry heretofore paramount to every other interest.

The less prevalence of the disease is also due in part to the increased facilities for selling to summer packers; the approach of the complaint in any given locality being the signal for the selling of every marketable animal.

In these hog-growing districts, the surface of the country is quite flat, affording very imperfect natural drainage, and as a consequence much stagnant water prevails. The soil is a mixture of clay and sand. The food is mainly corn, with some clover during the summer months, the animals often subsisting upon corn alone from the time of birth to that of slaughter.

In the county of Bartholomew there are several "grease factories," where they render dead animals, and it is estimated that during the year 1876 there were rendered at these several factories no less than one hundred thousand animals that died of the disease in that and adjacent counties.

It is the concurrent testimony of the leading and most intelligent observers, whose experience and observation have been most extensive, that while the disorder prevails more or less at all seasons of the year, it prevails to the greatest extent and with most fatal effect during the dry months of the fall season, and again during the last winter and first months of spring—February and March.

SYMPTOMS OF THE DISEASE.

A greater degree of uniformity was found to exist in the symptoms and character of the disease than was anticipated at the beginning of the investigation. The first symptoms that usually attract the attention of the farmer, indicating approaching disease, is a wheezing cough, coupled with a disposition to mope. During this period the animal stands about as if in a "brown study," with its ears dropped and its eyes inclined to water or matter.

Following in the usual succession of symptoms comes a failure in the appetite, with occasional vomiting and diarrhea, although the two last-named symptoms constitute an exception, to which constipation is the rule.

A complete failure in the appetite, intense thirst, with increased temperature of the body, indicates the supervention of the febrile and inflammatory stage of the disease. During this stage the temperature not infrequently rises as high as 107° F., as indicated by the introduction of the thermometer into the rectum of the animal. The cough increases; the breathing becomes more accelerated and laborious; the respiratory movements are scarcely observable in the walls of the chest, but become conspicuous at the flank, and range from 30 to 60 inspirations to the minute; the arterial circulation is increased in frequency and diminished in volume. Petechial eruption is often observed on the skin and is most distinctly observable on white animals. This is due to extravasated blood from the capillaries into the tissues, which, on undergoing decomposition, produces ulceration of the skin in the future course of the disease, particularly if the animal becomes convalescent.

In the last stage the animal becomes very weak; staggers in gait, if able to rise at all; refuses both food and drink; falls in temperature, sometimes as low as 60° F.; seeks the sunshine or a covering of litter, and speedily dies. Emaciation is a rapidly progressive symptom throughout the entire course of the disease.

DURATION OF THE DISEASE.

The disorder is by no means uniform in its duration, varying from a few hours to many days and even weeks. When death occurs only a few hours after the attack a complication of heart disease is usually the cause of the rapid termination of the case. Early fatality may occur also from rapid congestion of the lungs, producing hepatization of a large portion of that organ. The average duration of the disease can be, therefore, scarcely approximated. Perhaps five days would include the length of time consumed in most fatal cases, whereas a much greater length of time is required in cases that recover. In its most violent epidemic form a much less time than five days would include the course of the disease in all fatal cases.

PATHOLOGY OF THE DISEASE.

As before stated, all dead animals not too far advanced in decomposition were examined, and one or more sick animals were selected from each diseased herd, and after a careful study of their symptoms, as compared with the other sick stock of the herd, were slaughtered for examination.

Memoranda from thirty dissections made from fifteen separate and distinct herds fairly representing the disease as observed under all the varied circumstances as to food, soil, water, and general management, show the following results:

In every case, without exception, disease of the lungs was present, varying in degree from slight congestion to complete softening from suppuration and inflammation. In two cases the lung disease was tuberculous in character. In eight cases adhesion occurred between the costal pleura and lung. In six cases circumscribed spots of inflammation were found on the walls of the heart and its investment, with an effusion in the pericardial sack. In six cases were small patches of ulceration of mucous lining of large intestine. In six cases were congestion of mucous lining of the stomach. In all cases the liver presented a darker hue than natural, in four cases slightly, and in one greatly enlarged; but in all other cases in size and general appearance would compare favorably with that organ as usually observed in animals regarded sound and healthy. The spleen was in all cases discolored, as in case of the liver. In few cases there was slight congestion of the kidneys. In one case there was evidence of fatty degeneration, and in all others the organ indicated a healthy condition. The blood was always dark-colored, the muscles pale and relaxed.

The disease of the lungs was in all cases the leading pathological condition, to which all other diseased appearances were secondary in importance, constituting complications only.

A section of the lung of an animal slaughtered during the active inflammatory state of the disease shows, under the microscope, a complete solidification of lung-tissue, the air-cells being filled with epithetial exudation, no extravasated blood appearing. A section of the liver of the same animal shows a thickening of the septæ acini by a proliferation of epithetial cells, tending to or constituting fatty degeneration; other acini in the same section exhibit a perfectly healthy condition. A section of intestine from same animal shows a healthy condition. These three sections are transmitted with this report for verification. (See microscopic sections, Plate XV, Figs. 1, 2, and 3.)

The contents of the stomach and intestines were liquid in six cases, and dry, hard, and very dark colored in all others.

The gall-bladder usually contained a small quantity of thin, greenish fluid.

The trachea and bronchial tubes contained a large quantity of matter apparently consisting of mucus and broken-down epithelium.

DIAGNOSIS OF THE DISEASE.

Judging from the visible causes that appear most active in its development—the symptoms and pathology of the disease—we feel warranted in pronouncing it, in its milder manifestations, *bronchial catarrh*, and, in its most active and fatal form, *catarrhal pneumonia*.

There is no symptom uniformly present in the disease, as we have observed it, that bears any analogy to the symptoms of cholera as affecting the human subject, and the term "hog-cholera" is therefore a misnomer; and although there is, ordinarily, little or nothing in a name, in this instance the misnaming of the disease has been a source of incalculable loss, by suggesting a line of treatment irrationally administered and calculated to aggravate rather than cure it.

ITS CAUSE.

It is when seeking the cause of this wide-spread epidemic disease that the field of investigation takes widest range. As already stated, it prevails more or less at all seasons of the year, and under almost every conceivable condition and combination of conditions as to soil, food, water, locality, and general management; but the difference in its prevalence under certain circumstances is so marked and uniform that from these facts we may derive some definite information as to the causes *most* active in development.

The past history of the disease would indicate that it originated in this country at a time when the condition of swine was visibly altered from a comparative state of nature to one of more perfect domestication. When the country was new, affording almost unlimited range, the hogs bred, grew up, and roamed in the forest until maturity. Being allowed the free use of their noses, and being omnivorous in nature, they fed on worms, roots, mast, and such other food as was provided and given them by their owners; they exercised as their inclination or necessities inclined them; had free access to numerous springs and streams of running water; slept in storm-sheltered thickets on beds of clean leaves, and enjoyed under these circumstances a vigor of constitution and an immunity from disease unknown to the modern swine-breeders of the country. As the country became more densely populated, rendering it necessary to clear up and inclose the land for agricultural purposes, the lank, active, long-nosed animal of the pioneer age began to disappear in order to give place to a new and more advanced civilization in the history of his race. A close business calculation demonstrated that a hog fed to profit on food produced by manual labor must have an inbred tendency to take on flesh, and that tendency encouraged by close confinement and high feeding.

The hog of to-day is the result of persistent in-breeding for an obese habit, encouraged by want of exercise and over-feeding. An animal

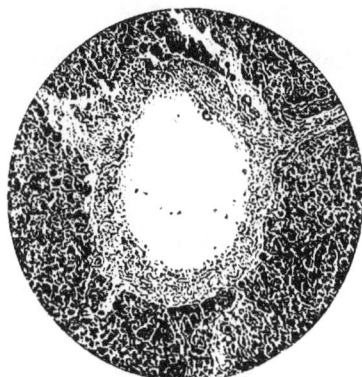

Fig. 1.
Microscopic section of diseased liver
in "Hog Cholera."

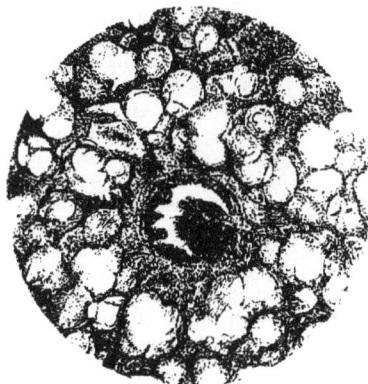

Fig 2.
Microscopic section of lung
in catarrhal pneumonia.

watering place. At the farm of Mr. Quinn, near Hartsville, Indiana, where the disease was prevailing, twelve head of sick animals were running in an inclosure, and when the proprietor was asked about the supply of water, he said, "There was plenty—a good spring." On personal examination the spring was found to issue from a hill-side, with but little incline; from the place where it issued to the point where it disappeared from exhaustion—a distance of some 40 feet—there was a long bed of thin mud, and no visible appearance of running water at any point. He was asked on our return when he last inspected the watering place, and answered, "This morning." He was then asked if he thought the supply of water at that spring would supply a few horses or cattle with water, if the hogs were taken out, and he replied promptly in the negative, and when asked by what process of reasoning he came to the conclusion that water of acknowledged unfitness for anything else was quite good enough for hogs, and sick ones at that, he replied, in substance, that hogs would not use water until they rendered it unfit for any other kind of stock!

We mention this case in detail because it fairly represents the views of the average farmer upon the subject of water for swine—"any water is good enough for a hog."

CLEANLINESS.

The domesticated animal does not approximate the habits of his pioneer ancestor in point of cleanliness. It is the instinctive habit of the animal to bathe in water and wallow in mud to counteract heat and as a protection against flies; but in a state of nature, when the mud has served its purpose, the animal cleanses himself by friction with the nearest tree; the filthy bed which the domestic animal becomes satisfied to occupy in a state of confinement is never occupied by animals running in the forest, and given opportunity to make and change their sleeping places at will—in short, when allowed to provide for his own existence, he exercises a more intelligent regard for his wants than is ordinarily exercised for him by his owner, who attempts to supersede instinct by reason.

The frequent allusions made to the native hog may provoke the inquiry, Are we to return to the ill-shapen and ungainly animal of forty years ago? Certainly not. In this age of high-priced corn, such an animal is unworthy of an existence. The only thing to be admired of him is his health and constitution; the only useful lesson to be derived from allusion to his history is the means by which these were acquired and maintained. Food, faulty in character and wanting in variety; water, deficient in quantity and purity; quarters, too limited in space and filthy in condition, are the three leading factors in the production of disease of swine.

Special attention was given to the examination of the surface land occupied by diseased animals, and while there were exceptional cases, in quite a large majority of instances they were running in fields producing quite a luxuriant growth of weeds which, during that season, were shedding their seed, bloom, and leaves. The earth was exceedingly dry and dusty. In traveling through the fields the animals created a dust from the earth and from the weeds also, which, together, were taken into the air-passages and lungs with the air breathed, constituting an active source of irritation. While pursuing this branch of the inquiry we were informed by some intelligent observers that they had noticed that animals running in such fields, particularly wheat and rye stubble, overgrown with weeds, were the most unhealthy; and under these circumstances the greatest amount of disease was observed. It is at this particular season of the year that hogs are most neglected. Having been turned out during the summer months to take care of themselves, while the grass is green and filled with nutritious qualities, they thrive and do well; but, at the approach of the dry season, green grass gives place to that which is mature and dry, in which state it is indigestible and constipating. The water at this particular season fails. It is also at this season that swine keep their skin clothed with mud as a protection against flies, seriously interfering with its healthy functions as auxiliary to the lungs and other depurating organs of the body. This is the season when the cold nights precipitate heavy dews, and while running through the grass and weeds, during the nights and early morning hours, the animals become wet and cold, to be dried off and scorched in heat and dust at the returning noonday. During the nights they are chilled, sending the blood from the surface to the internal organs of the body, and breathe a damp, cold atmosphere; during the day they are overcome with enervating heat, and breathe a dry atmosphere, loaded with dust and dry particles of decaying vegetation. Is not this an array of existing circumstances well calculated to excite catarrhal affections, and are not these conditions as universally present over a large area of country as the disease itself? It may be objected that the disease sometimes prevails where the conditions mentioned are wanting. That it does prevail in some instances where there is no visible cause for its production is true, but the instances are of rare occurrence. As before stated, it prevails again in an active and fatal form during the months of February and March. This is the season when bronchial and lung diseases prevail among the human family, due to the atmospherical changes, and exposure to the damp earth then in a state of alternate freezing and thawing. Swine are similarly affected during that period of the year from the same cause; and being more generally exposed to these causes than the human family, are more liable to such diseases in their epidemic form. The principal objection to this rational theory of the cause of the disease is that it is found to exist at other seasons of the year than those mentioned, and under circumstances where almost all the conditions named are wanting. In a few instances we observed it where there was no visible want of first-class care in the management of the swine as to food, water, cleanliness, and shelter, and when they were running on clean blue-grass pastures well shaded and watered; but the prevalence of the disease under such circumstances *was exceedingly rare*. It is the general opinion among farmers that the disease is due to some specific poison, and is contagious in character. This opinion was generally entertained by the farmers of Putnam county, where the disease prevailed this season for the first time as a general and widespread epidemic. Many claimed that the disease was communicated by a lot of diseased swine driven through that county from the county of Boone; but many cases occurred on farms entirely off the route traveled by the diseased animals, and entirely isolated from public highways, and upon which no new or strange animals had been introduced by purchase or otherwise. A toll-gate keeper living near the village of Bainbridge, in that county, had a few swine running at large, and coming in close contact with all the animals driven over the road, and still they had escaped the disease; while those occupying inclosures by the roadside generally had it. Numerous instances were reported by reliable and intelligent men, where the disease prevailed upon one farm with but a partition fence separating the sick animals from those of a neighbor, in an adjoining field, and the latter not be affected by it. No case of this kind was reported, where a stream of water led from the diseased herd to the opposite lot of animals, in which the latter escaped; which circumstance would indicate that while the disease may not be strictly contagious it becomes infectious, and can be transmitted by contact with diseased matter. Experimental operations conducted with a view to ascertain this fact were wanting, because of the lack of absolute knowledge that the animals operated upon would not have had disease without the introduction of diseased matter by inoculation; barring this doubt, the introduction of diseased matter into the system of a well animal produces the disease in four out of five cases. It is a safe practice to separate the sick from the well animals at the very first indication of approaching disease. The eating of the flesh of the dead animals, dying of the disease, by those surviving, is a very reprehensible practice, and should under no circumstance be allowed. The dead should be speedily removed and buried or cremated. Some farmers, however, claim that where they allowed the sick to eat the dead the animals seemed to recover faster by the practice—an observation, if correctly made, only demonstrating that the herd was suffering from want of animal food to such an extent that that furnished them in a diseased condition did them more good than harm. Those holding to the theory of contagion generally agree in the period of incubation as ranging from ten to twelve days.

Mr. William B. Taylor, of Martinsville, Ind., a gentleman of long experience as a feeder and packer, and an intelligent observer of the disease, states that when a herd of diseased animals were turned in a field with others not previously exposed, that the disease would almost invariably run through the entire diseased herd before attacking the others; and Mr. Joseph Goss, of Gosport, Ind., a feeder and packer of forty years' experience, and a most careful and intelligent observer, corroborates the statement of Mr. Taylor.

THE DISEASE AS AFFECTING DIFFERENT BREEDS.

This branch of the inquiry was forced upon our attention by certain parties who claimed in behalf of certain breeds of swine a partial or complete immunity from the disease. Unfortunately our field for observation in this regard was not good, since all the animals observed were grades in which the Poland-China and Berkshire blood largely predominated. The best information gained upon the subject was to the effect that the breeds for which such immunity was claimed were those not in general use, and that the absence of loss from such breeds is due to the small number of such animals existing in the diseased districts. Such claims were made in behalf of the Chester Whites and Jersey Reds. We saw none of either of these breeds in our travels, either sick or well. The latter breed may have a partial immunity from these considerations. It is an Eastern bred animal, developed in a section where in-breeding, close confinement, and over feeding and monotonous diet are not so generally practiced as in the West, and that breed has, therefore, *possibly* a better constitution with which to resist diseased tendency.

RECURRENCE OF THE DISEASE.

All experienced feeders agree in the opinion that animals having the disease and recovering from it seldom have a second attack, and state that in purchasing animals to feed preference is always given to those that have gone through with the disease. We are inclined to accept this opinion as of little consequence, for the reason that such as are fed for pork do not afford a sufficient lapse of time to clearly demonstrate this point; and, on the contrary, among breeding animals that are allowed to live older, in which timely opportunity is given, our information is that a second attack is not an unusual occurrence.

HEREDITARY EFFECT OF THE DISEASE.

Females having the disease when breeding almost invariably cast their young. If they escape that accident, the offspring usually die very soon after birth. Subsequent litters from the animal, after completely recovering from the disorder, do not appear to be wanting in vigor, and do not exhibit a greater aptitude for the disease than other animals.

PREVENTION OF THE DISEASE.

The widespread prevalence of the disease, its rapid course and dreadful fatality, warrant the opinion that measures of prevention, if discovered and applied, will be much more beneficial in result than the discovery of a successful line of treatment for the disease, unless that treatment shall consist of some specific remedy, a practical use of which can be made by the farmers in all stages of the complaint. That such a remedy will be discovered, we are of opinion, is not within the range of probability. The measures necessary to prevent disease in domestic animals embrace within their range a careful study of their natural habits and wants, and a strict observance of the laws of health that govern all animal life, the principles of which are the same in their application to the inferior animals as to man. Those errors alluded to when considering the cause of the disease, as, in our opinion, largely contributing to, if not wholly the cause of, its development, must be corrected. The idea that swine are exempt from the ordinary laws governing health, and will thrive under any and all circumstances, must be abandoned. Forced to keep pace in his superior development with the civilization of the age in which he lives, he requires additional care in his management in order to ward off the numerous ills to which he is liable, many of which were unknown to his race in its unimproved state of nature. The food of the animal should, at all times, consist of the greatest possible variety; the water drank should be strictly pure; too many animals should not be herded together; the young animals should be kept to themselves; frequent change of locality, by shifting from one field to another; the frequent plowing up or burning over of the lots usually denoted as hog-lots in order to disinfect them; frequent change of sleeping-places, and the removal and destruction of old, filthy bedding-material. During the dry fall months, when the swine are running at large, they should be daily inspected, and at the approach of that period when the succulent grass is giving place to the mature and dry, laxative food, such as bran-mash or oil-cake; or aperient medicine, as linseed-oil or Glauber salts, given to counteract the constipating effect of the dry grass; the watering-places daily inspected; if running in open fields with high weeds and grass, they should be taken out at night and kept from the cold, wet grass, and turned into woods, if there is such a place available; they should be kept from weedy and stubble fields during the dry dusty period of the fall season, both day and night. When confined in close pens, these pens should be cleaned daily, and disinfected when there is stench, by the use of copperas, chlorinated lime, or with dry, fresh dirt. The opinion that corn, almost alone, is sufficient food for swine, and contains all that is necessary for the growth and development of the animal, will not be abandoned by the average farmer until after many costly lessons from experience, while attempting to freight their corn crops to market through this uncertain medium of transportation. A judicious and intelligent system of in-breeding cannot be abandoned without a rapid reversion to the ill-shapen animal of forty years ago, and we do not insist that in-breeding, when judiciously and intelligently practiced, is materially deteriorating in its influence upon the health and constitution of swine; it is only by coupling animals near related, that have a constitutional defect or a diseased tendency, and where these defects and tendencies are duplicated, that such a course becomes positively injurious. In the natural state of swine, when running at large and growing up without man's intervention, in-breeding frequently occurs; and the bad tendencies are warded off by the more vigorous males fighting off or destroying the feeble ones and becoming the sires of the race. Thus nature provides for a "survival of the fittest." In artificial breeding, the selections made for breeding purposes are too often made with special reference to shape and beauty, and too little consideration is given to vigor and constitution. There is no practical test made in the prize-ring between the most comely male and his less handsome brother, as to which is by nature best entitled to become the sire; but the breeder makes the choice from other considerations than "might makes right." Good feeding is the counterpart of good breeding; but there is a marked difference between good feeding and overfeeding or stuffing. Good feeding consists in giving an amount of good healthy food in sufficient variety to provide for the waste of the body, and in quantity only sufficient to develop the future growth of the animal. Overfeeding or stuffing consists in pushing the amount of food to the full assimilative capacity of the animal, with a view to the greatest possible amount of excessive flesh. The first is essential to good breeding; the other is deteriorating to the constitutional vigor of the animal.

TREATMENT OF THE DISEASE.

This branch of the subject we might sum up in these few words: No remedy was discovered having any marked beneficial effect upon the disease when once fully established; no farmer was found who ever in his own experience tried any remedy or remedies that seemed to exert any well marked curative effect upon the disease. Many isolated cases were reported; one animal recovered by having the tip end of its tail cut off; two, by being saturated with coal-oil, and a few others of like absurdity.

The announcement of the names of the individual members of the commission appointed to conduct this examination brought to our notice by letter a large number of so-called hog "cholera cures," which their several proprietors asked us to test, or allow them to test in our presence. As the requests were coupled with the expressed or understood condition that in case said remedies proved efficient cures their proprietor should have the benefit, for his private use and gain, of an official indorsement of the remedy, we did not think the investigation of such remedies for such purpose came within the range of duties properly devolving upon a commission appointed to make an investigation at the public expense for the public good, and therefore declined to answer all communications relating to such subjects. What valuable discoveries left in temporary obscurity by our course in the matter time alone must disclose. We must say that in this matter we were not influenced by a strict regard to the observance of a high-toned professional code of medical ethics, but entirely from a sense of the proper discharge of a public duty. The sick herd of Mr. Quinn, previously alluded to, was taken as one offering a fair opportunity for treatment. The sick animals were all in the formative stage of the disease, and surrounding circumstances seemed favorable to their cure. They were confined to proper limits, in a pen well situated as to health and comfort, and were given a dose of purgative medicine as a starting point, consisting of Glauber salts. It was observed by all with whom we conversed that a larger per cent. of recoveries occurred from among those animals that at the commencement of the disease had vomiting and diarrhea than from others. The dry and hard condition of the fecal matter found in the animals dissected leads to the belief that purgatives at the commencement of disease would always be a judicious course. Bromide of ammonia was then given in solution in doses of 30 grains every six hours. This remedy we tested at the suggestion of the Agricultural Department, at the instance of a gentleman who insisted that inasmuch as it exerted a salutary effect in the disease of cholera as affecting the human subject, it might prove equally beneficial in such disease in swine. So it might, but we did not find that an analogous disease, and therefore the remedy having no properties calculated to meet the character of the disease that *we did find*, proved of no practical benefit in its treatment, the animals dying in about the same proportion as when not subjected to any plan of treatment, but left entirely to themselves. Mr. Stadda's herd, in the same county, was subjected to the same plan of treatment with the same results. The herd of Mr. Thomas, in Harrison county, was treated under our direction by giving a mild purgative at the commencement of the disease, and during the acute inflammatory state of the complaint administered antimonials as a sedative to the circulation, and in the second stage tonics and nutritious food of milk, mill-feed, and vegetables, but the per cent. of deaths remained much the same as when not treated. Other isolated cases occurred under circumstances where extra care and effort was made in trying to effect a cure by several different lines of treatment, but candor compels the admission that as far as relates to the discovery of any plan of treatment proving sufficiently efficient to entitle it to respectable consideration, our efforts were without good results. And, lest our speculations and theories as to the proper line of treatment may be wrong, and present further obstacles in the way of the discovery of a successful remedy, we will refrain from giving them, preferring to present such points only as we fully believe will be of practical value.

I remain, very respectfully, your obedient servant,

D. W. VOYLES, M. D.

NEW ALBANY, IND., *November* 23, 1878.

REPORT OF D. E. SALMON, V. S.

Hon. WILLIAM G. LE DUC,
    Commissioner of Agriculture:

SIR: In my investigations of the contagious hog-fever as it exists in North Carolina, it has been my endeavor to decide those points which it was indispensable for me to know before adopting preventive measures, rather than others which might be equally interesting from a scientific standpoint. What is the percentage of loss from swine disease in this State? Is it one and the same disease from which the hogs are dying in the different parts of it? If but one, what are its symptoms, *post-mortem* appearances, nature, and cause? And what are the means by which such losses may be diminished or entirely prevented? These are the questions which it seemed most important to answer; they are those to which my time has been entirely devoted.

It was found very difficult to obtain information of localities in which the disease existed; for although requests were made through our newspapers for such information, and although, as I have since learned, swine were dying largely in every section of the State, I received during the whole time but three letters naming such localities. If to this we add that a large part of this State is without railroads; that the farms are large, and, consequently, the country is thinly settled; that usually but few hogs are kept on each place, it is seen that a great part of the time must have been spent in unproductive work in searching out infected localities, and, when these were found, in traveling from farm to farm to find herds suitable for experiment, or dead animals for examination. These facts must explain the small number of experiments which I was able to carry out.

To give a connected view of the subject, and one convenient for reference, the report is presented under the following headings:

I.

THE LOSSES OF SWINE.

a. Extent of disease, number and percentage of deaths.
b. Are the great bulk of these losses caused by one disease, or are they more equally distributed among all those to which these animals are subject?

II.

THE CONTAGIOUS HOG-FEVER.

a. Symptoms.
b. Post-mortem appearances.
c. Nature.
d. Cause.

III.

MEANS OF PREVENTION.

a. Hygienic and medical treatment.
b. Sanitary regulations.

EXTENT OF DISEASE, NUMBER AND PERCENTAGE OF DEATHS.

North Carolina is a State with a great diversity of soil and climate. In the western or mountainous part the summers are not excessively hot nor the winters extremely cold, and, with the exception of river bottoms which are of comparatively small extent, the soil is rolling and naturally well drained; the water is good; there is no malaria, and the country is rightfully considered a very healthy one. Extending from the mountains for two hundred miles eastward is a strip of country much of which is not sufficiently rolling for good drainage through the compact subsoil, and in a large part of which intermittent fever prevails to a considerable extent among people. Still farther east is a strip of sandy and swampy country, extremely malarious, and very subject to intermittent fever and other diseases of malarial origin.

Now, if our hogs were dying of unhealthy surroundings; if their disease or diseases originate to any extent from malarious emanations, it is certainly in this eastern belt that we should expect to find by far the largest percentage of losses. We should not be disappointed in finding a few in the central belt, but in the healthy, elevated west, where the hogs roam in vast mountain forests, we should certainly expect an unusual freedom from disease, especially in summer. Viewing the matter from this standpoint, I visited the western and central sections, and would have gone to the seaboard if my own health had not failed me at this point.

Fortunately statistics have been collected of the number of deaths among swine in the different parts of the State for the year ending April 1. 1878, and these, as far as can be obtained (twenty-three counties only out of ninety-four), are as follows:

Counties.	Total number of swine.	Number of deaths.	Counties.	Total number of swine.	Number of deaths.
Bertie	22,286	5,151	Lenoir	16,604	3,853
Buncombe	12,076	3,194	McDowell	6,011	2,363
Burk	6,341	1,940	Martin	12,755	3,670
Camden	5,586	2,158	Mitchell	8,972	1,380
Chatham	27,858	9,103	Pender	14,964	1,977
Cherokee	5,183	538	Person	12,789	3,084
Clay	4,998	1,286	Richmond	10,630	1,192
Craven	11,446	3,493	Robeson	27,411	3,764
Cumberland	13,466	2,006	Rowan	14,409	1,943
Currituck	7,064	2,451	Wake	17,448	4,112
Franklin	16,045	6,359			
Guilford	22,392	1,041	Total	304,492	66,946
Hyde	8,358	888			

That is to say, hogs have died to an alarming extent from Cherokee, Mitchell, and Buncombe counties in the mountains, to Camden, Currituck, and Craven on the seaboard. Nor was the year above reported an exceptional one, as these losses are now being repeated in Haywood and Yancy in the west, and from thence in localities eastward to the sea. Speaking in round numbers we have reports here from one-fourth of the counties in the State, and these counties in 1870 contained about one-fourth of the hogs in the State, and contain now very nearly the same number as then. We may, therefore, estimate the losses in the entire State at four times the number in these counties, say 260,000. Taking the counties mentioned, the loss amounts to 21½ per cent. of the whole stock, and ranges from 38½ per cent. in Camden to only 4½ per cent. in Guilford.

ARE THESE LOSSES THE RESULT OF A SINGLE DISEASE?

This question has been raised again and again, whenever any measure has been proposed for diminishing the death-rate of these animals, and notwithstanding investigators in widely different localities have observed similar symptoms and similar *post-mortem* appearances, the great objection to sanitary laws has always been the uncertainty in regard to the affection or affections from which death occurred. It, therefore, seemed advisable to visit a large part of the State in order to decide this question of primary importance. The disease was seen by the writer in Haywood, Buncombe, and McDowell counties, in the mountain district, in Rowan, Mecklenburg, Lincoln, Gaston, and Alamance, in the central belt, and particular inquiries were made of those who had observed it in the counties bordering on the coast. Several counties not enumerated above were visited, but I was not successful in finding infected localities. My greatest regret is that I was not able to make personal observations in every part of the State.

In each of the counties mentioned a considerable number of herds were visited and examined, and without exception the living animals presented similar symptoms, and the dead ones showed similar changes in the different organs of the body. Slight variations were of course observed, as is always the case in any disease, but these were as great between different individuals of the same herd, sick at the same time, as between different herds, even in different counties. And, what is of great importance, I did not find a single case in which it could possibly be supposed that death resulted from a local disease; but in every case a variety of organs, belonging to different apparatus, were found diseased; the blood often showed marked changes; there were extravasations in various parts of the body, and always inflammation of the lungs and large intestines, generally, also, of the heart, and often of the eyes; the skin, too, was often plainly affected, and the temperature was found to be increased before any other symptoms of disease were in the least apparent.

Considering all these facts, there can be no doubt that these animals all died of a general disease—a disease not caused by changes in any single organ; but, on the contrary, a disease which caused the various organic changes observed. Again, from the similarity of symptoms in all these cases which I saw, and in those reported to me from other parts of the State, and from the correspondence in *post-mortem* appearances, there can scarcely remain a shadow of doubt that the great mass of the hogs dying in North Carolina are affected by one and the same disease.

SYMPTOMS.

An increase of temperature precedes for an undetermined and probably variable length of time the appearance of all other symptoms. In one lot of seven ten-months-old pigs, only one of which showed symptoms of disease, the six remaining had a temperature varying from 103.6° F. to 106° F., and this temperature was preserved unaltered for six days, with no other changes in the condition of the animals than increased dullness of the eyes, a general unthrifty condition and a disinclination to search for food, although the appetite was still good. The pig first affected died about this time, and a *post-mortem* examination left no doubt of the disease.

In another lot of ten three-months-old pigs, but one of which was plainly sick, six had a temperature ranging from 104½° F. to 107° F.; with one this was 103½° F., with two 101° F. and 102° respectively, while with the sick one it reached 107.4° F.

In a herd of twelve, from which one had just died, and one was plainly sick, four others showed a temperature from 103½° F. to 107° F.

In a lot of fourteen animals, one had died, one was plainly sick, and three others had a temperature from 103° F. to 104° F.

Of five pigs, one had just died, three had a temperature of 105° F. to 106° F., and the remaining one 103° F.

Of eleven hogs, two had died, one was plainly sick, and five had a temperature ranging from 103° F. to 106° F.

From these and similar cases it has seemed probable that a high temperature may exist several weeks before other symptoms are manifested, or even that the disease may in some cases be confined to, and run its course in, the blood, without a localization in any organ or organs. Such a view is also sustained by the often-observed fact that when the cholera exists in a herd, animals, which show no positive signs of sickness, are found in an unhealthy condition, and cannot be made to thrive and fatten. This point, however, remains to be cleared up by future investigations. An objection may be brought to the lower temperature here recorded, that according to other observers it is common to find a temperature of 103° F. to 104° F. in healthy animals. This, however, does not agree with the observations which I have been able to make. In one herd of ten, the last of a much larger number which had been reduced by this disease, all of which appeared healthy and thriving, not one showed a temperature by my thermometer as high as 103° F. In several other herds of healthy animals which I examined, but notes of which were not preserved, the temperature was found to range from 96° F. to 102½° F. In nearly all these cases the animals were called up from fields where they were running at liberty, and were immediately examined. So that, although there may be differences in thermometers, I think there can be little doubt from these observations that an increase of temperature precedes other symptoms by a number of days.

The first symptoms apparent externally are a dullness of the eyes, the lids of which are kept nearer closed than in health, with an accumulation of secretion in the corners; there is hanging of the head with lopped ears, an inclination to hide in the litter, to lie on the belly, and keep quiet; as the disease advances there is considerable thirst, more or less cough, a pink blush, rose-colored spots, and papular eruption on the skin, particularly along the belly, inside of thighs and fore-legs, and about the ears. There is accelerated respiration and circulation, increased action of the flanks in breathing, tucked-up abdomen, arched back, swelling of the vulva in the female, as if in heat; sometimes, also, of the sheath in the male; loss of appetite, and tenderness of the abdomen; occasionally there was persistent diarrhea, but generally obstinate constipation. In some cases large abraded spots are observed at the projecting parts of the body, caused by separation and loss of the epidermis; in these cases a slight blow or friction on the skin is sufficient to produce such abrasions. In many cases the eruption, blush, and spots are entirely absent; petechiæ were formed in about one-third of the cases; in one outbreak, chiefly confined to pigs in which the eruption was remarkably plain, there was considerable inflammation of and discharge from the eyes. Some animals have a very disagreeable odor even before death. In nearly all cases there is weakness or partial paralysis of the posterior extremities, and occasionally this paralysis is so complete in the first stages of the disease as to prevent walking or standing.

The percentage of animals affected and the violence of the symptoms vary greatly, according to the time the disease has existed in a locality. In the early part of an outbreak from 70 to 90 per cent. die, and most of these in the first stages of the disease, from deterioration of the blood or apoplexy. In one case there was a loss of 102 out of 107 head; in other cases whole herds of 30 or 40 succumbed; later, many of the animals linger for weeks, and finally die from persistent lesions of the lungs or bowels. In some instances a considerable number of those affected—20 to 25 per cent.—recover; many of these lose all their hair, and often the epidermis as well. Of those recovering, a very few fatten rapidly and do well, but by far the greater part cannot be fattened, and are always unthrifty and profitless animals.

POST-MORTEM APPEARANCES.

In about one-third of the cases petechiæ and larger blood extravasations are seen on the thinner parts of the skin; in a somewhat larger proportion of cases the abraded spots, already mentioned, are present; making a section through these, the skin appears thickened and of a very high color, but the sub-cutaneous tissue is not appreciably altered. In one or two cases there was no effusion in the abdomen, but in all the rest this cavity contained a variable quantity of liquid—sometimes of a bright yellow color and clear, sometimes of a straw color, and very often turbid and mixed with the coloring matter of the blood. In every case the colon and cæcum were plainly affected, reddened externally, and internally showed changes varying from simply a deep coloration to inflammation and great thickening; in some cases they were studded with petechiæ, in others there were none; ulcers of various sizes were frequently found, and also thickened fibrous, concentric patches, occupying sometimes nearly the entire walls of these organs. In one case there were large blood extravasations in the walls of both colon and cæcum, distending them to a thickness of half to three-fourths of an inch; on section, these spots had the appearance of a clot of black blood; they were firm and tough and did not yield to scraping with a knife. Round, firm nodules, one-half inch in diameter, were frequently found in the walls of these bowels, which, on section, were of a grayish-white color, and appeared to be composed of compact fibrous tissue, with the exception of one case in which they were less firm, and presented the appearances of the extravasated-blood patches already described. With the exception of petechiæ the small intestine was nearly always normal; in one case there were two or three patches of inflammation one to two inches in diameter. The rectum was congested or inflamed in spots only; there were occasionally the nodular masses mentioned above, but in a majority of cases this part of the intestine showed little or no change.

The stomach in one-third of the cases was unchanged; in the remainder there were patches of inflammation from the size of the palm of the hand to the involving of half of the surface of this organ. Sometimes this was confined to the mucous coat, but often implicated the whole thickness of the walls.

The cavity of the thorax in every case contained a considerable quantity of a turbid, bloody liquid, in some cases nearly black in color; the pleuræ were generally thickened and covered with false membranes; the lungs were constantly found inflamed, occasionally in a few small spots only, but generally the greater part of the lung tissue was involved. Often these organs were greatly congested throughout, and would break down under the slightest pressure. The bronchial tubes were also found congested or inflamed, and contained considerable frothy mucus, which in some cases entirely filled them. The pericardium was in nearly every case distended with a turbid, blood-colored liquid, but no false membranes were discovered, and only in one case a piece of coagulated lymph the size of a hen's egg was found floating in this liquid. The heart seemed to be congested throughout in most of the cases, and had patches of a deeper hue than the rest on its external surface. These patches were very suggestive of inflammation, but in the absence of coagulated lymph this may be considered doubtful. This organ at times contained clots of blood of different consistency, and always of dark color, and at other times all the cavities would be found empty. In all cases the blood was very dark, and generally formed an imperfect clot, and the lymphatic glands were enlarged and greatly congested. The larynx and pharynx were found normal in all the *post-mortem* examinations, but in some of the living cases there was considerable swelling about the larynx and ulcers on the posterior part of the tongue. The liver was generally as in health, though in some cases it was congested, spotted, and softened, and once was found smaller and more dense than natural. The bile was at times very thick and dark, and again very thin and of a bright yellow color. The spleen was normal in two-thirds of the cases; in the remainder it was slightly enlarged and softened. In two cases the interior was almost of a fluid consistency, while in one the organ was smaller and firmer than in health. The bladder was generally normal, but in two or three cases was inflamed and covered with blood extravasations about the neck, and contained in these cases bloody or very turbid urine. The kidneys were seldom more than slightly hyperæmic, but in a few cases there was considerable extravasated blood in the tissues about the hilum, and on section the substance about the pelvis was found infiltrated with perfectly black blood.

We have here a considerable variety of pathological changes, the only constant ones being congestion and inflammation of the lungs, colon, and cæcum, and congestion of the lymphatic glands. To mention any single peculiarities of these lesions as characteristic of this disease would not be possible from this investigation. Neither the thickened fibrous patches, the ulcerations, gray elevations of the intestines, the cuticular eruption, nor petechiæ were constant.

NATURE OF THE DISEASE.

In studying the nature of an unclassified disease the first question that occurs to us is: Is the affection a general or a local one? In other words, does the disease originate from functional or organic disorder of any particular organ or apparatus, or are the anatomical lesions developed secondarily as the consequence of a general affection? And this question, as regards the disease under consideration, can now be answered in a definite and satisfactory manner. Indeed, when we consider that the first symptom, and one preceding all others by several days at least, is an increase of temperature; that when localized a great variety of organs belonging to different systems and apparatus are involved, as, for instance, the nervous system, as shown by occasional paralysis and apoplexy, the lungs, pleura, bronchial tubes, heart, liver, stomach, intestines, spleen, kidneys, bladder, and skin; that there are considerable changes in the blood, as shown by imperfect coagulation, solution of the coloring matter, and blood extravasations, there can scarcely remain a shadow of doubt that the trouble is not a local but a general one.

The next question in logical succession relates to the contagiousness of the disease. Is its extension due to a principle which is multiplied in the bodies of sick animals, and which is of itself sufficient to cause the disease in healthy ones? In answering this question I will merely mention the experiments of Professors Axe, Klein, and Osler, which prove that the disease may be inoculated without detailing their facts; and I will only allude in like manner to the instances already recorded by Dr. Sutton, Professor Axe, and others, which seem to prove its highly contagious character. Most of these facts have been published in recent reports of the Department of Agriculture, and there is no need of repeating them. In my own investigations I have met with facts which entirely confirm the opinion of these observers in regard to this latter point. Thus I have found the disease to start at some point and spread slowly in different directions—not rapidly, as though depending on atmospheric conditions—and the rapidity of this extension depends to a very great degree on whether these animals are allowed entire liberty

9 SW

or whether they are kept on the premises of the owner. In Mecklenburg county no stock is allowed to run at large, and the disease existed during the present year, in some localities, from early in the summer, and up to October first by far the greater part of the country was free from it; while in Alamance county, where no restraint is put on the animals, the disease spread from one extremity of the county to the opposite in a few weeks. In each of these outbreaks, and, indeed, in every one I have observed, it is no difficult matter to find one locality where the hogs have nearly all died and the disease has finished its work some weeks or even months before, while in almost every direction, at a distance of five, ten, or fifteen miles, these animals are just taking the affection; that is, the disease has extended and is extending, and it has required this length of time to travel this short distance. Can it be possible that an atmospheric or climatic change would travel no faster than this? Again, if dependent on such conditions, why do we find one township devastated by it and another not many miles distant entirely free from it? Such instances are very apparent in Haywood, Mecklenburg, Lincoln, and Gaston counties at this writing, and were not less so in Buncombe county in 1877. If it is claimed that this depends on the condition of the soil, it is only neccessary to reply that in the outbreak just mentioned, in Buncombe county, there are no facts to justify such a theory. In Swannanoa township, which is high, rolling land, with very few bottoms, no swamps or malaria, and which cannot be surpassed for healthfulness, the loss was 60 per cent. of the whole stock; while in Upper Hominy, which has no advantage over Swannanoa in healthful location, but which is more remote from thoroughfares traveled by western droves, the loss was only 2 per cent. It was probably entirely free from this disease.

A large number of instances could be produced of outbreaks in this State, particularly in the western part of it, clearly traceable to infected droves, and this is, above all, the cause with the first introduction of the disease. It is difficult to establish exact dates, but all accurate testimony points to 1859 as the first appearance of this trouble. Some think the earliest outbreaks might have been a few years before that date, but of this I have been able to get no evidence. Mr. Morris, of Polk county, remembers that a drove stopped at his place in 1859; that some of the hogs died there of the disease, and that soon afterward this malady spread among most of the hogs in that locality. This was the first appearance of the trouble in that county. Mrs. Davidson, of Buncombe county, remembers that during the life of her father, who was a large hog-raiser, and who lived on the route followed by the droves, no hogs were lost by this disease, but that about the time of his death (1858) droves came through with sick animals, and that this was the first appearance of the disease in that locality. Many other people who cannot remember dates are positive in the opinion that the disease was introduced by droves from Tennessee and Kentucky. One man remembers that he was employed by the drovers to kill the animals that were sick and cure the meat. He also remembers that these animals had diseased lungs, and such a bad odor that they could scarcely be dressed. This was his first experience with the disease known as "hog-cholera." Colonel Polk, our present commissioner of agriculture, informs me that the first appearance of this disease in Anson county was in 1859; that it was undoubtedly brought there by western droves, and that these animals died to such an extent that the drovers took them secretly to the woods and buried them under brush and rails to conceal them. A drover who sold his hogs in Georgia at that time informed me that the disease was first introduced in that State in 1859, and that he had no doubt it was carried there by the droves. Indeed, I have found but one opinion among those best informed on this matter, and that is, that the disease was never known in this section till introduced by animals driven from Western States; and in some sections of this State, a part of Alamance county for instance, the disease never existed till the present year.

Judging from all these facts, therefore, we cannot escape the conclusions that this disease is a contagious fever.

In this connection there is one more question that is generally raised by those discussing the nature of this fever, and that is, does the disease always originate from pre-existing contagious germs, or is it often or generally developed *de novo* as a result of improper hygienic surroundings? In the consideration of this question I shall confine myself to the facts brought out by the investigation in this State, simply premising that most of these facts are as true of the Middle States and probably of most of the Southern States as of North Carolina. The first point that attracts attention is the fact that this State was free from the disease till about 1859, certainly till it was introduced by droves from other States, whatever the date may be; hogs had been kept in th s State from the time of its first settlement undoubtedly under similar hygienic conditions, and yet the disease had not appeared up to that time, when it was brought by imported animals, just as England was free from contagious pleuro-pneumonia up to 1842, when it was imported with animals from the Continent. It is claimed that in the west the disease is produced by overcrowding and filth, but I doubt if these animals are crowded any more now than forty years ago; indeed, I was surprised at the results of my investigations on this point, for, in all the time I have been visiting infected localities, I have not found a case of overcrowding, and not more than two or three where there was anything like filthy surroundings. In the western part of the State most of the hogs are kept in the large mountain forests, or are at least allowed the run of the highways and commons; in the east they either run in the highways and old fields or have ample pastures. If it originates from restricted range and unhealthful climatic conditions, it is certainly in the east that we should expect to hear of its originating and proving most disastrous; but it was known in the mountains as early as in the other parts of the State. And if we examine the list of counties which I have given above, we shall find it as fatal in the elevated and heathful west, with its immense mountain ranges, as in the malarious east. I append some conspicuous examples of this:

Loss in eastern counties.	Per cent.	*Loss in western counties.*	Per cent.
Camden	38	McDowell	37
Lenoir	21	Buncombe	25½
Robeson	14	Mitchell	15½
Hyde	10½	Cherokee	10½

We find here, then, just as large losses in the west as in the east, and just as small ones in the east as in the west; in other words, the disease rages irrespective of these climatic and hygienic extremes; and this becomes still plainer when we add that in Swannanoa township of Buncombe county the loss reached 60 per cent.

Of course, at the present time, as with all contagious diseases which have existed for several years in a country, there are some outbreaks which it is impossible to trace to their source; and it seems probable that the contagion may be preserved over winter in manure, straw, litter, or in the remains of unburied animals which died the preceding year. There are some outbreaks that cannot well be explained otherwise, and, indeed, there is no reason to doubt that this may be the case; contagious germs may also undoubtedly be carried a considerable distance by other animals or birds, and it is for this reason that many farmers have concluded that pasturing hogs on wheat-fields produces the disease; but hogs were pastured on wheat-fields as well thirty years ago as now; why did not the same result follow then?

I have concluded, therefore, after a careful study of these facts, that this contagious disease does not originate *de novo* in North Carolina; and that if the contagious germs now in the State can be destroyed and their importation prevented, we shall be as free from it in the future as we were before its first importation, about the year 1859.

HYGIENIC AND MEDICAL TREATMENT AS PREVENTIVES.

It was one object of this investigation to determine if the best hygienic conditions, clover pasture, large range, and variety of food have any preservative influence against this contagion; and while a large number of cases where these conditions seemed perfect could not be collected, the few that were observed prove that these alone are absolutely powerless to keep off the disease. Thus, Mr. Wadsworth, of Charlotte, lost 117 animals, nearly his whole stock, which had the run of a clover-pasture and large wood lot, which had in addition slops from the city hotels, and grain. In this case disinfectants were freely used. Mr. Davidson, of Hopewell, lost 50 per cent. of his herd under similar conditions. A herd kept at a slaughter-house, in Charlotte, which had other food as well as the refuse, was the first to take the disease, and suffered to the same extent as others. Indeed I met with hundreds of cases where animals had large pastures and other food in addition daily, where such popular preventives as salt and ashes, sulphur, tar, oil of turpentine, charcoal, and copperas were freely and regularly given, where the majority of the animals were neither too fat to be vigorous nor so poor as to be wanting in this respect, and yet from 50 to 90 per cent. succumbed to this affection. In one case where I had the tincture of chloride of iron given regularly as a preventive, commencing before any of the animals showed even an elevation of temperature, and where they were in a large pasture at a considerable distance from any others, the disease has appeared: two have died and others will probably follow.

Some experiments were made with bisulphite of soda, salycilic acid, bichromate of potassa, and bromide of ammonia to determine if these have any power to arrest the disease when given before any symptom but increased temperature had appeared; the results of these were as follows:

Agents.	Number of animals.	Beginning of temperature.	Dose per day.	Length of experiment.	Final temperature.
Bisulphite of soda.				*Days.*	
Experiment No. 1..............	6	103. 6° to 106° F...	4 drachms	7	96° to 99° F.
Experiment No. 2..............	4	104½° to 107° F...	1 ounce	4	102½° to 105° F.
Experiment No. 3..............	3	103° to 104° F...	1 to ½ ounce.......	7	103° to 106° F.
Salycilic acid.					
Experiment No. 1..............	4	104½° to 107° F...	30 grains..........	7	100° to 101° F.
Experiment No. 2..............	8	103° to 106° F...	45 grains..........	6	103° to 105° F.
Bichromate of potassa.					
Experiment No. 1..............	3	103½° to 107° F...	½ grain............	7	103° to 105° F.
Bromide of ammonia.					
Experiment No. 1..............	4	103° to 106° F...	23 grains..........	7	103° to 106° F.

These experiments show that none of these agents can be depended on to stop the changes going on in the blood as a consequence of this disease. Although both bisulphite of soda and salycilic acid in one experiment each appeared to accomplish this they failed in other cases where given in larger doses for an equal length of time; and when we consider that in no contagious fever has a remedy been discovered capable of arresting the course of the malady, the doubt in regard to the efficacy of these agents in this disease must increase.

SANITARY REGULATIONS,

We are finally brought to the irresistible conclusion that sanitary regulations properly framed and enforced are the only means at our command for checking the ravages of this disease and relieving our farmers from the enormous losses at present occasioned by it. We cannot expect, however, that this desirable object will be accomplished without considerable expense, especially in the first years of the attempt. We must expect outbreaks in all parts of the country where the disease has previously existed, caused by contagious germs which have been preserved in some of the ways already mentioned; but we should be encouraged by the fact that in most parts of the country, at least, these germs, unless especially preserved in straw, manure, remains of dead animals, &c., are entirely destroyed during winter. Thus, in Swannanoa township, where 60 per cent. of the hogs died in 1877, there has been no outbreak up to October 30, 1878. Above all must we realize the *necessity* of thoroughly destroying every particle of contagion wherever it appears. Although this would undoubtedly be very expensive, it would certainly be a great saving, even at the start, on the great losses which we are now annually experiencing; and if the work is thoroughly done we may expect that this expense will be reduced to a comparatively small item in the course of a few years. At the worst such expense would be much less than the use of a specific by individual farmers, even if such a remedy were discovered. In regard to such regulations I would suggest the following points as necessary according to what is now known of the disease:

1. The regulations should go into effect in winter or early spring when fewest animals are affected, or when, as my experience indicates, the disease is entirely extinct.

2. People living in localities where the disease has prevailed within two years should keep their hogs in an inclosure free from accumulations of manure, straw, litter of any kind, or remains of dead animals in which the contagion might possibly be preserved, and in which there were no sick hogs the preceding year.

3. That in such localities, *i. e.*, where the disease has existed within two years, it should be made obligatory for persons owning hogs to report each and every death occurring in their herds promptly (within forty-eight hours if but one, or twenty-four hours if more than one, or if others are sick), to a designated person to be located in every township or county, unless such deaths were plainly caused by mechanical injuries, drowning, maternity, &c. And that there should be districts established of convenient size, in each of which a competent veterinarian (or physician in case the veterinarian could not be obtained), should be appointed, to whom the above township or county officer should report whenever two or more such deaths have occurred in the same herd within a fortnight; whenever an unusual number of deaths have occurred in any locality, or whenever there is any reason to suspect the presence of this disease.

4. On receipt of such report the veterinarian should visit the locality and make a careful investigation into the nature of the disease, using the clinical thermometer and making *post-mortem* examinations.

5. If the contagious fever is indicated the whole herd should be slaughtered, the animals deeply buried, the place thoroughly disinfected, and no more hogs allowed there till after a succeeding winter.

6. When the disease exists to any considerable extent in a locality, those owning hogs in adjoining townships or even counties, according to the extent of the outbreak, should be required to keep them in small inclosures or pens, at a distance from roads or streams of water coming from infected localities. This is necessary to lessen the danger of infection and to allow more thorough disinfection in case the disease appears.

7. A certain compensation should be allowed for slaughtered animals—say 25 per cent. on a fair valuation for those plainly sick, 50 per cent. for those which simply show a rise of temperature above $103\frac{1}{2}°$ F., and full value for the healthy ones.

8. In case a hog-owner fails to comply with above regulations a penalty might be fixed, or at least such a person should receive no compensation for slaughtered animals.

These are the regulations that seem to me most necessary, but there may undoubtedly be circumstances in which these may be advantageously modified. Thus in case of a herd of several hundred animals, in which but few are affected and the remainder show a healthy temperature, it might be advisable to simply kill and bury the *affected* ones, to thoroughly disinfect the premises and to kill others as soon as a high temperature becomes apparent. Or in case all were killed the meat of the healthy ones might be preserved and marketed. It is also possible that, through negligence in making reports or an improper diagnosis of the disease, such a large territory may become infected as to make it advisable to establish a sanitary cordon, isolating the locality as much as possible; and leave the disease to run its natural course. In such cases no live hogs should be allowed to leave the infected section till after a succeeding winter, nor any carcasses of hogs till after freezing weather; people living within this district should be prohibited from going near swine outside of it, nor should drovers or others from outside be allowed to visit the infected swine. All dead animals should be promptly and deeply buried, and disinfectants freely used. All hogs in such district, and for twenty miles distance from it in all directions, should be kept in small inclosures at a distance from roads, in order to lessen the chances of extension and to allow thorough disinfection.

If such regulations are thoroughly carried out there can be no doubt that the ravages of the disease will be greatly diminished at once, and in a few years many States which now suffer terribly from it will be completely exempt: while in those where it now proves most disastrous there is reason to believe it would never cause serious losses. Sanitary regulations similar to these are the only means that have ever been successful in combating the contagious diseases of animals, and while we would not be understood as discouraging the search for specific remedies we cannot disguise our opinion that it is extremely irrational and absurd to delay action in this disease till such specific shall have been discovered; in other words to neglect those measures which have alone succeeded and cling to those which have always failed.

Respectfully submitted.

D. E. SALMON, V. S.

SWANNANOA, N. C., *November* 15, 1878.

REPORT OF DR. ALBERT DUNLAP.

Hon. WM. G. LE DUC,
    Commissioner of Agriculture:

SIR: On the last day of July, 1878, I received from you a "commission to act for the Department of Agriculture in the examination of diseased animals," accompanied with printed instructions directing me to particularly examine into causes of the disease known as "hog cholera." I interpreted my instructions as follows: Find out what disease or diseases are destroying the swine and the symptoms of the same; the causes, both predisposing and exciting; the stage of incubation, morbid anatomy, &c., and to discover how far attention to hygienic care will prevent the spread of the disease in infected herds and its inception in healthy droves; and in addition to test the value of various medicinal remedies for curing the sick and preventing the spread of the disease. Recognizing the primal fact that the hog is an animal of short life, low vitality, and of comparatively little pecuniary value, singly, as compared with other domestic animals, and that they are kept in large droves by most Western farmers, I considered it of little profit to attempt to meet each special symptom with its appropriate remedy; but rather, after having fully diagnosed the disease or diseases, their nature, causes and lesions, and the predisposing causes which had assisted in the spread of the same, to try and devise a system of treatment, both hygienic and medicinal, which could be used in the treatment of large droves already infected, and reduce the liability of healthy droves contracting the disease. I do not claim for this report any degree of perfection. The limited time allowed only permitted the examination of the disease under certain climatic influences, and not through the various seasons of the year. I am, therefore, only able to report on the diseases which came directly under my own observation in this State (Iowa) during the two months of investigation, briefly referring to cases of diphtheria which I carefully observed last winter, and of which I have seen no cases during this investigation.

The medical literature upon the subject of the diseases of swine was very limited, and I could find no strictly scientific work treating upon the topic. I was, therefore, forced to fall back upon my knowledge of the diseases of man as a foundation, and after having fully examined the symptoms and morbid lesions in a series of cases selected out of an infected drove, I compared those symptoms and lesions with like symptoms and lesions found in man, and thus arrived, I think, at correct conclusions as to the proper name of the diseases under consideration. I was thus materially assisted in tracing out both the predisposing and exciting causes of these ailments. To the casual observer it may seem absurd to form conclusions in regard to diseases of swine from a previous knowledge of the diseases of man, but when we consider that the hog resembles his two-footed brother in many respects, has a similar alimentary canal, like viscera, the same system of blood-vessels and nervous structure, is also omnivorous, and that the diseases under consideration are caused by specific blood poisons, which act in like manner on man and brute through the process of inflammation, we can but conclude that if we find a set of certain classified symptoms in a hog with a distinctly marked uniform set of pathological lesions, and a similar set of symptoms in man with like morbid lesions, that these two are one and the same disease, and should bear the same title, especially when we can trace the cause in both cases to the same exciting agent. I have been forced to depend entirely upon my own observations for the material of this essay, and I will say in defense of the position or theories I advance, that they are my conclusions after inspecting over three hundred herds of diseased swine in various counties of this State, and after a careful dissection of nearly one hundred diseased animals. In justice to the farmers of Iowa, it is my duty to state that I received much valuable assistance from their hands. During the progress of my investigations prominent symptoms were pointed out by farmers who had made the disease a study, and I am only sorry that I cannot give each one credit for his particular contribution. I made my "headquarters in the field," and strived to obtain a thorough knowledge of the subject in all its details. I was forced to abandon the use of the microscope after a few days' trial.

HOG-CHOLERA.

Definition.—Any contagious or infectious disease attacking swine with usually fatal results. This definition will include all fatal diseases that are contracted by one hog from another, either by direct contact or by contact with the discharges or exhalation of any diseased animal, or the gases arising from any contaminated matter. Under this head can be properly included the three diseases I have discovered during my investigation, viz., *diphtheria, typhus,* and *typhoid fever.* The definition will exclude worms, lung-fever, pneumonia, pleurisy, or any special inflammation of internal viscera which are the results of climatic influences, vicissitudes of weather, or improper food. I am led thus accurately to define the disease and draw the line of distinction, because I have repeatedly found droves of swine suffering with so-called hog-cholera, when, in reality, there was no contagious disease whatever prevailing, but they were sick and dying because the rules of common sense had not been observed in their care. Because a number of hogs in a drove are taken sick at one time and with like symptoms, it does not follow that they are suffering from any contagious disease, and the sooner the fact is impressed upon the farmers the better it will be for their pockets. Often it is not medicine that is needed but a change of food. I will give a few cases which will best illustrate the ideas I wish to convey. Mr. B. kept his swine in a lot of one acre, more or less, where they had but little exercise, regular food, and sheltered bed. After gathering his corn he turned his entire drove into the field to glean. They also had the range of a forty-acre wood lot. Two days after he found a number of his shoats sick, five of which soon died. The disease was pneumonia or lung-fever. Morbid anatomy in each case showed at least one lung hepatized and inflammation of pleura.

Cause.—The hogs were previously confined without exercise and had regular food and sheltered bed. They were then turned out in large range, exercised fully (especially the shoats), slept on bare ground at a time when the weather changed suddenly colder, and the result was lung-fever and death. No medication was needed to prevent the healthy shoats contracting the disease, and a little care and simple medication would have probably cured the sick.

Mr. M. kept his hogs on a clover and grass range. They had stagnant water for drinking, and sour, fermented swill was fed freely twice a day. The land was flat river bottom, with black soil; ringers were used to prevent rooting; no roots, vegetables, or corn were given. The natural result of such errors in diet was sickness, emaciation, and death. First, the young pigs pined away; sudamina appeared upon the eyelids, nose, and ears, and one animal after another was attacked with convulsions and died. The brood sows and stock hogs soon followed in the same way, and when I visited the farm fifty out of eighty head had been cut off in this useless and unprofitable way. Three sick hogs were killed and dissected. The lungs were white, but showed no signs of organic disease. The kidneys were light colored and showed some irritation in tubules; all internal viscera without organic disease. There was a lack of red corpuscles in muscular tissue, which appeared almost white. The disease, in this case, was simply starvation. As yet no contagious disease has appeared in the herd, but the hogs were in such a condition that if exposed to the slightest miasma they would inevitably contract any contagious disease, and, with the debilitated blood to begin with, would rapidly succumb to it. Now, I assert that although the drove was supplied with abundance of food in kind, yet it was not the nourishment demanded. There was an excess of certain constituents and absence of others necessary to health. Every article of food furnished this drove contained acid. This was the case with the clover, grass, and slops given them. The water was poisonous also, and they were deprived of the alkaline salts necessary to life. The small quantity they might have obtained from the ground was made inaccessible by the rings in their noses.

In this drove the tongues of the hogs were large, white, and flabby, indicating plainly the need of change of diet. There are many other errors in diet which will be alluded to when we come to speak of the predisposing causes—errors which do not cause death, but which render the hog peculiarly liable to contract contagious diseases, and also increase the expense of feeding.

I will now give an illustration of a case where too much care, misdirected, caused disease and death: Mr. C. builds a so-called model pig-pen. It is low and tight; the sun and air are excluded; the floor is of boards, and is raised above the ground. To prevent dampness, straw is furnished liberally to keep the hogs warm. The feed-lot is exposed to the north and west winds. The hogs, sleeping in this damp place, with cold boards under them, pack closely together in the damp straw, for, no matter how dry the straw may be when put in, in the course of a few hours it will be wet and loaded with ammonia. Mark the results. At reveille they come from their sheltered house wet and heated, pass into the feed-lot exposed to the bleak north wind or cold rain from the west, and the natural consequence is coughs, colds, bronchitis, pleurisy, lung-fever, inflammation or irritation of some internal viscera from the sudden check given to perspiration, or sudden change of temperature by the inhaled atmosphere. If the exposure is not sufficient to cause a fatal inflammation, it will cause a bronchial irritation, as shown by cough. The system is vitiated, and any contagious disease prevailing in the vicinity is liable to attack the drove. The owner reports the cough as existing for one or two months as the first symptoms. In this case the cough was caused by errors in care, and was but a symptom telling the farmer that his swine had contracted a cold, and that this disorder of the system would debilitate and render them more liable to contract any contagious disease to which they were exposed.

We will now take up in their order the three diseases which come properly under the title of "hog-cholera," that is the diseases which answer to the definition we have given of hog-cholera. We do not claim that these are the only contagious diseases which are known to cause death. There may have been others in past years, or even in this year, but they did not come under my observation, and having accurate reports from many prominent and intelligent stockmen in all the Western States, detailing the symptoms in their infected hogs, I can but conclude that these three diseases are the only contagious diseases which have attacked hogs in the last two years. After describing each disease, its symptoms, course, stage of incubation, pathological lesions, causes of death, and exciting causes, we shall take up the subject of predisposing causes toward the contraction of these diseases. Then we shall point out the best plan of treatment, both hygienic and medical, for curing the sick and preventing the spread of any contagious disease among healthy animals.

TYPHUS FEVER.

Definition.—A specific continued fever, attended with increased temperature, usually above 105° F.; stupor; congestion of brain; swelling of forehead; stiffness of joints; excessive soreness of all tissues: a profuse eruption on the belly and inside of thighs, with costive bowels during the first few days, and usually terminating in death within fourteen days.

Symptoms.—Headache, as shown by wrinkled forehead; partially shut eyes; nose held near the ground; loss of appetite; stupor; indisposition to move: excessive soreness of all tissues, the slightest pressure causing excessive pain; swelling of forehead between the eyes; tongue generally large, white, and flabby, especially if the disease is complicated with malarial poisoning. There is also great restlessness, shortness of breath, and cough. The sick hogs are frequently lame in one limb, and cannot even put it to the ground. The heat of the body is excessive, the temperature rarely ranging below 105° F., and generally reaching as high as 108° to 109° F.; and if the hog is not carried off from the fifth to the seventh day a copious eruption appears on the bowels and on the inside of the thighs and other soft parts. The bowels are almost always costive during the first week and the discharges hard and dark colored. Thirst is excessive, and the hog will often drink until it falls over dead. During the second week we have increase in the severity of symptoms. Sordes collect in mouth; small watery pimples appear on nose, eyelids, and ears; there is great prostration of strength, with staggering gait when forced to walk. Costiveness may now give place to diarrhea; urine is passed while lying down, and convulsions or fatal stupor intervenes; enlargement of glandular structure, especially in the neck, is a common symptom, but in no case have I found abscess with healthy pus, but rather thin sanious fluid. A common symptom during the second week is thumps, and I have never known a case to recover when this symptom was present. The thumps appear to be nothing more than a spasmodic action of the nerves, like hiccough in man, and denotes great prostration and approaching death. In advanced stages of the fever these are the main symptoms, and this alone is a common course of the disease, as I have observed it. But there are many exceptional cases. Many hogs, especially those debilitated by errors in food or from the effects of malaria, will succumb to the influence of accumulated poison acting on the brain and nervous system, and die within twenty-four hours. This is of frequent occurrence, especially in young pigs and shoats. Others will die from obstinate constipation, the impacted feces causing ulceration and rupture of descending colon and rectum. In some herds convulsions, from congestion of the brain, occur during the first day, and unless relieved the case terminates in death in a few hours. Tubercular deposit in the lungs and in the mesenteric glands is very common. In this disease, as also in typhoid fever, the smouldering spark of scrofula is fanned into a flame by the fever, and the tubercular matter is deposited in the lungs and glands, and the patient that might have recovered from the fever is carried off with consumption. The odors of the exhalations are peculiar, and will at once diagnose the disease from any other. To describe this peculiar smell would be impossible in words.

Duration.—The duration of the disease is variable. Many animals die within a few hours, but if the bowels are emptied by saline cathartics or injections, the animal generally lingers into the second or third week before the crisis will occur. The prognosis is very unfavorable, especially in large droves, where little can be done to relieve symptoms. Our advice is. in all large herds where this disease obtains access, to destroy at once the sick animals, burn or bury the carcasses of the dead, and labor to check the progress of the disease by prompt hygienic measures. In small droves, or where the stock is of peculiar value, an effort may be made by the use of medicinal agents and care to relieve the symptoms and guide the case to health. But in large herds this effort will be found unprofitable. When we remember that in man, with all the advantages of a thorough knowledge of the disease, with skilled physicians and competent nurses to care for the sick, many of those attacked in crowded armies succumb to the influence of the disease, we certainly cannot advise farmers having large herds to attempt remedial measures. Another argument against attempting to cure those having well-marked symptoms of the disease is that, if there is the slightest taint of the scrofulous diathesis in the blood, the spark will almost certainly be fanned into a flame, and the patient, reacting from the specific fever, will be carried off by deposit of tubercular matter in the lungs or mesenteric glands. Now, we know that the hog is an animal of low vitality, and, in a majority of cases, of scrofulous habits, hence we need not be surprised to find consumption a very frequent sequence in this disease.

Pathological lesions.—During the first three days after the appearance of the outward symptoms of the disease dissection will show but little, if any, change in the viscera. The bowels will be found loaded with hard fecal matter, and careful examination will disclose some thickness of the inner coat of cæcum and ascending colon. A hog which has been sick a week or ten days will still disclose no disorganization of internal organs sufficient to account for the severe outward symptoms. The blood is blacker and less coagulable than in health; a general irritated condition of all mucous membranes will be noticed. The lungs will show no organic change, unless tubercular matter has already been deposited. In a majority of cases dissected, I have found the liver, kidneys, and spleen healthy—at least showing no signs of disorganization. I have never, in this disease, found abscess of any internal viscera, but have frequently found a low form of inflammation in the glands of the neck, which discharged a thin, sanious matter, but not true pus. In all cases examined I have found certain uniform morbid lesions, invariable thickening, and deposit in certain portions of alimentary canal, particularly at opening of small bowels into large bowel. This increase of tissue may take place in stomach or in any portion of alimentary canal, but will always be found in the cæcum around the ilio-cæcal valve. In a large number of cases I have found at this point that "peculiar bearded appearance" spoken of by Flint. But these black specks were only found during the first few days of the disease. At a later stage there was invariably great increase of tissue, thickening, and hard deposit. During the investigation I dissected over fifty hogs, all presenting the peculiar symptoms of typhus fever, and in every case I found thickening or deposit around the ilio-cæcal valve; in several cases where the disease was recent I found the minute black specks, and my own opinion is that the bearded appearance or black specks are the commencing lesions of the disease, and that this is followed by thickening or deposit. Anomalous lesions were found in many cases. In one the entire mass of bowels were found agglutinated. In several others were found enormous thickening or deposit in coat of stomach ; but in all cases, as before mentioned, there was one lesion always present, a deposit or thickening around the ilio-cæcal valve where the solitary glands of cæcum are situated.

The cause.—The exciting cause of this disease is a specific poison in the blood. an infectious, miasmatic poison, and the disease cannot be generated by any excess of filth, by want of care, or any errors in food. The specific poison must be there. The hog, to contract the disease. must be exposed to the specific miasma arising from another animal suffering from the disease. This disease is very contagious, and if it once obtains access to a drove of swine, prompt measures only can prevent its spread to the entire lot. The rapidity of its spread depends upon the condition of the drove and the ventilation. When the hogs are allowed an extensive range, and are not crowded together, it will spread slowly : but where they are cooped up in a contracted pen it will spread very rapidly. Although, as I have before said. this is the most contagious and fatal of any disease that has attacked swine, yet it has one redeeming feature, it is more easy to prevent its access to a drove, as the miasm cannot be carried as long distances by wind and other methods of conveyance as can the poisons of diphtheria and typhoid fever.

It may be well for me to explain the statement that filth cannot generate the disease. No amount of filth, no confinement in close quarters, no errors in food can produce the disease, but filth, want of ventilation, and improper food can deprave the system. disorder the stomach and render the animal more liable to the inception of the malady. Hence the disease often obtains access to a drove by means of one or two animals whose systems are disordered, and having once obtained a foothold spreads to the healthy ones, the contagious influence being now nearer and stronger.

Incubation.—From the few cases where the stage of incubation could be accurately determined. that is, the period of time elapsing from the time of exposure until the outward manifestations of the disease, I would place the period of incubation at fourteen days. I have but two instances to report where the time of exposure could be exactly determined. To verify this statement. in each of these cases, the exact time of exposure (by arrival of strange hogs suffering from the disease.) and the first outward symptoms of the disease were noted, and in each case it was fourteen days from time of exposure until the symptoms of disease appeared. [See notes on Homestead (Amana Society) Colony.] We shall speak of the predisposing causes when we come to consider the three diseases collectively, as the same causes will promote the spread of either one of them, but in different ratio.

Typhoid fever—Definition.—A specific continued fever, attended with great prostration of strength, stupor, tympanites, diarrhea, showing specific anatomical lesions, namely, ulceration of the solitary glands of cæcum and colon. The disease, when uncomplicated, runs its course in nine days. During the first month of my investigation I made no separate classification of those two diseases—typhus and typhoid fever. My course was as follows : From each infected drove inspected, I selected from two to five diseased hogs of various ages and at different stages of the disease. After carefully noting the age, history, and morbid symptoms in each case, the animal was killed, and exact notes taken of the condition of the blood, and also of each of the internal organs. From the start I noticed that the symptoms varied greatly in different droves, particularly in condition of bowels, the amount of eruption on skin and the duration of the disease and its fatality. I also noticed the morbid lesions varied greatly in different droves. On my return to my office, after three weeks' inspection, I made a careful review of my notes taken in the field, and found I could separate the symptoms into two distinct classes, only resembling each other in the one peculiarity of being low or typhoid in their character. I also found that the pathological lesions could be separated into two distinct classes, and that each of the two classes of symptoms were accompanied with one of the two classes of lesions. I was also impressed with the results obtained from treatment by those whose swine presented the peculiar symptoms and lesions which I now call typhoid fever, who reported that the disease had been promptly checked in their droves and a large part of the sick hogs cured by following my instructions in regard to hygienic care and medical treatment. Those whose hogs presented the typhus symptoms and lesions almost invariably reported that all the sick had died, and in most cases the disease was still continuing its ravages. I had also noticed that the peculiar odor spoken of was present in some droves and absent in others, and on examining my notes found that this odor was confined to the typhus cases. There was, of course, more or less smell wherever there was any disease among the swine, but the odor in the typhus form had a certain difference that could be noticed by any one. Thus finding I had two distinct diseases to deal with, the one resembling very nearly typhoid in man, the other presenting symptoms and lesions with which I was not particularly familiar, I turned to my medical library for information, and found a disease described as occurring in man with symptoms and lesions exactly resembling those I had classified in swine. I, therefore, called this second disease typhus fever. I placed typhus first in my list, because I found it the most frequent and most fatal of the three, and the one which has caused the greatest pecuniary loss to the farmers.

Typhoid fever symptoms.—Loss of appetite; headache; avoidance of light; standing with its head in a fence corner, or lying in such a position as to keep the light from its face; will only move when urged, and then but a short distance to resume its former attitude; a hot, dry skin; high fever; thermometer often showing 105° to 109° F.; increased urine; diarrhea; tympanitis; cough; shortness of breath, or quick breathing; stiffness of hind quarters. The hog moves his back from side to side as he moves his hind legs. Bleeding at the nose is a common symptom. These symptoms continue with remarkable uniformity during the nine days. There is an entire loathing of food, and as the disease progresses great weakness is manifest. The hog cannot be forced up, but lies for hours in a semi-stupid condition, but still restless and showing signs of nervous excitement. If the case is of a severe type the symptoms will be aggravated. The bowels will be enormously distended; urine scanty and high colored; fecal matter will be passed while lying down, and the urine will pass every time the hog is moved; more or less petecchiæ will be found on the abdomen, but in limited numbers. The first sign of improvement is inclination for food and disposition to move around, and in this disease this is the most critical period. Improper or over-abundance of food is liable to cause rupture of bowels and death, and it is at this time that many swine which have passed through the disease to the crisis are killed by incautious feeding. The nose bleeding is seldom severe, but hardly ever absent. The cough is of no importance as a symptom, as it is present in all inflammatory diseases depending upon a specific blood poison. Many farmers point to a cough lasting from one to three months as a preceding symptom of the disease, but this is a mistake, as the cough is due to the climatic changes or sudden exposure as set forth in another part of my report, and has no connection with the specific fever. This preceding cough should have told the farmer that there was some error in his management, which, unless corrected, would render his drove more liable to contract any infectious disease if exposed to its influence. The tympanitis is a prominent symptom in this disease, and if the hog is lying down a gentle tap on its distended flanks will show the presence of wind. Thumps or hiccough occurring during the second week is a fatal symptom. During convalescence small abscesses or boils often appear, and also sloughing of ears; in many cases the entire ears rot off. This condition is due to depraved blood, and demands tonics. Tubercular disease in lungs often makes its appearance during convalescence, and the hog is carried off by what is known as galloping consumption. Malarial complications often render the dangers of the disease more difficult, and have a material influence upon the rate of mortality. In most cases the malarial debility or fever is the primary disease, and the typhoid fever the secondary. The causes which lead to this fever we shall speak of under the head of predisposing causes of the specific fever. Enlargement of the glands of the neck does not often occur in typhoid fever. The duration of the disease may be set down at from nine to fourteen days when uncomplicated. I have no data from which I can give any information on the period of incubation.

Morbid anatomy.—I can best illustrate the lesions by quoting a few cases from my field-notes: Visited the farm of G. W. Davis, near Frank Pierce post-office, Johnson county, Iowa; breed of hogs, Poland-China; range, rolling prairie with clay subsoil; about ten acres in lot; lot covered with grass and brush; hogs also had a run of rye stubble; water running through lot; feed, raw sound corn and sour slop regularly. There was no disease in vicinity, and could trace cause to no contagious influence; disease had appeared two weeks previously, and nine head had died, three large animals and six shoats. There were twelve animals sick, five large brood-sows and seven shoats. This man complained, like many others, of losing his pigs. Symptoms, loss of appetite, high fever, diarrhea, emaciation, general stupor.

Dissection, No. 1.—Shoat three months old, sick one day; lungs white and showing no organic disease; some inflammation of stomach; liver and bowels appeared healthy. Thermometer showed 106° F.

No. 2.—Shoat two months old, sick one week; heat 107° F.; hepatization of one lung; liver, spleen, and kidneys appeared normal; some inflammation of inner coat of stomach. On opening the cæcum there were found deep ulcers scattered around the ilio-cæcal valve. These ulcers had only the peritoneal coat for a floor, and were in position of solitary glands.

No. 3.—Age two months, sick twelve days; thermometer showed 102° F.; lungs, liver, spleen, and kidneys showed no organic disease; lungs lighter colored than normal; considerable enlargement and inflammation of mesenteric glands; ulceration of solitary glands, but ulcers small and evidently healing.

Here we have an illustration of the disease in three different stages. In the first case ulceration had not yet commenced in the bowels. In the second it had eaten through all the coats of the bowels except the peritoneal. In the third case the ulcers were healing, and in a few days the pig would have been well again. In some cases I have found small abscesses of solitary glands, each one discharging matter on pressure, and I think this will usually be found the primary stage of the ulcer. Whatever other morbid lesions may be found, the farmer should carefully inspect the inner coat of the bowels and determine the nature of the lesions of the ilio-cæcal valve in order that he may accurately diagnose the disease. I would impress this particularly, because in several instances I opened swine and found no morbid appearances whatever to account for the severe symptoms until the bowels were examined and the inner coat exposed. In one case a farmer in Hamilton county, who had made a specialty of doctoring hog cholera, after making what he called a thorough examination, declared that a diseased hog was healthy, but I opened the bowels and showed the signs of specific disease on the inner coat. Although we find the lesions in both typhus and typhoid fever at this point, we can only look upon it as an effect of a certain poison in the blood, but why it uniformly develops morbid lesions at this one point has not yet been determined, even in man.

Diphtheria.—A specific septic blood-poison, contagious in its character, with inflammation of mucous membrane of pharynx (throat), and exudation of lymph; inflammation and abscess of kidneys; constipation and fever. Symptoms: Loss of appetite; fever; swelling of glands of neck; discharge of blood and matter from nose and mouth; weakness; the bowels are casually constipated; the urine is at first increased in quantity, but afterwards decreases in amount. The hog may try to eat, but there seems to be a difficulty in swallowing the food. As the disease advances all the symptoms are aggravated. The hog becomes stupid, and only moves when forced to do so. The glands of the neck are enormously enlarged, the urine diminished, and is at last entirely suppressed. The animal strains to evacuate its bowels every time it gets up, but passes only a few hard lumps, and, unless relieved, it dies within from two to five days from suffocation, caused by swelling of the throat or accumulated poison in the blood acting on the brain. The primary disease may be either constitutional or local, but in either case both general and local effects are soon manifest. This disease is a contagious blood-poison received into the blood, and passing through the stage of incubation, manifests its presence first when the system strives to rid itself of the poison through the four great waste-gates of the body—the lungs, kidneys, bowels, and skin. The expired air, loaded with the poisonous excretion, passes from the lungs, and as it obtains exit from the windpipe, is thrown with force against the posterior fauces. There the poison is deposited, and diphtheritic inflammation and exudation is the result. The kidneys, also, strive to eject the foreign matter and are at first stimulated to increased work; hence the increased flow of urine; but as the labor increases the kidneys become irritated from overwork, then inflamed, and the septic poison, instead of being eliminated, is deposited in the kidneys, and abscess in the same is the result. If the free egress of air from the lungs is prevented, either by swelling of glands externally, or swelling and exudation internally, either in fauces or wind-pipe, the poison cannot be thrown off as freely as it passes into the lungs, and abscess of the lungs is the result. In fact, wherever this poison is deposited an abscess at once forms. The skin is hot and dry, and there is often an eruption or rash apparent on the surface. Abscess of the liver is also a common sequence of this disease if it has continued for any length of time. The bowels are invariably costive, and, unless relieved by injection or brisk cathartics, the hog will die, either in convulsions or coma, from the united depressing influence of the septic and uræmic poisons acting on the brain, as the costive bowels causes increased labor for the kidneys and hastens inflammation of those organs, resulting in abscess. The duration of this disease is from. one to six days. Death is caused either by suffocation or from accumulated poison acting on the brain. I saw no case of this disease while making my investigation for the department, and my account of it is taken from my record of the disease as it appeared during the winter of 1877–'78. At that time it spread rapidly, and I had no means of testing the period of incubation. Dissection showed the following morbid lesions : In the first stage, inflammation of throat with diphtheritic exudations on fauces, and inflammation of all internal viscera. In the second stage, all pathological appearances were more positive. The glands of the neck were enlarged, and often contained pus; throat often a mass of ulceration, with diphtheritic membrane extending to windpipe; lungs inflamed and kidneys containing extravasated blood, and showing signs of commencing abscesses. In every case where the symptoms were severe and had continued for several days, abscesses of lungs, kidneys, liver, and spleen were observed, and putrefaction set in very rapidly, rendering examinations very dangerous. The specific cause of the disease, as stated in definition, is a septic poison, specific in type, and very contagious. It spreads more rapidly than either of the other fevers, and usually within two weeks after it obtains access to a drove it spreads to the entire herd, unless prompt and thorough means are adopted to check its progress. Although, probably, the most contagious of the specific fevers, it yields more rapidly to treatment and care, but if neglected it is more rapidly fatal, few that are attacked escaping with life.

Having treated of the three diseases I have found in swine, I will now glance at the symptoms and lesions which assist us in a diagnosis of the disease. In typhoid we have diarrhea, tympanitis (wind in bowels), very little eruption, and entire loathing of food. There is seldom much swelling about the neck, but there is ulceration of the bowels and loss of substance. In typhus we find costive bowels and lank flanks, except when filled out with solid feces; profuse eruption; except in few cases, considerable swelling of glands of neck, but not containing true pus. Dissection shows increase of tissue and deposit, frequently in coats of stomach and invariably around the ilio-cæcal valve; also accumulation of feces in bowels if they have not been relieved by purgatives before death. In either disease there is seldom much disorganization of internal viscera, unless in advanced stages, when tubercular deposits may be found in the lungs. In diphtheria we found constipation. There may be eruption, but this is not a uniform symptom; discharge of matter and blood from the nose and mouth, swelling of glands of neck. appetite not entirely absent, but, although the hog tries to eat, soon turns away from food. Dissection shows ulceration of throat, exudation and inflammation of lungs and kidneys, and in advanced cases inflammation, disorganization of kidneys, lungs, liver, and spleen. In diphtheria also the disease spreads more rapidly and is of shorter duration, except in cases of constipation in typhus, where death often occurs in a few hours. In all three diseases we have cough, rapidity of breathing, and fever.

Predisposing causes.—Included under this head are any causes which have a tendency to reduce the vital strength of the hog, disorder the stomach, or deprave the blood in any way. These causes are foul air, food improper in quantity or quality, bad water, filth, malaria, atmospheric influences, scrofulous diathesis, unusual exercise and over-suckling. All of these causes combined cannot generate the disease, but any one of them, by reducing the vitality or disordering the system in some way, may be the cause of the disease obtaining access to the drove. We will consider each cause and how it can be avoided. One of the common causes of disease among swine is confinement in a pen where the air does not circulate freely enough to carry off the carbonic acid expelled by the hog. The result is that from dark to daylight the hogs are forced to breathe an impure atmosphere. Many farmers build luxurious pens, tight and warm, and with an abundant ventilation only above, and abundance of straw below, forgetting that in such a house there is no ventilation, that, in fact, the breath, loaded with exhalation from the hog, is heavier than the air and sinks to the bottom of the pen. Even if, by reason of increased heat, the expired air attempts to rise, the cold air from above congeals the moisture and it falls as minute rain or snow. Other farmers build tight pens without any thought of ventilation and let the hogs pack in as they choose. In this case the air becomes very foul before morning with noxious gases, and if the owner would but put his hand within he would hardly find the air with sufficient power to sustain life. Now, it follows that we must have the pens so constructed that the swine can have pure air, at the same time the intense cold of our northern winters must be avoided, and either artificial heat must be provided or the heat of the hog utilized to increase the temperature where the surrounding atmosphere is below zero. We must remember that the natural haunts of the species in a wild state are in the torrid zone, and that swine are never found in a northern climate in a wild state except where they have escaped from domestication and become wild—that they are not provided with fur to protect them from extreme cold. Now, common sense teaches that when attempting to domesticate any wild animal his natural habits—food, climate, and mode of life—should be carefully studied. Again, effort has been made by careful breeding and feed to change the natural form and development of the hog—to raise a breed of swine with small bone, little muscle, and capacity for taking on fat while young, and these changes have been made at the expense of natural strength and endurance. It is a common remark among farmers that wild hogs do not have cholera, and acting upon this idea many farmers keep their hogs in large timber lots without shelter, and are disappointed to find disease appear and carry off a large proportion of the drove. In these cases, where the hog is not confined and forced to breathe foul air, but is exposed to the vicissitudes of weather, with loss of vital force by so-called improvement of breed, he becomes weakened and succumbs. I have noticed this particularly in regard to diphtheria; several large droves were almost swept away in a few days, although they had large range, pure water, and good food. This is true of diphtheria poison, but I have never known the other fevers to attack any isolated drove having pure air, clay soil in range, and good food, unless hogs having the disease were allowed in the same lot. The confinement of swine in close pens has another danger. The animal, heated by the confined atmosphere and damp straw bed, goes out at feed-call on a cold or rainy morning with its skin and hair damp from the accumulation of the gases which have congealed during the night. The cold, frosty air is a sudden change from the heated atmosphere of the pen, and bronchial lung irritation is the result. It is also wet, and this moisture, if it is a very cold day, is congealed, and the skin is chilled; and thus, from this error in care, the animal is exposed to a double danger. To avoid these dangers the pen should be so constructed that free ventilation can take place at the top, as it is absolutely necessary in a cold climate to utilize the natural heat of the hog to keep the pen at a moderate temperature. It will not do in winter to have any openings below to admit cold air,

10 sw

hence we must use some absorbent for the poisonous gases constantly being exhaled by the hog, and the best and cheapest yet known to man is dry clay, which will take up a large amount of gas in proportion to its bulk. The dry clay will also assist in keeping the hog dry and clean, and with reasonable ventilation above the air will remain quite pure. The plan for a pig-pen annexed I have furnished to many prominent stock men, and all have united in stating that it is the most perfect plan they have seen. (See drawing of pig-pen.)

The lot should, if possible, have a clay soil surface, and the feeding floor should have a slope of two inches to carry off the rain that falls upon it. By having the floor open to sun, rain, and wind it is kept clean and pure; by having the lot sloping away from the pen, the rain will assist in keeping it clean by removing refuse matter from the surface. In this way nature assists the farmer in keeping his pens clean and healthy. No straw or other litter should be allowed in sleeping rooms, as it will accumulate moisture and give forth noxious air at all times. Straw should not be allowed in the lot, as it will absorb any poisonous vapors passing over, and birds coming from herds infected with septic disease will bring the matter on their feet, and it will retain its life in the straw. But on dry ground, even if it finds lodgment, it will soon be disinfected. The hogs should be furnished with pure fresh water in abundance, not only because it is necessary to health, but because water assists materially in producing fat. On the subject of food supply there has been much difference of opinion, and I can only give my own views and the scientific reasons for them. The prime object in feeding swine is to accumulate fat as rapidly as possible on those intended for market. to keep stock hogs in healthy growing condition, and to have brood-sows in the best condition for bearing and suckling young. Of course, to accomplish these objects the stomach must be kept in healthy condition and not overloaded; the food must be of due variety and in suitable quantities, and its character and quality must be considered. For stock hogs, of course, green food is absolutely necessary. The hog cannot thrive upon an exclusive diet of dry corn and water; but the green food must not be the exclusive diet any more than dry corn. If the hogs are kept on a clover lot, sour fermented slop should not be fed at the same time, but rather roots and vegetables, as potatoes, turnips, rutabagas, and beets, which contain large quantities of the soda salts. which the clover lacks. Hogs fed or corn may have sour slop to advantage, as this will assist digestion, and in this case prevent an undue acid condition of the stomach and blood. The hog's natural instinct will lead him to seek just what his system demands, and he will root in the ground not for the mere pleasure of destroying the clover-field, but to find certain salts necessary to health that cannot be obtained except from the ground. Then if you deprive him of the means nature has furnished for obtaining these necessaries of life, you must furnish him with them in some other way.

Observing farmers have learned by experience that sickness in swine shows error in feed, and at once change to the opposite extreme. If feeding clover they change to dry corn, and if dry corn to clover. This rule has saved many droves from being swept off by infectious diseases. But I will give a rule which I have adopted in my investigations which is simple, but which at once tells the farmer what general course to pursue. If the herd is not doing well, if they do not eat well and appear less active than usual, at once examine the tongues of a few and notice the color; if the tongues are red and contracted give sour slop or turn them on clover pasture or on green food, and they will at once improve. If their tongues are large, pale, and flabby, give corn, corn-meal, cooked

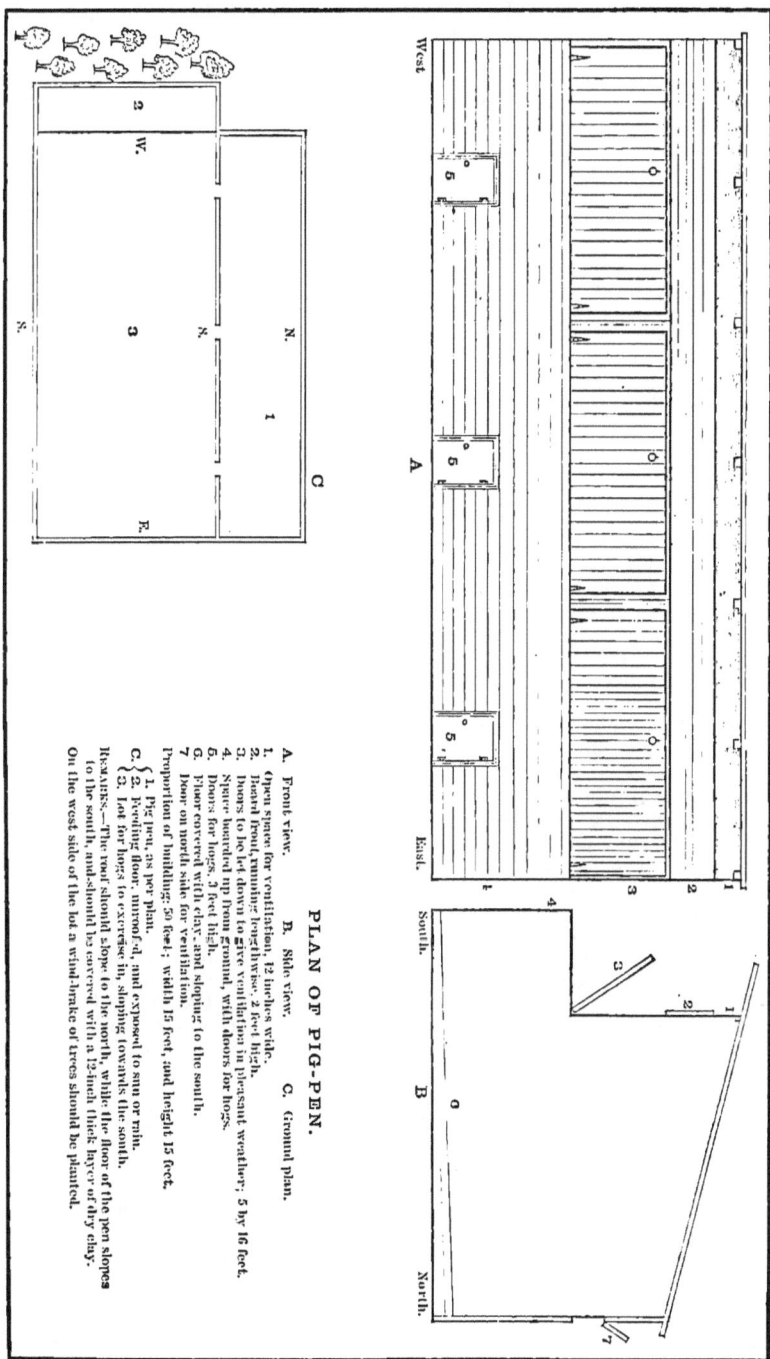

PLAN OF PIG-PEN.

A. Front view.　　B. Side view.　　C. Ground plan.

1. Open space for ventilation, 12 inches wide.
2. Board front, running lengthwise, 2 feet high.
3. Doors to be let down to give ventilation in pleasant weather; 5 by 16 feet.
4. Space boarded up from ground, with doors for hogs.
5. Doors for hogs, 3 feet high.
6. Floor covered with clay, and sloping to the south.
7. Door on north side for ventilation.

Proportion of buildings: 50 feet; width 15 feet, and height 15 feet.

C. { 1. Pig pen, as per plan.
    { 2. Feeding floor, unroofed, and exposed to sun or rain.
    { 3. Lot for hogs to exercise in, sloping towards the south.

REMARKS.—The roof should slope to the north, while the floor of the pen slopes to the south, and should be covered with a 12-inch thick layer of dry clay.

On the west side of the lot a wind-brake of trees should be planted.

root vegetables, and add soda to the feed, or soda and milk, but give no
sour slop. The large, white, pale tongue shows that the stomach and
blood are in acid condition and need alkalies; the contracted red tongue
shows a subacid condition, and that acids or sour remedies are needed.
For years the farmers' journals have lauded clover-fields and advised
keeping swine upon a clover range during the entire summer, on the score
of economy of feed and health. As far as it goes this is good advice,
and yet following this advice has been the chief cause of the spread
of the contagious diseases among swine. When the clover range is on
clay soil, or the hogs have access to clay banks, and the use of rings is
avoided, all will go well, but if rings are used in the nose, or the soil is
exclusively black loam and no other food is furnished but the grass and
water, or, perhaps what is worse, sour slop in addition, an acid condition of the blood is engendered. The hog becomes debilitated and
peculiarly liable to any contagious disease which may appear in the
vicinity. Of course farmers must keep their swine on grass and clover,
and, as a matter of economy, must use rings to prevent the clover from
being rooted up when the range is limited, but they must at the same
time study the natural habits and food of the species and supply that
food or its constituent elements in some form. The natural food of this
class is not a vegetable diet, but they were designed by the Almighty
so that they could obtain those roots from the ground. When, therefore, they cannot obtain them, they should be furnished in kind. As a
rule, the constituents of all grasses and annual plants are acid—have an
acid reaction. Especially is this the case with clover. Root vegetables
have an alkaline reaction, and are composed largely of phosphates and
soda salts. In clay soils the hogs can probably supply themselves from
the ground with phosphates, but when confined to a black, loamy soil
they can obtain but little of these necessary salts from the earth. A
noticeable fact is, that no matter how wide the range the swine will
select the bare points to root in rather than the soft loam. Where root
vegetables cannot be obtained and hogs are kept on clover range, soda
and lime or sulphate of iron should be given regularly. Dry corn as
an exclusive diet is not a natural food for hogs, and some additions
should be made to the bill of fare. Turnips, potatoes, or some other
cheap vegetable must be added to insure good health. I know there is
a bitter feeling among many farmers against cooking or grinding corn
for food, on the score of extra expense and trouble, but I have never yet
known a farmer abandon the practice when once thoroughly tried. It
will pay any farmer to grind and cook the corn fed to his hogs, even if
that staple is worth but 13 cents per bushel. Practical farmers, who
have made the profitable feeding of hogs a study, report that one pound
of cooked corn-meal is equal to one and one-half pounds of raw meal,
and to three of whole corn, in fat-producing power. One advantage in
feeding cooked feed is that root vegetables can be combined with corn-
meal and cooked at the same time. Where raw corn is used as a steady
diet sour slop will assist in its digestion, and should be given regularly
to prevent as far as possible the evil results of error in diet. The use
of coal, charcoal, ashes, and rotten logs in the pen assists in keeping the
hogs in health by supplying certain chemicals needed by the animals.
I have been thus particular in speaking of errors in diet because I believe that this cause more than any other has helped to spread the fatal
diseases among swine. A single hog with diseased stomach may be
the cause of imparting the malady to a herd, and having thus obtained
a foothold it may, unless prompt measures are taken, spread to the well

hogs, which would, if it had not been for the one or two unhealthy ones, have escaped infection.

The water should be clean, pure, running water, and should be within reach of the hogs at all times. Stagnant water, covered with green scum and loaded with organic impurities, is unfit for hogs to drink, yet many farmers furnish only such to their swine. Foul air, by vitiating the blood, is one of the common predisposing causes of disease. I have already spoken of the influence of heated air on the health of swine, and the evil effects of sudden changes, but I did not mention the depressing influence of the foul air itself upon the animals. Swine breathing air loaded with carbonic acid and ammoniacal gas for half of each day cannot remain healthy any more than man can, and the same natural results will follow—impure blood, disease of lungs, and other viscera. A pen erected on the plan set forth in diagram will remove this cause of disease. The dry clay is the best and cheapest disinfectant yet discovered, and will absorb the poisonous gases and render the air pure. Even though a large number of hogs are confined in a limited space, by opening the large doors on the south side on a clear day the sun's rays will dry the clay and renew its absorbing powers.

Scrofula is another common predisposing cause, and one of the principal causes of the large mortality in diseases of swine. The two chief causes of the scrofulous diathesis are breeding young sows and in-breeding. In order to avoid these causes sows should not be allowed to become pregnant until one year old. By that time she has matured and is fitted to bear young. Before that time she is growing and is immature. Not only the mother may be injured by early breeding, but the progeny will inherit disease. In-breeding has been largely practiced in the Western States, and whenever practiced it is easy to pick out the young resulting from this management. They were the first of the pigs attacked, and the *post-mortem* examinations disclosed tubercular disease in every case. Before the close of the investigation I became so thoroughly convinced on this subject, that, whenever I detected tubercular disease in lungs or mesentery, I sought out parentage of the pig. In several droves where a portion of the diseased swine were the offspring of in-bred sows and part cross-breed, the tubercular disease was found in the former and not in the latter. In-breeding is often practiced through the effort to obtain a perfectly pure breed of any particular species. With but few exceptions, and those among the imported stock, the pedigree does not extend back more than one or two generations, and often unwittingly the same blood is infused into a drove of sows, although the male may have come from a distance. To avoid this grave error, I would advise crossing breeds, selecting carefully the male from some special breed, as Poland-China, and crossing with an opposite breed in shape and habits, as the Essex. The finest drove I saw this year was the result of such a cross. Mr. Pendroy, of Monroe, Jasper county, bred two years to Essex boar and two to Poland-China, making a special effort to obtain as different blood as possible from that in his own herd. The herd of nearly three hundred head were in fine health, except some brood-sows which had been suckled down and were poor. These sows contracted the disease, but it was promptly checked by proper measures, and did not spread to any extent in his drove.

And this illustrates another very frequent cause of the contagious diseases obtaining a foothold in a herd of animals. The brood-sows become worn down with oversuckling and want of suitable food during pig-bearing and nursing, and with systems thus disordered are very liable to contract any disease in the vicinity. See that brood-sows have root-vegetables and milk, and that soda is furnished them liberally with green food, and they will not become so emaciated and debilitated.

Malarial influences can affect swine as well as man, and is one of the most troublesome and fatal of complications in infectious fevers in their first stages. The low sloughs, covered with green mold and surrounded with rank vegetation, are not the most healthy resting places for swine or any animal, biped or quadruped, especially between the hours of sunset and sunrise. The plants are giving off carbonic acid gas during the night, and the wet ground, loaded with organic matter, is giving off malaria. If the hogs are allowed to breathe this poisonous air their blood becomes vitiated and health is impaired, as in man. Typhus and typhoid fever find a favorable location for incubation here. To avoid these two causes combined, impure water and malaria, let the drove be gathered in at sundown into a large pen on high ground with sloping surface prepared, and kept there until the morning sun has dispelled the perceptible mists. If the day is inclement the drove may be allowed to range two hours after the hour for sunrise. They should be furnished pure water to drink before leaving, if they are to be confined in a range with stagnant water. Many will say that all this trouble will entail increased expense; but it has not been found more expensive where tried. Swine, like any other domesticated animal, can be trained to regular habits, and a drove can be trained to return to its sleeping place, if a small quantity of food is furnished them each night until the habit is formed.

Unusual exercise, which debilitates the hog and weakens his vital force, is another cause of the inception of contagious diseases. In several cases which have come under my observation, choice hogs for breeding purposes were purchased from apparently healthy herds, taken on cars and wagons a considerable distance, and after their arrival showed signs of disease and eventually died. In a few days others in the herd to which they had been taken were affected, and thus the disease was spread from a new focus to a large number of droves. I, of course, could not state where these hogs contracted the disease. When they started from their first home they were probably in perfect health, but confined in a close box and jolted around in a wagon, or confined in cars with irregular or unusual feed, and nervous excitement as additional causes, brought on a gastric irritation, and during these travels they were exposed to a contagious illness more or less intense, and their system being in a condition to receive and take up the poison, it found a lodgment, and after a stage of incubation showed itself by outward symptoms. Hogs brought from strange droves should invariably be kept in strict quarantine for at least fourteen days, no matter how perfect the bill of health they bring from their former owners. Neglect of this precaution has been the cause of the spread of the disease from new points, and many counties could trace the disease which had carried off thousands of hogs to a single imported animal. In one county visited in Western Iowa, which had previously had no swine disease, an estimated loss of over $100,000 worth of hogs was claimed to have been sustained during the past year, and this disease started from a central point—a single imported hog. (I use the term imported as meaning from a distant county, or another State.) The disease spread to the drove in which it was placed, and from that drove to adjoining herds. Several expensive lawsuits for damages and much ill-feeling between stock-men might have been avoided by attention to this point.

We will now take up the subject of treatment, which naturally divides itself under two heads—*Preventive* and *Curative*. Each of these can be divided into two classes, hygienic and medicinal. The whole secret of success in preventing the inception of contagious diseases by hygienic care, as has been already pointed out, can be included under two rules, viz., keep the system of the animal in a healthy state, and avoid exposing it to poisonous, contagious influences. We have already shown how the first rule can be followed with success—by fresh, uncontaminated air, suitable food, fresh water, seasonable exercise, and avoidance of low, damp places for sleeping quarters; also avoidance of those causes in breeding which are known to engender the scrofulous diathesis. The second requires that all dead organic matter, such as straw, hay, litter, and other matter, which is liable to catch the poisonous fungi floating in the air or carried along by the wind, should be kept away from the animal. All strange hogs must be kept in quarantine for fourteen days before being allowed to run with healthy herds. If there is any disease in the vicinity, especial care must be taken that no man, vehicle, or animal from infected localities be allowed to pass over meadows where healthy hogs are allowed to range; and if any stream passes through your range from an infected district, the stock must be kept from the water, as water will hold the poison and keep it alive for a considerable time. The yards and pens where the swine stay at night must be kept clean of cobs or other organic matter, so that the rains can wash the surface clean. All swine, either brood-sows, shoats, or pigs, not in general health, or showing evidences of debility, should be kept away from the drove and carefully treated, the causes of sickness removed and effects remedied. No medical treatment can be positively recommended as a preventive for contagious diseases. Remedies may be used to correct any derangements of system, as has already been recommended—soda, if the tongue is broad, flabby, and pale; acids, if the tongue is narrow, red, and contracted. In sows worn down with nursing, nothing can have a better effect and improve their condition more rapidly than soda and sweet milk or buttermilk. If the bowels are constipated, Glauber salts may be given in doses of one-half to one ounce to each hog, or one pound to every thirty hogs, once a day, until the bowels are acted upon. Salt should be furnished to all swine, in small quantities, every day. If any contagious disease is in the near vicinity, hyposulphite of soda in milk or fresh slop, given every morning on an empty stomach, offers the most reasonable hope as a preventive, and if the disease is diphtheria or typhoid, belladonna should be added. There is much difference of opinion in regard to the power of belladonna to prevent the spread of the septic diseases, diphtheria and scarlet fever. From my own observation I base the belief that it is a positive preventive or prophylactic, and on that account I extend its use to swine, and have recommended its regular use in small doses whenever diphtheria or typhoid was prevailing. As a preventive, the following would be a good formula for general use: Saturated solution of hyposulphite of soda, one gallon; tincture of belladonna, one fluid ounce. Of this mixture, give one gill to every twenty hogs in slop every morning on an empty stomach. Believing that all the contagious diseases are received into the system through the mucous membrane, and that any agent having power to destroy these minute fungi before their absorption will prevent the disease, I have for years recommended the use of chlorate of potash or sulphate of soda as preventives when persons are exposed to any contagious diseases. As typhoid has but a limited power of contagion, I cannot say positively that the remedy has prevented the spread of that disease; but I have never had a second case occur in a family where the remedies I recommended were used regularly. I would therefore recommend this formula to be used once a day where contagious diseases are in the near vicinity to diminish the chances to the lowest point. If the tongue is pale and broad, bicarbonate of soda must be added to neutralize the acid in the stomach, also sulphate of iron in doses of five grains will be found useful.

The curative treatment like the preventive must be both hygienic and medical. If disease has appeared in a herd prompt measures must be at once taken to prevent its spread. The sick must be immediately separated from the well. All organic matter, such as hay, straw, and litter, to which the hogs have access, must be burned, the lots cleaned up, and every possible effort made to destroy contagion. The well hogs, if possible, should be at once placed upon fresh ground; that is, on ground over which the sick hogs have not passed since a heavy rain cleansed the surface. Any disorder of stomach or general system should be at once corrected, and at least once a day the remedy before mentioned should be given in slop. Each day all hogs in well herds showing symptoms of disease should be at once separated from the others. Where the season will permit, especially in cases of typhoid fever, keep the entire drove on plowed ground, and have the ground harrowed every day to insure thorough mixture of fecal matter with the soil. Keep the sick hogs on a dry clay floor, with free ventilation, and protected from cold wind and rain; feed nothing but cooked slop and milk, and these only in limited quantities, adding the medicines recommended with the slop. In typhus and diphtheria the important point is to relieve the bowels as speedily as possible, and for this purpose castor-oil or saline cathartics must be freely given until the object is accomplished. In typhoid, diarrhea is a prominent symptom, and cathartics should be avoided. When the animal is a valuable one and will repay the trouble. injections of warm soft water into the bowels will be found the best plan for moving the same. The injections should be repeated until the bowels are well acted upon. In diphtheria the important point is to neutralize the poison as rapidly as possible, and eliminate it from the system. This can be effected with the sulphite and belladonna. The following will be found a useful formula, viz: Saturated solution of sulphite or hyposulphite of soda, one quart; fluid extract belladonna, three drachms; fluid extract aconite, two drachms. Of this mixture give one gill to every sixteen hogs five times a day, in a limited amount of milk or cooked slop. If the glands of neck are swollen to such an extent as to threaten danger from suffocation, oil of turpentine and sweet-oil may be freely applied externally. By following these directions in treatment few, if any, of the hogs suffering from diphtheria will die, and recovery will be rapid and permanent. When a good article of the powdered herbs can be obtained, the following will be found preferable to the tinctures and fluid extracts : Sulphite or hyposulphite of soda, five pounds; sulphur, two pounds: powdered belladonna leaves, four ounces; powdered aconite root. two ounces; powdered elecampane, a half-pound; powdered ginger, two ounces, and mix thoroughly. On one pound of the powder pour three quarts of water (boiling); add a quart of molasses, stir and cover. Of this mixture give one gill to every fifteen hogs, or one tablespoonful to every hog, in a little milk, four or five times a day. The medicine should be kept in a stone crock or wooden bucket—not in a tin vessel. In typhoid fever the condition of tongue is our principal guide to determine treatment. The sick will, as a rule, utterly refuse food, and very little medicine will be needed. Carbolic acid in milk, in doses of two to five drops in one pint of milk, as often as the hog will drink, or three times a day if given by force, will accomplish a good purpose if added to medicine, and oil of turpentine may be added as a useful adjunct. If the tongue is red, muriatic acid, diluted in doses of ten drops in a little slop, can be given as often as the hog will drink, or the water may be acidulated with the acid. Where there is a large number of hogs sick it will be impossible to attend each one. I would therefore advise the use of these remedies to all the sick, not attempting to treat special symptoms in each case. If the disease is promptly treated as above, the first symptoms of typhoid may be destroyed, and the hog will improve at once, but if treatment is delayed the case must run at least a nine days' course. Great care must be exercised in returning to solid food, as this error may render a hog worthless that might have entirely recovered from the effects of the disease.

Under the above course of treatment I have succeeded in checking the spread of the disease, and a large majority of the sick hogs have recovered.

Treatment of typhus fever.—I must confess I have not had any very flattering success in the treatment of this disease, and can only give my views and recommendations and the reasons therefor, hoping that some of my colleagues may have been more successful. As mentioned before, the bowels must be relieved either by saline purgatives or by injections. This is an important point, as impacted fecal matter is a frequent cause of death. Another important point is to keep the hogs on a large range, scattered as much as possible, as crowding together only increases the intensity of the poison. Internally give as follows: Bromide of potassium, ½ ounce; bromide of ammonia, ½ ounce; gelseminum (fluid ex.), 2 ounces; aconite (fluid ex.), 2 ounces; capsicum (tr.), ½ ounce; water sufficient to make 4 ounces. Of this mixture give one teaspoonful to each hog three to six times a day, in milk or slops. After the bowels have been freely moved the amount of podophyllin (may-apple) must be reduced. The same remedies can be obtained in powdered form and given in infusions: Bromide of potassium and ammonia, of each one-half ounce; powdered gelsemini and powdered aconite root each one-half ounce; powdered capsicum (cayenne), two drachms; powdered elecampane, one-half ounce; powdered podophyllin, two to four drachms. Upon this powder pour one quart of boiling water, stir and cover, and give a tablespoonful to each hog twice a day, or oftener, in a little slop. The same medicine may be given to the well animals as soon as they are separated from the sick. It should be given on an empty stomach every morning. The great difficulty in obtaining powdered drugs is that most of the powdered vegetable drugs have been kept so long in stock that the medicinal properties are lost, and are perfectly inert. I would, therefore, advise the use of fluid extracts in preference to powdered medicines, unless a reliable article can be procured.

The following may be considered the best general treatment for a drove of hogs attacked with contagious disease: Separate the sick from the well animals; keep the sick on bare and fresh ground, not having been passed over by diseased hogs since a heavy rain. If constipated, see that the bowels are moved either by using salts, oil, or injections. Protect them from inclement weather, and give internally, if the tongues are large, white, and flabby, soda, hyposulphite and bicarbonate, each one-half drachm, sulphite iron five grains, belladonna leaves two quarts, powdered aconite root two grains, elecampane (powdered), twenty grains, once a day to well and three times a day to sick hogs, in milk or fresh, rich slop. If tongues are red and contracted, give water and slop acidulated with muriatic acid to all, and to sick hogs give bromide, gelseminum, and mandrake, in regular and free doses. I would particularly caution the farmer not to rely upon medical treatment to the exclusion of hygienic care, but rather to follow carefully the directions set forth for the case of swine, and make the medical treatment an auxiliary.

A few words may be proper in regard to worms in alimentary canal. I have found no species of worms which could be strictly included under the head of contagious diseases, or could in any way be called a cause of the disease to which swine are subject. I have seldom examined a hog in any stage of the disease without finding worms in some form. The long, round worm in the stomach, and frequently the small thread-worm in the cæcum, have been found. These worms are natural to the swine and to all domestic animals. They may increase in numbers and cause trouble, but they are not the disease or the cause of the disease, but rather an effect of the condition of weakened mucous membrane which has increased the parasite. Oil of turpentine, in milk or slop, given once a day (preferably on an empty stomach), will expel those worms when so numerous as to affect health. I have received many letters from farmers and proprietors claiming that the worms were the specific cause, in fact the disease itself, and approving remedies to meet their single indications. I will therefore state emphatically that, in the dissections I have made, numbering over one hundred, I have found no form of worms which are not frequent in health, and have found no foreign parasites of any kind that could be detected with the naked eye that could possibly be a cause of the disease of swine. A careful examination of the liver, lungs, spleen, and kidneys with a powerful microscope may disclose some minute animalcula or parasite (as I said before, I found I could make no practical use of the microscope in field); but even these minute objects are but an effect, and the poison germ lies behind as the cause of the depraved system which has permitted the parasite to find a home. There is one disease known as kidney-worm, of which I have heard almost every farmer speak, but I have not seen a specimen of the parasites, although I have dissected a number of hogs which farmers claimed were suffering with this affection. I invariably found inflammation of kidneys, but no worm visible to the naked eye.

There seems to be a general belief among farmers that rings are a strong predisposing cause of disease, and, instead of meeting the opposition to this theory which I expected, I find that careful observers are willing to admit the truth of the statement, and either abandon the rings or furnish the food which is cut off by their use. In Jasper county, which has a rolling, clay soil, and the hogs generally have extensive ranges, I particularly noticed the fact that, in a ride of fifteen miles through a thickly-settled country, the droves in which rings were used were invariably sick, and in those in which they were not used there were no sick animals. In one drove only the brood-sows were rung, and these alone were attacked at the time of my visit. Although this is but one isolated county, yet it furnishes food for reflection. I do not claim that clover does not contain potash and soda (sodium) in a neutral form. The claim I set forth is that this food has an acid reaction in a green state; that it contains an excess of vegetable acid; and that confinement to this diet will induce flatulency and dyspepsia in any omnivorous animal. It is the natural food of horses, cattle, sheep, deer, and buffaloes, but not of swine; and the anatomy of the hog proves the statement. My claim is that no omnivorous animal can remain in health on an exclusive diet of green clover. I append the following notes from my daily journal: John Nimick, near Washington, Iowa, had a breed of Chester Whites, mixed with Berkshire. Pens, filthy; range, timber, with clay soil and grass; food, soaked oats after they had been taken sick; good water. Had ninety-four head; sixty-seven died and seven recovered; seven now sick. Symptoms: diarrhea, prostration of strength, tympanitis. Dissection of one hog one year old showed great emaciation, tympanitis, little change in liver, lungs, spleen, and kidneys. Bowels were expanded with gas, and there were a number of ulcers situated at seat of solitary glands. The disease was typhoid fever. No treatment was attempted, as the man refused to follow instructions; but the slowness of progress under the rather unfavorable circumstances showed the slight contagion there is in typhoid, as merely turning the drove out on a different pasture had alone checked the rapid progress of the disease, although sick and well remained together.

John V. Anderson, Washington, Iowa. Poland-China herd; on grass and away from straw or manure when attacked. Soil, black loam; water, open ditch; intense heat, 99° F., at time of attack in July; disease in near vicinity, and had been in this herd three weeks. The owner had lost seventeen hogs, fifty-eight shoats and pigs, and had remaining thirty-eight hogs and three pigs. The disease had spread gradually, and was killing two or three per day. The symptoms were the same as in last drove—pale, large, and flabby tongue; diarrhea; tympanitis. All sick. I ordered milk and lime-water, and ground cooked feed made into slop, and limited quantities of soda bicarbonate and hyposulphite (each five pounds), sulphite iron (one pound), given at the rate of one pound to drove of thirty-eight hogs twice a day. Mr. Anderson reported that all the sick animals recovered except three pigs, two of which were sacrificed in the cause of science. This man did not allow in-breeding; no scrofulous taint was detected in dissection; he did, however, use rings, which I consider were the predisposing cause of the disease in this herd.

W. J. Hamilton, Washington, Iowa. Breed, Poland-China; feed, growing rye and dry whole corn, with slough water for drinking. Pens clean, large range, but soil black loam, and rings used in nose. The disease had continued ten days, during which time ten had died and thirty-five had been attacked. Dissection of a few of the sick showed tubercular disease, and the gentleman stated that he had been in-breeding for some time. The same treatment was adopted as in the last case, but report showed no beneficial results when used with sick hogs, or in preventing the spread of disease.

Amana Colony, Homestead, Iowa county, Iowa. Breed, Poland-China; pens in poor order; feed, corn and slop. No disease had been known in the colony for twelve years. In July the agent purchased in Iowa City five boar pigs, and they were hauled to depot and forwarded on cars to Homestead when it was intensely warm. On arrival one of the pigs refused to eat, and was put in a small pen in breeding-house, where it died a few days afterward. On the fourteenth day the hogs in pens on each side of the one occupied by infected hog were taken sick and died, and the disease gradually extended until over two hundred had died. Two of the five boars were sent to the North Amana Colony, and in five days refused food, sickened, and died; and in nineteen days from their arrival the pigs in pens on each side of the one containing sick boars were taken sick and over five hundred died in a few months. In each of these cases the disease appeared first only in the pens immediately adjoining the infected pen, and afterwards spread to the other pens. The other boars were carried to the other colonies of the society, remained well, and no disease appeared in any of the seven settlements except the two mentioned. I visited the man from whom the boars were purchased, but could elicit but very little information. He stated that his " hogs had cough, as all hogs had, and that he had lost about thirty-five head by the intense heat, they being very fat, but that no disease had appeared in his drove; and, further, that he lost no hogs for some weeks after selling those to the colony." It may have been that these hogs had the poison germ in their system before starting from home, and might have succeeded in throwing off the poison if they had been retained at home; but worry, fatigue, and confinement during excessively hot weather, in a close box in a tight car, was enough of itself to reduce the animal vitality to a low ebb, and give the most favorable encouragement for the disease.

Respectfully submitted.

ALBERT DUNLAP, M. D.

Iowa City, Iowa, *December* 3, 1878.

REPORT OF REUBEN F. DYER, M. D.

Hon. WILLIAM G. LE DUC,
   Commissioner of Agriculture:

SIR: Having been appointed by you to investigate the diseases of swine in this locality, I entered upon that duty August 1st, which duty was to extend over a period of two months. Having performed that duty to the best of my ability, I now proceed to make a detailed report of my investigations.

Having carefully noted the origin and spread of the epidemic among swine in this county, which first made its appearance on the farm of Mr. William O'Mera in May, 1877, the report I am now to make will commence at the time when, from that starting point, the disease has become quite universal in this locality.

In order to thoroughly understand the cause of the disease, I will commence at Mr. O'Mera's farm. He is situated on the bottom-lands of the Illinois River, close to the bluff, which rises some 60 or 70 feet. His hog-yard, which comprises about one acre, is close to the Chicago, Rock Island and Pacific Railroad, so that his herd was exposed to any contagion that might be transmitted by moving stock-trains. An instance of this kind occurred in the case of Mr. A. Holderman's herd, which was attacked about one month ago. There was no diseased herd within several miles of his place, but the same railroad passes through his farm.

The same condition is seen again in this town near the stock-yards of this railroad. Pigs confined in pens near the stock-yards have been infected in the same manner. Also on the Chicago, Burlington and Quincy Railroad, where the railroad crosses the Illinois and Michigan Canal, a Mr. Loudergrau had some pigs confined in a pen close to the railroad. The trains stopped directly opposite his pen to take in water, and his pigs became diseased. As it is a well-known fact that these roads have been shipping diseased hogs, it appears quite evident that these points became infected by disease transmitted by the railroads, and also by wagons transporting hogs to market.

Owners of hogs, as soon as the disease attack their herds, and sometimes before, sell all fat animals, hauling them to market in wagons. All along the road thus traveled herds will take the disease, and it is probable that the herd so attacked is infected by hogs thus transported. This is evidenced in the manner in which it is distributed, as one herd will take it, and then it may pass two or three farms before another one is infected, and this peculiarity of attack is only observed on roads over which diseased as well as dead hogs are hauled. When not carried in this or some similar manner, but left to its own natural course, as a rule it moves steadily along, taking in each farm in turn. There are but few exceptions to this rule.

In the northern part of this county it is particularly observed on roads over which dead hogs have been transported that hundreds of animals are suffering all along the line of these roads with the same peculiarity of attack as is witnessed by the live diseased hogs passing. In view of these facts, it is fair to presume that Mr. O'Mera's herd contracted the disease from the stock-trains on the Chicago, Rock Island and Pacific Railroad. From this herd it began to spread to the adjoining farms, going up the bluffs to herds on farms situated along the bluff.

In June, 1877, it struck Mr. A. Strawn's herd, and he lost very heavily. West of Mr. Strawn's it attacked Mrs. Hardy's herd, and she lost nearly all. The next farm west, which was only separated by a common board fence, on each side of which hogs were confined, it did not attack, and the owner attributed his immunity to adding sulphur to the swill fed his hogs; but it went east, taking several farms, and was only arrested for want of material to prey upon.

From Mr. O'Mera's it crossed the Illinois and Michigan Canal, and extended east and west up and down the Illinois River.

Mr. J. Delbridge had a herd of young hogs, which he sold late in the fall at an auction sale. At the time of sale it was not supposed that his herd was affected, but the heard adjoining his had been dying for some time. The sale was made, and different parties purchased the pigs, took them home, and placed them with their own hogs. In a few days after it was noticed that these pigs were diseased, and every herd in which they were placed, without a single exception, was attacked by the disease in question. In the herds thus contaminated the disease lingered until the spring, but it did not spread much until warm weather, and since the growth of vegetation became rank it has spread all over the southern part of the county, destroying not less than $50,000 to $70,000 worth of hogs up to this time, and it is still raging. One great source of spreading the disease is observed by the small pigs wandering to the herds of adjoining farms, and thus importing the malady. Farmers usually confine their hogs in lots only sufficiently fenced to keep in the large ones, hence the small pigs readily escape and gain access to other herds. Many farmers tell me that when their herds are sick they do not know what becomes of the small pigs, as they all disappear and seldom return. When asked if they know how their herd contracted the disease, they very frequently answer, "Well, one morning I noticed a strange pig in my herd which was sick, and in about ten days or two weeks mine began to die." Another instance proving that the disease is transmitted by those infected occurred only a few days ago. Mr. Dunlavy, who lives north of the Illinois River, in Ottawa township, purchased five pigs from a Mr. Poundstone, who lives in the infected district south of the river. Soon after Mr. Dunlavy placed those pigs in his herd he noticed they were sick. Two of them soon died, and this morning he tells me he has lost seventy of the remainder of his herd and all his small pigs; also lost eleven of his fat hogs. He had one hundred and ten head, all told. Mr. Poundstone tells me he has lost his own since selling those to Mr. Dunlavy.

The same rule holds true by placing well pigs in a diseased herd. In March last three well pigs were placed in a diseased herd, and in a short time they were taken sick. This shows that the disease retained sufficient vitality through the winter to impart itself in the spring. I carefully examined three of these cases, and found the disease a typical case.

Only one of them had then died. I might go on and illustrate by a good many examples to prove the contagiousness of the so-called hog-cholera.

The mycetic theory, which is now so popular among scientific men. and which ascribes the disease to parasites of the lowest form and smallest size. would seemingly offer the only explanation for this disease. It cannot be a toxic poison, as no one has ever been able to demonstrate an organized poison as a cause of any contagious disease. The lowest forms of organisms live in the air and in water as well as when attached to solid bodies. A specific germ, a favorable medium of development, and contact with the animal to be infected are fundamental conditions for the development of the disease and its diffusion; and every purturbation, every solution of continuity in the chain of these factors of development may prevent or lessen its destructive action.

From numerous observations I am convinced that the moving of dead animals does not import the disease as readily as do the live ones. I am led to believe that putrefaction diminishes the capacity for infection, and that the bacteria of decomposition is destructive to the germs of the disease. It is a well-known fact that one low form of organism is destructive to another low form. Climatic influences have but little control. I think that warm weather acts more favorably to the formation of the infecting germ. Along belts of timber it readily spreads; it also extends out on the prairie where the growth of vegetation is luxuriant. Contact of diseased with well animals imports it under all circumstances, climate. having no influence to prevent its spread. As to diet and care, it matters not how well or how poorly fed, or how cleanly kept, if such well-fed hogs come in contact with the disease, they are as sure to contract it as those that have no care. Where not caused by other means, the prevailing wind. gives the direction or march of the disease. The greatest distance that it has been carried by the wind, in any well-authenticated case that has come under my observation, is two miles. As a rule, a greater or less number of animals in every herd will escape the disease. or have it so lightly as not to interfere with their doing well.

It appears that quantity as well as quality of the germ, and aptitude of the animal to receive it, are the conditions which influence contagion. Some animals possess an absolute power of resistance. Trousseau says that "there are individuals who pass unharmed through every kind of an epidemic, be it influenza or cholera, scarlet fever or measles. small-pox or typhoid fever. There are individuals whom it is impossible to affect with the vaccine virus; inoculate them twenty times, and you will obtain no result. If I may use the expression. ·the soil is barren.' and in it the seed cannot germinate. There are others again in whom the power of resistance is only temporary. It is in general difficult to find out the condition upon which this power of resistance depends. It is known that the ability to resist contagion varies with the age of the individual. There is less power of resistance in the youth than in the old man. One attack of a contagious disease generally confers complete immunity from any subsequent contamination. Occasionally it may be repeated, but these exceptional cases do not at all invalidate the general rule."

The same writer still further says: "It would appear that virus or morbific matter, upon its entering the economy for the first time, puts in motion all therein that is fermentable, and so thoroughly destroys it that the leaven—the contagion—when introduced again, finds nothing whereupon to exert its action."

Wilson says: "That in every epidemic there is always a great variety in the gravity of the disease, some cases being very serious, others very slight, without any apparent cause for such difference. Sometimes an epidemic begins with moderation and closes with severity, and *vice versa*."

Trousseau holds: "That every contageous disease must have a spontaneous development, as contagion necessarily implies the presence of two individuals, one the giver, the other the receiver, of the morbific germ." This remark he follows by another which modifies it: "While there is every reason to believe," he says, "that at present there are some diseases, such as syphilis, small-pox, and measles, that are always reproduced by contagion, there are other maladies which we see arise spontaneously."

I believe it is now generally conceded that all diseases that pass through a regular period of incubation are contagious or infectious, and that they depend upon a morbific germ for their development. In several of the contagious diseases the morbific germ has been discovered by the microscope, and in all probability the morbific germ in all contagious diseases will yet be discovered, as has already been the case in the measles, small-pox, whooping-cough, scarlatina, typhus and typhoid fevers.

Lubermeister, in his introductory remarks on acute infectious diseases, says "that a peculiarity of infectious diseases, which they have in common with the poisons proper, or intoxications, but by which they also differ in the most marked manner from all other diseases in their specificness, which shows itself in the fact that always and under all circumstances a given kind of disease is solely due to a given kind of morbid agent or cause. There is no such constancy between cause and manifestations in other diseases. Exposure to different degrees of cold will produce different affections. * * * On the other hand, vaccination with the virus of variola only produces variola, if any disease at all is produced by it; vaccination with the vaccine matter only produces vaccinia; the infection from a patient with measles only produces measles, and never anything else, and *vice versa*. Whoever, therefore, is affected with small-pox, measles, syphilis, &c., is certain that he has taken the disease by becoming infected with small-pox, measles, syphilis, &c., and of no other disease. In infectious diseases the predisposing cause, which in most other diseases plays a more important part than the exciting cause, is to be considered only in so far as it may determine the severity of the disease. The kind of disease is entirely independent of it. Various physiological conditions may induce other pre-existing affections, and are influential in so far as they may increase or diminish the susceptibility, but the kind of disease will not be determined by it.

"Through the longest series of generations diseases preserve their specific character with the utmost persistency, and if at times some of these characteristics are not brought into complete maturity, owing to an unfavorable field for their development, they assume them again as soon as they are planted in favorable soil. The weather, the period of the year, the climate, the conditions of the soil, &c., conduce to, or prevent the spread of, an infectious disease, but they never change the nature of the disease. The kind of diet and all other physio-chemical influences act indifferently with regard to the nature of the affection, and one infectious disease is never changed into another. The doctrine of specificness would arise, as a necessary consequence, from the hypothesis of a contagion vivum, even if it were not already proved by the facts. From the specificness of infectious diseases we naturally conclude that they never arise spontaneously, but are dependent upon a transmission or continued propagation of the diseased person."

When a hog is attacked by the disease in question, the first thing that is usually noticed by the owner will be that it has refused its food; it walks slowly along with its nose to the ground. The attack may or may not be preceded by a cough, but a cough is usually noticed in starting the animal from its resting place. It is inclined to hide itself in its bedding. Sometimes a distinct chill will be noticed, the animal shivering or shaking like one with ague. There may be bleeding at the nose, also bloody urine. The bowels may be loose or costive. Usually in small pigs a diarrhea will be observed, sometimes quite severe and producing pains. Vomiting is often present, and many cases, especially among old hogs, where this is the case, they recover, while others in the same herd that do not vomit or have diarrhea die. In many herds quite a percentage of all that have an active diarrhea recover, while in other herds that are not thus affected, nearly all die. A swelling of the face, ears, watering of the eyes, increased saliva, and also increased discharge from the nose, are all symptoms of the disease. The genitals in sows will be frequently swollen; an eruption over the entire body; in some cases quite red, in others dark discolored spots appear. Some limp off as if lame in all the feet; others only in one foot. Some are attacked by convulsions. The fever runs high for four or five days, if the animal is not sooner destroyed. In fact, all the tissues of the animal suffer more or less as though the poison affects all. The mouth and throat often have a diphtheritic appearance, and bronchitis and inflammation of the lungs supervene with pleurisy. On *post-mortem* examination during the period of incubation you will notice the capillaries of the lungs already inflamed and bursting. Later, a circumscribed interlobular inflammation; still later, gangrene of the lungs. The liver may be inflamed, also the mucous membrane of the stomach and intestines. The kidneys sometimes present traces of inflammation; in some the peritoneum with slight effusion into the abdominal cavity. The temperature during the fever often runs very high, from 107° to 108° F., but some time before death it decreases. The same or nearly the same temperature will be observed morning and evening. There are exceptional cases that have come under my observation.

Among the affections of the nervous system is an inflammation of the meninges with rigidity of limbs, spinal meningitis, muscular paralysis, and convulsions with eclampsia.

Among inflammations may be mentioned that of the pericardium, gangrene of the lungs, interlobular inflammation of lungs, abscess of lungs, peritonitis and inflammation of mucous membrane of the stomach and intestines, liver, and spleen. The inflammation of the stomach and intestines is of a catarrhal character, sometimes moderate and sometimes severe: diarrhea with intense pain: bleeding from the kidneys: abortions by sows with pig; also abscesses in subcutaneous tissue. A hemorrhagic condition manifests itself by bleeding about the ears: inflammation of pleura with adhesions of a fibrinous character, but no effusion into the pleural cavity.

Aggregating a large number of cases in the same herd, you will find all the tissues diseased, but more particularly the lung tissues and the mucous membrane of the intestines.

I saw one case that had survived the acute attack that in two months terminated by tuberculosis and ascite; gangrene of tissues in hams and about the face; inflammation of fetlock or ankle joints, involving ligament and bone. In observing a diseased herd of several hundred head, you are impressed with the fact that the infectious poison invades all the tissues to a greater or less extent. In one hog it will be noticed that the brain or spinal cord is the point most severely attacked; in another, the muscular and ligamentous tissues suffer; another, the bowels receive the attack, but all ending alike, with a destruction of lung tissue. The whole course of the attack very much resembles the effect produced by an epidemic of measles, and quite similar to typhus fever in man.

The first herd that I visited after receiving my appointment was Mr. J. Follet's, of Deer Park. Mr. Follet had a herd of six hundred head, large and small. They had been dying for three weeks. He had been giving kerosene and lime in their drinking-water. The herd was a mixed breed of Berkshires, Poland-Chinas, and Chester Whites. Two years ago he lost nearly his whole herd. His pasture was woodland prairie, traversed by ravines, so that every rain washed the ground, especially his feeding-ground. The water to drink was from a spring, pumped into a trough by a windmill, and the trough was so constructed that they could not get their feet into the water. This herd was well sheltered from storms and sun, and their sleeping places were scattering out-buildings, so that there was no crowding together.

I advised him to continue lime in water, and to disinfect thoroughly with carbolic acid and chloride of lime, and to give sulphur, soda, bicarbonate, and salt, which he did; also turpentine in swill. The animals soon ceased to die, and he saved nearly all of his older hogs which he had wintered over and a few of this year's pigs. One hog, whenever it found a dead pig, would at once eat into its entrails and devour the whole internal viscera. This hog thrived finely.

Joseph Watts, who had a large herd, lost a great many hogs. They had been dying for about the same length of time. I advised the same course as with Mr. Follet's, but I cannot say that any very satisfactory results followed. His herd nearly all died, and out of one hundred and fifty head he saved only thirty.

Mr. Henry Green's herd had, since May, been running on a timothy and clover pasture, through which ran a creek. They had no corn. His year-old hogs began to die first, then the breeding sows, and lastly the pigs. He disinfected very thoroughly with carbolic acid, chloride of lime, and lime. As he had a very choice lot of Poland-China hogs, he was very anxious to save them. He sold what pigs would do to go to market, but with all his care by changing lots, turning into his cornfields, &c., he saved only four or five head.

In this herd I separated a few sick ones and placed them by themselves and gave fluid extract aconite to control the fever; but the results were unfavorable, as those thus treated finally died. A few others I gave a physic of mandrake with like results, losing all or nearly all the small pigs. I will here remark that few of the farmers that have large herds know anywhere near how many small pigs they have, as they only count the larger hogs. Mr. Watts thinks he has lost a hundred small pigs.

Mr. Rockwood's herd is confined on an adjoining farm to Mr. Green. He also had a very choice herd of Poland China hogs, numbering one hundred and sixty-five, ninety large ones, seventy-five spring pigs. He sold twenty-two large ones after his herd was taken sick, lost thirty large animals, and has only five or six small pigs and thirty-eight large ones left. He used soda, turpentine, sulphur, and kerosene after the herd was taken sick. Fumigated once with sulphur, and regrets he did not repeat this process, as, he says, "after doing that they appeared so much more lively." I made several *post-mortem* examinations in all these herds with like results.

Talman and Ed. Libby's herds were in a woodland pasture, with

11 SW

plenty of good water. Previous to turning out to pasture this spring he fed salt, sulphur, and wood-ashes combined. As soon as he discovered the herd was sick he took them from the woodland pasture and divided up the herd, placing some in a yard and some in an orchard, and others in an open field with straw stacks in it, and upon my advice gave salt, sulphur, soda, and turpentine, disinfecting with carbolic acid.

On the 6th of October I visited his herd and found he had only lost a few of his hogs, and these were mostly small pigs. He said he "never had hogs do any better than they are now doing." He continues the sulphur treatment.

Michael Ryan's herd consisted of only six shoats, which he had wintered. They were running in a pasture of timothy and clover; grass tall; clear stream of water; hedge fence for shelter. When I visited the lot I found them lying in tall grass, and all sick. His farm adjoins that of Mr. Rockwood. One half died. No treatment.

Mrs. David Strawn has a large herd, which she fed sulphur, copperas, and salt up to three months ago. She has commenced this treatment again. This herd lost heavily. The surroundings in the way of sleeping places were rather bad, being old straw stacks and dirty sheds; but they had a good pasture with plenty of spring-water for drinking. Mrs. Strawn's hogs being in very fair condition, she shipped all that were not sick. She lost most of her small pigs. Just in this neighborhood the disease appeared to be more fatal than in any other locality in this section.

John Craig Morr's herd consisted of thirty large and twenty small animals, and were confined in woodland pasture. He lost three large and six small hogs. He gave sulphur, copperas, and wood-ashes.

Isaac Reed's herd was confined in an orchard and open-lot pasture. He had five old hogs and seventeen young pigs. Once a week he gave fine soft coal, wood-ashes, and salt, with occasionally a little sulphur. He lost both large and small animals; has only two left.

John Goss had a herd of seventeen and lost twelve; the remainder had the disease, but got well. He bought seven more and put them in the pen two months after, and they did not take the disease.

Joseph Black's herd is situated just across the road south of Mr. Henry Green's. Mr. B. put sulphur and asafetida in his swill-barrel, and disinfected with chloride of lime, and saved a large number of his pigs and nearly all the older hogs, while Mr. Green lost severely, and the only difference in care and situation consisted in Mr. Black commencing treatment before his herd was taken sick. I saw no reason why Mr. Black should not have lost as many as Green or Rockwood under the same conditions.

Mr. Black's herd was in a timber and prairie pasture, cut up by ravines. He had seventy-five head, and lost five old and half his young pigs. He gave lime, sulphur, and wood-ashes.

Richard Smith, living on the south bluff of Illinois River, had seventeen hogs, a year old, and thirty young pigs. An old animal and a young pig were the first to die. The pig weighed from 75 to 100 pounds. The old animal was a sow with sucking pigs. All the pigs died, and in ten days more other pigs began to die. After he had lost four he gave one sow nitrate potash in water and she recovered. I advised asafetida, sulphur, and soda, with turpentine, in swill. After he commenced this treatment he lost no more hogs. Mr. Smith says, "Every time I give turpentine I can see that that cough gets better."

Mr. Gentlemen's herd was treated with a secret remedy by a Mr. Sutton. Mr. Sutton claimed specific treatment. He also treated some of

Mr. Watts's and E. C. Lewis's herd, but they report no particular success. Mr. Dunlavy also employed a patent-medicine man to treat some of his hogs, but he says "His medicine does not amount to a row of pins, if the government did give him a patent."

Mr. Newell's herd, at Deer Park, was treated with bi-sulphite soda, but without success. He then changed to sulphur in swill, and there was marked improvement. On October 11th Mr. Newell reported that this last treatment succeeded well. In all cases where carbolic acid has been used for disinfecting purposes, parties so using it have added some to the swill in trough. One litter of pigs which I treated entirely with carbolic acid passed the acute attack, but finally wasted away and died. On *post mortem* examination I could not discern any immediate cause of death.

Cornelius Sullivan, living in the outskirts of the city of Ottawa, had three large and six small pigs taken with the disease. At the time I saw the lot he had lost two large and one small one. I gave him bromide ammonium, but have not yet heard how it acted after the second day of administration. He said then that he could see no difference. I gave the same remedy to Mr. Thomas Toombs and a Mr. John Hickey, but have not yet received any report from them.

Mr. Hunt tried a remedy administered by Dr. Dunlap, of Iowa. At last accounts they were still dying, but he says he thinks it helped them some.

Many have used tar as a preventive quite freely with more or less apparent advantage. While nothing gives entire immunity, yet herds in which this disinfectant has been used do not suffer so severely as others not so treated.

Abner Strawn had a very fine herd of Berkshires. He is largely engaged in raising fine stock, and is fitted up with every convenience for feeding and sheltering it. Still he lost very heavily. The widow Hardy directly west of him lost all but one or two of her hogs, but in the next herd west of widow Hardy's, owned by Mr. Duffy, which was only separated by a common board fence, not one died. He fed sulphur mixed in swill. This was in the summer of 1877. This year the disease is not in that locality, and what few animals Mr. Strawn had left have done well, and he has raised some very fine pigs from a sow and boar that had the disease last year. A Mr. Degan has also raised a fine litter of pigs from a sow and boar that came very near dying last year. I have seen several instances where those that had passed through the disease and were used for breeding purposes have done well. I met with one case, that of Mr. Goss, who says that he did not succeed in raising pigs from parents that had been affected, but the cause may have been in the boar, as he made no further test.

Peter Donlavy, situated north of the Illinois River, imported five sows and introduced them into his herd the latter part of August. He purchased of a Mr. Poundstone, whose herd it has since been proven was infected at the time, as they subsequently died. As Mr. Donlavy was situated in a neighborhood where there was no disease pending, I desired to make an effort to quarantine the disease and confine it to his herd. Now, at the present writing (October 8th) it has not spread to any adjoining farms. His nearest neighbor is eighty rods away. Mr. D. has disinfected thoroughly and continuously with a solution of crude carbolic acid, a tea-cupful to a pail of water, using a sprinkling pot to sprinkle his hogs and yards, sleeping and feeding places.

If it can be established that the disease can be quarantined, then I think we have made a move in the only direction with which I have any

knowledge by which we can prevent its spread, unless the government
will do as England did with the cattle plague, kill every infected hog
and pay the owners a part of the loss, and thus stamp it out. Certain
it is that some stringent measure should be used to prevent transporting diseased animals. As long as railroads are allowed to ship,
or owners to sell, diseased animals, just so long will we have the disease
spreading over the country. The loss, starting from one contaminated spot
in this country by transportation by rail of diseased hogs, has cost this
county this year already not less than seventy-five to one hundred
thousand dollars. Some place the figures much higher. The loss is not
only to the owners immediately, but in the future. When it shall become
universally known that diseased animals are being continually slaughtered
and packed for shipment, when Europe shall learn that we are sending
them cholera hog-meat to eat, then one of the greatest sources of rev-
enue to this country will be seriously damaged. It is a notorious fact
that the stock-yards in Chicago are full of diseased animals. Commission men say that they are selling that class of hogs for slaughtering and packing, and think nothing of it. I know that in the yards in
this town hogs die from this disease, and as well hogs are put into the
yards preparatory for shipment, they will, of necessity, contract the malady. They are sent to market, and about the time they should be
slaughtered are taken sick. I know this is not a very pleasant picture
for those that like a steak of ham with eggs, but it is a true one, and
when Congress can only appropriate the paltry sum of ten thousand dollars to aid in trying to stop this annual loss of twenty or thirty millions
of dollars' worth of property, I want every Congressman to just reflect
that almost everything he eats has a little lard in it, and that every time
he calls for ham he may be eating a piece of cholera hog. I do not feel
competent to present this subject in the light it ought and deserves to
be presented. If we wish to preserve this industry the matter must be
grappled with vigorously and with no stinted hand, and prosecuted
until the last vestige of this disease is swept from this country.

I have used by way of experiment nearly all the articles recommended
in your circular, but the time of observation is so limited I cannot yet
report results that would be of any practical information to the government. Owners of hogs were willing to pay the expense of medicines
themselves, and I have to thank those gentlemen who have kindly and
earnestly seconded my efforts to arrest the disease, and at the same time
try to obtain information in regard to this terrible scourge. In sum-
ming up I do not deem it necessary to give a history of each individual
herd that I have seen, as those mentioned are types of them all.

As to treatment, I am led to the conclusion that the use of disinfectants offers the best field for success. The use of turpentine for the cough
acts better than anything I have tried, and when given early, I think,
very much mitigates the severity of the disease. A mild laxative like
sulphur also acts well; besides, it has the additional advantage of being
destructive to low forms of organisms. Alkalies during the attack are
certainly beneficial. Frequent changing of the location of the herd and
stamping out every sick pig will, in the end, save money to the owners.

I hope, now a beginning has been made, that Congressmen will see the
importance and real necessity of following up this small beginning until
it is thoroughly ascertained what must be done. If it proves, like most
contagious diseases, largely uncontrollable after the animal has once
been attacked, and must have its own run, then we must turn our attention to eradicating the plague by more expensive and radical means.
Such legislation in regard to transporting diseased animals, or the

sale of them by owners, or the killing of all animals that have been exposed to the disease, must be enacted as will effectually put a stop to the spread of it over this country.

I am, very respectfully, your obedient servant,

REUBEN F. DYER, M. D.

OTTAWA, ILL., *October* 1, 1878.

REPORT OF DR. ALBAN S. PAYNE.

Hon. WM. G. LE DUC,
Commissioner of Agriculture:

SIR: My description of this disease (so-called hog-cholera) will be confined to its history as it invaded that beautiful section of country lying between the Blue Ridge and the Catoctin chain of mountains, in Virginia, during the summers of 1869–'77–'78.

GENERAL CONSIDERATIONS ON CONTAGION.

Before speaking of the endemic and epidemic disease under consideration, generally known as hog-cholera, although a palpable misnomer, I will offer a few remarks upon the subject of contagion. This is always a question of paramount importance, not only to the investigator of diseases, but to the people at large. One great difficulty in arriving at a definite conclusion as to the contagion or non-contagion of a disease, I am persuaded, arises from the too great latitude given to the definition of the word *contagion* by the older and more systematic writers. In the sense in which this term is used at the present time it strikes my mind as being too vague and indefinite. The same objection may be urged against the term *infection*. For if you mean to signify by the term contagion a disease that transmits disease from one subject to another by direct contact, without the assistance of any susceptibility or predisposing cause on the part of the patient, I should then contend that very few epidemic or endemic diseases were so, strictly speaking. But if you mean by contagion to signify a disease from which exhalations or emanations may arise during its progress, capable of exciting a similar disease in those exposed to the influence of the noxious exhalations, or rather deoxygenizing emanations, then I will say that most of these epidemic and endemic diseases to which man and the domesticated animals are equally liable are more or less contagious. For here you have an exciting cause furnished by a foul deoxygenized atmosphere and a predisposing cause furnished by a weakened, impoverished system from improper food, bad water, or from the want of proper protection from inclement weather, or from sudden climatic alternations, causes sufficient of themselves, under certain circumstances (which we call epidemic influences), to produce disease in man or domestic animals. Infection is as unfortunate and indefinite a term; nor are the terms "specific" contagion and "contingent" contagion, as defined at the present day, by any means explicit. In my humble opinion fevers are a unit, varied in their character by surrounding circumstances; that is, in a temperate climate a remittent bilious fever becomes yellow fever in a hot climate when the temperature of the atmosphere is at its acme of power. The theories of ozone, "disease germs," micrococci, &c., are very plausible in theory, but they have yet to be proven.

Contagious diseases are produced either by a virus capable of causing them by inoculation, as in small-pox, or by miasma proceeding from the sick, as in the plague, measles, and scarlet fever. No two physicians agree as to which diseases are contagious and which are not. The contagia of the plague and typhus, especially the latter, is denied by many. It seems probable that a disease may be contagious under certain circumstances and not so under others. That is, a case of ephemeral fever, fever of acclimation, the mildest form of fever known to the medical profession, arising from cold superinduced by sudden and decided climatic alternations, may, if the patient is kept in a close, foul condition, be converted into a disease capable of producing emanations which will reproduce a similar disease in those exposed to them, and with great virulence. Ephemeral or camp fever is almost sure to manifest itself in cases where large bodies of healthy men are brought into camp from different sections of the country. This is equally apt to be the case when you bring together healthy young animals from different parts of a country, even if from different parts of the same county. We know this much; but how much this *materia morbi* weighs, what its color is, how it smells, are to us secrets yet hidden from our view. We know that if a man has fever and it intermits he becomes cold and shakes; we say he has "intermittent fever," "chills and fever," "ague and fever," and we know if he has a long continuance of this kind of fever, one of the organs of his system (the spleen) is apt to become enlarged, and this is about all we really do know as yet, because no one has seen, weighed, or smelled the peculiar miasma which causes intermittent fever.

I noticed two facts which threw important light upon this subject of hog-cholera in this Piedmont country, viz., that recently the larger portion of the sick hogs were under twelve months of age (shoats), and the larger portion of them were taken sick while eating the corn after cattle which were being fattened for market. The popular name given this disease is, as I have before said, a palpable misnomer. If I am correct in my diagnosis—and I think I am—it is *Rotheln*, or Dutch measles, and should be classed with the exanthemata, along with erythema, erysipelas, rubeola (measles), roseola, scarlatina, nettle-rash, and the artificial exanthemata. The young hogs being mostly the ones affected, strengthens the hypothesis of its being an eruptive fever. As far back as 1852 I recorded the fact that I considered epidemic tonsilitis (Rotheln) as the most frequent epidemic disease to which Piedmont, Va., was liable, and that this arose from the moist and variable character of the climate. I have since seen nothing to make me change this opinion, but much to strengthen and confirm me in this theory. Horses, hogs, cattle, and sheep are as susceptible to disease from exposure to cold, rainy weather, and to sudden climatic alternations, as the human family; probably more so. They suffer from exposure to cold as easily, and are as much given to catarrh or cold as the human race.

A disease peculiarly liable to be felt by the young of both the human and animal race, yet no age, sex, or color affords any certain protection from this epidemic disease, called Rotheln, or German measles. In my opinion, then, this so-called cholera is no cholera at all—has not a single choleroid symptom, as the bowels are invariably constipated until moved by medicines, or give way under the last throes of speedy dissolution; but that it is rather a fever prevailing in an endemic and epidemic form, subject to all the natural laws governing fevers, from its inception to its termination, in restoration or in death, and more closely resembling scarlatina and scarlet fever than any other of the varieties of the anginose exanthemata, and is now known to some of the medical profession as Rotheln, or German measles.

I will now proceed to give you a short history of the so-called hog-cholera as it appeared in that section of country known as Piedmont, Virginia, during the fall of 1877 and during the spring of 1878. In the fall of 1877 hog-cholera, so called, made its apppearance in that section of country lying south and east of the Bull Run Mountains, and the losses by death reached an aggregate of 85 per cent., mostly young animals, as I learned from Messrs. John and Ludwell Hutchison, intelligent farmers living near the old Braddock road, four miles below the village of Aldie. The people were much divided in opinion, some believing the improved stock of hogs most liable to the disease, others that they proved to be more exempt from its fearful ravages. The care which a farmer took with his hogs, I presume, had more to do with lessening the bill of mortality than the difference in breeds. Hogs feeding after cattle, and young hogs, were generally the first to show symptoms of the disease. No remedy so far as they knew seemed to be of any benefit. Dr. Ewell recommended calomel, and some persons thought it of service. So far as I could learn no case occurred north or west of Catoctin Mountains until October of 1877. The section of country where it occurred as early as February, 1877, is at an average altitude of 400 feet above tide-water. On the 13th day of October, 1877, J. Milton McVeigh first noticed that one of his hogs, feeding after his fat cattle, appeared stupid, dull, droopy, mopy. He very soon noticed others appearing to be affected in the same way. This farm is located just above the little village of Aldie (the William Berkley farm), at an average altitude of 550 feet above tide-water. He had on his farm at this time fifteen home-raised hogs, but having some large cattle that he thought would justify him in corn-feeding he determined to purchase some hogs to follow after the cattle and eat up the waste corn. Accordingly he bought, about the 1st of August, 1878, of Mr. Cox twenty-two fine, healthy shoats, of Mr. C. B. Rogers twenty healthy shoats, and of Jack Simpson ten more. These fifty-two animals were turned into a field to run after his cattle. The field was high and dry, rolling, and at an altitude of 600 feet above tide-water. The hogs had good, comfortable, dry, warm shelter to go to, and in the field there was an abundance of fresh running water from a large, fine mountain-spring. About the middle of November the disease commenced in earnest, first with shoats purchased of Mr. Cox, then with those bought of Mr. Rogers, and lastly with those procured from Mr. Simpson. He lost fifteen head between the middle of November and the 1st of December. One or two would be taken at a time and die, and about the time he would flatter himself that the disease had subsided, one or two more would be taken. This continued until the 1st of February, 1878, and during this time he lost thirty-nine out of the fifty-two shoats. After this, no other cases occurred. None of his *home-raised* hogs took the disease until he had sold his cattle and disposed of the remaining shoats, when, supposing the disease killed out by frosts and the cold weather, he turned a fine large sow and eleven pigs into this field where the sick shoats had run. The sow escaped the disease, but the pigs soon became sick, and he lost seven out of eleven of them. About the 1st of January following, the remainder of these shoats having become fat, and being apparently healthy, he killed five, and after dressing them he found the skin purplish, red to pale black; little pustules or pimples covered the shoulders, and by pressure pus would spin out. The throat gave unmistakable evidence of disease, and the lungs were in a condition of decay. The lower bowels were full of black, hard, dry balls (scybalæ) the color of tar, and very dry and hard. These animals had never been in the barn-yard, and there were no marshy places in the field in which they ranged. This history, as it occurred on Mr. McVeigh's place, militates strongly towards the theory of ephemeral fever (fever of acclimation) as the exciting cause. The weather was rainy, warm, alternating with damp, raw, chilly weather. The hogs of his neighbors, John Carl, William Tiffany, and Samuel Simpson, living in a southeasterly direction, were dying at the same time. They gave signs of great thirst, would eat mud and soft soap avariciously. As a general thing they had a cough, and occasionally vomiting; appearance of eye not noticed. E. C. Brown's hogs, of Middleburg, began to show signs of disease; would mope about and look dull and stupid. About the 20th of June, 1878, all his hogs had a cough ; bowels very much constipated; discharges from calomel sticky and tarry, black as tar itself; great thirst; would eat mud, soft soap, and their own excrements. All had more or less eruption upon the skin ; skin had scarlet blush. Hogs had plenty of good feed, grass, grain, slop. He tried every remedy, almost everything; thought calomel the only thing of service that was tried; lost about 50 per cent. of his hogs. Shoats proved to be most liable to the disease. The hogs of Mr. A. B. Moore, proprietor of Aldie Mills, commenced to show symptoms of disease about the middle of June, 1878. The disease was not as fatal with his hogs as it generally was with those of his neighbors. Attributed this fact to good clean shelters, good food, mill-feed, apples, and slop. Gave no medicines. Altitude of his place 400 feet above tide-water. About this time, advancing from the northeast and traveling south (in direction of prevailing winds and fog), it began to be felt at all the farm-houses along the road leading from Middleburg, in Loudoun county, to Salem, in Fauquier county, playing sad havoc with the young hogs of A. B. Rector, Mr. Hathoway, John Middleton, Howell Brothers, Maj. T. B. Hutchison, &c. Mr. A. B. Rector thought the plant known in some neighborhoods as barrowroot, in others as burvine, in strong infusion, was beneficial. This region of country is mostly 600 feet above tide-water. Here also the hogs running after cattle were those most affected. About this time the disease passed up the pike leading from Aldie to Upperville and Paris, never halting until it reached near to the summit of the Blue Ridge, above the village of Paris, in Fauquier county, at an altitude of 1,100 feet above tide-water. From Salem it passed up the main road, leading from Salem to Markham, Mr. T. A Rector's hogs being among the first affected. His nearest neighbor, Mr. Wilford Utterback, living between Mr. Rector and Salem, was unusually fortunate with his hogs. He did not lose many ; thinks they need good attention ; knows of no remedy. Altitude of Mr. Rector's and Mr. Utterback's farms. 550 feet above tide-water. F. W. Maddox, proprietor of Oak Hill farm, lost about one hundred hogs. Mr. Charles Brown lost all he had, except five shoats. The disease was very fatal at Maj. S. B. Barley's farm, near Delaplane Station. At A. J. Chunn's, John R. Strother's and others, on the west side of the Little Cobbler Mountain, the disease was very fatal. These farms all lie at an average altitude of 600 feet above tide-water. No remedy seemed of any avail in stopping its ravages on any of these farms. Above Markham, at Mr. George Strother's, Mr. Conner's, and Mr. Charles Trussel's, the disease was quite fatal. At Mrs. Palmer's, above Petersburg, at an altitude of 1,150 feet, it prevailed with violence. The altitude at Mr. Strother's, Mr. Trussel's. and Mr. A. Conner's is about 550 feet above tide-water. Mr. Trussel's hogs were fed upon mill-stuff, corn, and slop. He lost sixteen out of twenty. Mr. A. Conner lost eighteen head out of twenty. Young hogs were the ones that suffered most. Mr. Charles Trussel thought his hogs had some kind of a fever. He tried no remedies. I think I can safely set down the loss by disease this season in hogs in this rich productive country at 75 per cent. In my travels through this section of the State I saw many hogs, partially recovered, but still in a low state of health, that had lost their hair and their hoofs. The tegumentary tissue (skin) looked as if it came off in fine bran patches, instead of coming off in large flakes. This I considered unmistakable evidence of tegumentary excitement. The internal mucous membrane being a continuation of the external tegumentary tissue (skin), we may reasonably expect to find the internal mucous membrane likewise in a state of phlegmhymenitis. Add to this symptom the significant fact of such great thirst, and we raise a strong presumption that the disease is a fever, and one of the eruptive fevers, beyond peradventure. The instinct of the hog tells him what is cooling to him, therefore you find him eating mud, soft soap, his own excrements, rotten wood, ashes, and the like. I met no intelligent man who did not believe that either the hog's lungs or his throat were affected.

Mrs. Simpson's hogs, running in the common just below the village of Aldie, within fifty yards of Ish's tan-yard, were among the first to take the disease. Ish's hogs ran regularly in the common, yet none of them took the disease, while almost every one of Mrs. Simpson's hogs died. Ish gave his hogs chamber-lye in their slop. Mrs. Simpson did not use this remedy with her hogs. J. Milton McVeigh tried the same remedy, but without apparent effect. B. F. Carter, sr., gave his hogs coal-oil, and lost none. B. F. Carter, jr., gave his hogs the oil in same quantity and lost all. D. Mount and Daniel Lee used asafetida one year, with supposed good effect; another year it had no effect at all. Thomas A. Rector gave his hogs soap-suds and soda in their slop one year, according to advice of the writer, with marked success; persuaded by others to give turpentine and sulphur in the present epidemic, his loss was large. I found many persons who had come to the conclusion that during some period of the disease the hog's throat was sore, and that the disease was the putrid sore throat, which was so fatal to swine some forty years ago in this Piedmont region of Virginia. I find most of them agree that there is swelling about the face and eyes, eruption on the skin, great thirst, often cough, occasional vomiting, constipated bowels, a thumping in the side or sides, lower bowels full of hard, dry balls of fecal matter, with a rapid loss of flesh. Other farmers seem to notice sequelæ of the disease more, and speak of swelling of the forelegs; that they shed their hair and hoofs; skin peels off, and new skin becomes scurfy.

I gave for publication a short history of the so-called hog-cholera as it prevailed in this section of Virginia in 1868 or 1869. I have no notes left, and I am not morally certain in which year the disease prevailed. I remember, however, to have remarked that the first indication of sickness in the hog noticed by me was closing the eye in the bright sunshine of morning. Now, this symptom may have been from swelling of the face, but I then attributed it to contraction of the pupil of the eye and from intolerance of light. The next one had a ticking in the side, and then a rapid loss of flesh, so much so that a large fat hog would become so thin in a few days that you could almost read a newspaper through him. I will remark that the only symptom at all like cholera is this rapid loss of flesh. But then there is no purging, no loss of fluid by urination, but it seems rather that the heat in the internal organs of the hog is so intense that all the fluids in his system are dried up. To satisfy myself on this point I placed them in pens, with clean, dry plank for flooring, overnight, and in the morning the large hogs would be

almost living skeletons; but you never could discern any urinary or
other discharges on the clean dry floor of their pens. I made some
post-mortem examinations, and generally found inflammation in various
stages in the posterior portion of the lungs, and the glands and throat
in a gangrenous condition—blood thick and black as tar and disinclined
to flow; indeed, in some cases it was black, hard, and as dry as a chip.
Any one who carefully reads the reports of the Department of Agriculture for 1877 will perceive that some of the writers describe the disease
as attended by a fever; others, again, speak of the peculiar eruption
attending it. Now, I submit that if there is a fever accompanying hog-
cholera, and an eruption also, it is *prima facie* evidence that it is a disease which rightfully belongs to that class of maladies known as eruptive fevers, and it only remains for us to establish to which species of the
exanthemata it belongs for us to place its treatment on solid and well-
established grounds.

The description I gave in 1872 and the account given by Dr. Gillespie
in 1877, goes very far to identify *rotheln* with the hog disease that prevailed in Piedmont region of Virginia in 1877–'78. Fortunately the remedy
I shall recommend as a preventive, as well as a curative, agent during
its prevalence is equally beneficial in scarlet fever, diphtheria, and erysipelas in some forms. It is a trite saying but a true one that an ounce
of *prevention* is worth a pound of *cure*. If this is true in regard to diseases in the human family, it becomes eminently more so in the diseases
incident to domestic animals.

Etiology.—The causes of disease are, unfortunately, frequently obscure,
although they are sometimes evident enough. The causes of disease
resolve into several varieties. As writers divide them differently, a short
explanation may not be out of place. As a general thing the predisposing and occasional causes are the only ones on which much stress is
laid by medical writers. Causes accessory are those which have only a
secondary influence in the production of disease, as the want of proper
shelter for domestic animals in inclement weather may be indirectly the
means of producing disease among them. Accidental causes are those
which act only on certain given conditions and which do not always
produce the same disease. Cold may be an accidental cause of acute
pneumonia, inflammatory rheumatism, &c. Proximate cause is the disease itself; superabundance of blood is the cause of plethora, &c.; external causes are such as act externally to the patient, as cold, &c.; these
causes are such as determine the form of the disease; internal causes
are those which arise within the body; mechanical causes are those
which act mechanically upon the windpipe in producing suffocation;
negative causes comprise all those things the privation of which may
derange the functions, as want of food, water, &c. They are opposed
to positive causes which of themselves directly induce disease, as the
use of crude, rotten, indigestible food, &c; occasional or exciting
causes (actual causes) are those which immediately produce the disease.
Occult, hidden, or obscure causes, any causes with which we are unac-
quainted; also certain inappreciable conditions of the atmosphere—
if I may use such a word, "distemperature"—which we believe gives
rise to endemic and epidemic diseases. Physiological causes are those
which act only on living matter, as narcotics; predisposing or remote
causes are those which render the body liable to disease, as previous
low, depressed condition of system, bad health, &c.; principal causes
are those which exert the chief influence in the production of disease as
distinguished from the accessory causes; specific or asserted causes are
those which always produce a determinate disease, contagia, for example.

The deaths, in many instances, in this hog-disease arose from a mechanical cause. Throwing him down on his back to "drench him" with some remedy produced suffocation, the wind-pipe or the swollen tonsils were tilted back by pressure upon the epiglottis, and the glottis being thus mechanically closed no air could penetrate the lungs, and the result was death. When drenching is resorted to, the animal should be made to stand up on its hind feet, and sudden deaths will not so often occur from the administration of such remedies. The treatment of *rotheln* and epidemic diseases generally resolves itself into prophylactic (preventive) and curative. Among the most valuable remedial agents to prevent epidemic diseases among domestic animals, especially the hog, may be enumerated a good, clean, dry bed of leaves or straw often renewed, protected by a good shelter and with a plank floor; a good supply of pure running water to drink; plenty of good, strong, generous food, made up of corn, buckwheat, or oats, vegetables, fruits, and slop. Give them regularly a little dry salt, all the "soapsuds" you can, and let them have a bank of hickory ashes to run to. By this means the hog would be better able to withstand the sudden climatic alternations of from heat to cold, for these climatic alternations are, in my opinion, the most prolific source of all epidemic diseases to which the human race as well as domestic animals are liable. It is an admitted fact, I believe, that domestic animals, in fact all animals, breathe more through the pores of the skin than the human family do. By this the internal organs are relieved of a considerable burden. Hence arises the importance of keeping the pores of the skin open and in a healthy working condition. To effectually do this you must provide your hogs with frequent new beds; burn up the old ones, which, when worn down to dust, become moistened and the whole tegumentary tissue of the hog is agglutinated, as it were, by a paste-like substance, and is rendered totally unfit to perform the functions necessary in the animal economy. We can see why this should strongly predispose to disease. To further prevent this undesirable condition of the hog's skin, I would recommend washing with strong soapsuds and then scrubbing them dry with a clean corn-cob until their skin presented a red, healthy glow. See that the pores in the fore legs are open (*the little safety-valves*); give them plenty of chlorate of potash of the strength of two drams to a pint of water, and the chances of disease will be greatly lessened. Timothy, orchard, and other grasses incline them to constipation, which cannot be relieved except by the strongest remedial agents. Green plantain and purslane are good for hogs.

For a long time a great many German physicians, and a number of the profession in our own country to-day, believe that the extract of belladonna (deadly nightshade) given beforehand will prevent children from catching scarlet fever. Now, as *rotheln* is a kindred eruptive fever, might not some herb be found that would prove a preventive in this disease? I am more inclined to recommend *Veratrum viride* (American hellebore) as a prophylactic in this disease, because I am satisfied that venesection (bleeding) in the early stages of the malady is demanded. I remember that all hogs not castrated, and those castrated early in the disease of 1868 or 1869, recovered, and not only recovered, but made good recoveries. So did all the hogs I saw in those years early enough to get blood from them. After the first and second stage of the disease in those years the blood was very dark, black, thick, and could not be made to flow. From this condition of the blood, and from the low temperature I found in many hogs, I suspected congestive chills, or more probably dumb chills, of a very severe character. I am still disposed to cling to this opinion. In all those cases where the hog is mopy and chilly looking, I would, after the first stage of lowering the pulse has passed, recommend a teacupful of a strong infusion of the leaves of dogwood or the same quantity of a strong cold infusion of boneset. In either case add a teaspoonful of powdered ginger or thirty drops of the oil of black pepper, to be given morning, noon, and night regularly. Chlorate of potash, two drams to a pint of water, for drink at will.

I think the hog is peculiarly susceptible to the influence of malaria, therefore they had better be kept in the woods, or in a pen, or on high and dry places where there is not much grass, and fed on corn, oats, and buckwheat, with a proportionate admixture of fruits, vegetables, and slops. Soapsuds, all the preparations of potash, hickory ashes, soda, saleratus, &c., are anti-febrile, and will be found very beneficial when given in slops. In my opinion the throat and the adjacent parts, the upper and the posterior portions of the lungs, are the only really vulnerable portions in the animal economy of the hog. Protect these and you thereby protect the whole hog. I have no doubt that in one epidemic in this hog disease you may have it so dressed in the livery of pneumonia that the most accurate observer might diagnose the disease to be primarily pneumonia. In another case you may have an exudation of membrane, thereby simulating very closely diphtheria. Again, you may have *rotheln*, but the disease spreading to the parenchymatus portion of the lung and on to the pleura, producing *rotheln* complicated with pleuro-pneumonia, and so on. To show that the stomach of the hog is not very susceptible to the action of poison, I will state a fact known to almost every one in this region of country, that the hog can feed sumptuously on the rattlesnake, moccasin, and the poisonous copperhead with perfect impunity. Again, unless the snake bites the hog about the throat, and on the jugular vein and carotid artery, there is no harm done, but if over either of these blood-vessels the bite is speedily fatal. The internal remedy upon which I most rely, both as a preventive and curative agent, is that invaluable remedial agent, chlorate of potash. Dr. L. P. Dodge, in Georgia Medical Companion, December number, 1872, page 717, says:

The therapeutical effects of this agent are obtained by direct application and by absorption. When taken into the stomach it imparts a cooling sensation to the mouth and throat; the circulation is somewhat depressed. Hence it has been classed by authors as refrigerant, and from increased action of kidneys diuretic. By some it has been supposed to exert hepatic action. Without doubt it does, but to what extent we are not prepared to state. When applied locally to ulcerated surfaces of the mucous membrane, as in ulcerated stromatitis and many other diseases of the mucous membrane, and also to ulcers of the integuments, it has a stimulating action, as shown by increased sensation of the parts and excited vascular action, which becomes alterative, and, therefore, salutary. Its most decided effects are obtained when taken into the system. Chlorate of potash, we think, has a specific action on the mucous membrane—the glandular and cutaneous systems. In scarlatina it is universally recognized as the best remedy. In diseases of the mouth and throat, whether ulcerative or inflammatory, chlorate of potash has a salutary effect. In diphtheria it is one of the most reliable remedies for lesions of the throat. In no disease is its alterative action better shown. Given to an adult in tablespoonful doses of the saturated solution every hour for twenty-four hours, and there will be a marked change in the general appearance of the diseased parts. The exudation will be diminished, the fever removed, the surface paler, the swelling diminished, the vascular action less, the sensation ameliorated; the skin becomes cool, the pulse less frequent; in fact, a large per cent. of the incipient form of diphtheria requires no other remedy.

You can, then, safely give the hog one good dose of calomel in this disease, and then rely with an abiding confidence on the chlorate of potash.

Respectfully submitted.

ALBAN S. PAYNE, M. D.

MARKHAM, VA., *November* 25, 1878.

REPORT OF DR. J. N. McNUTT.

Hon. WM. G. LE DUC,
Commissioner of Agriculture:

SIR : I have the honor to report that, in obedience to instructions from your department, I have devoted the past two months to the investigation of diseases of swine. Though my labor has been confined to one county (Jefferson), I had abundance of material, and have examined several hundred diseased hogs, and made thirty *post mortem* examinations.

While the results of my experiments and examinations may not be as satisfactory as could be wished, I am convinced, first, of the nature of the disease, and, secondly, that if it cannot be cured in all cases, it can by proper hygienic measures be with much certainty prevented.

I have aimed to have the results of my examinations as practical as possible, and will endeavor to present them devoid of any scientific theories.

The disease has, in this county, as in other portions of the State, prevailed in different localities for a number of years. It usually begins in early spring, and increases in extent and severity until the late summer and fall months, disappearing toward the approach of winter, only to appear in another locality with the return of spring. Although in different seasons and localities it presents different symptoms, it is evidently the same fatal enemy to the pig raiser, only in another garb. Unfortunately, as the name of a disease should convey some idea of its nature, this dreaded scourge is called " hog-cholera," why we know not, unless from its rapid and almost certain fatality.

While the pathological conditions found in my examinations were many and varied, yet the main lesions pointed to the intestinal mucous membrane and lungs, with sufficient uniformity to clearly indicate the nature of the disease; and as it is clearly shown that the disease, while contagious, is not communicable to other animals nor to man, it is evidently a specific contagious disease *sui generis*—typhoid fever of swine. The disease occasionally begins suddenly with symptoms of a chill, the pig standing drawn up and shivering on the sunny side of a barn or fence. But the disease generally begins more insidiously, and the first thing noticed is, in a previously healthy pig, a dull appearance with a wrinkled, drawn look about the head and neck. It stands with back humped, head and shoulders drooping, eyes listless and watery; loss of appetite, or perhaps eats for a few moments and then stands over its food with an appearance of loathing; sometimes it show a disposition to nausea, great and constant thirst, increased temperature, first about breast and belly, and after one or two days extending over body and limbs. Fever at first of a remittent character; temperature in rectum 102°–104° F., in morning; in the evening rises to 106°–109° F. Has hacking cough, which is increased on exertion; sometimes attended with frothy (white or yellowish) and in last stage offensive discharge from the nose. Breathing rapid and labored, with drawing in of the flanks; panting. Bowels usually, at first, constipated; in some continue so; in others become lax after a few days, to be frequently followed, especially in protracted cases, by very dark fetid diarrhea. Kidneys usually act well, though urine is generally scanty and high colored. In very malignant cases it is suppressed. As the disease progresses the patient shows a disposition to get away from the herd; lies on its belly under straw, brush, or any place for a shade; is stirred up with difficulty ; walks with a staggering, painful gait. Some, if they attempt to run, go sidewise, and carry their head to one side. In white hogs, rose-colored spots appear on belly and inside of arms and breast, effaceable by pressure, but return immediately. On dark hogs, the spots are of a petechia or hemorrhagic character, with elevation of the cuticle, especially behind the shoulders and on the neck and back of the ears. In one case, sick three weeks, I found sloughs one inch or more in diameter, thickly scattered over belly, neck, and snout. Large abscesses are occasionally seen in parotid glands (behind the ears), and in a few malignant cases the legs swell until the skin bursts, discharging a thick, yellow serum. In some cases the hoofs fall off. If the case does not end fatally, as it often does in a few days, the symptoms increase in severity. The animals rapidly lose flesh, get lousy, refuse to eat or take note of their surroundings; if possible to arouse them, they immediately relapse into a stupor. Some pass off in this way; in others, convulsions close the scene. When one occasionally gets well, it is after a very protracted convalescence. Abscesses, ulcers, &c., form on different parts of the body. The hair all falls off, and it seldom makes much hog anyway.

My subjects for *post mortem* examinations were taken, some of them a few hours after death, and others were killed during various stages of the disease, from the first day to the third, and in the fourth week, by bleeding. The subjects that had died were usually very much emaciated, lousy, offensive; snout and ears a dark purple; eyes shrunken, sometimes ulcerated, and body covered with dark spots of extravasated blood.

The principal lesions found were in the alimentary mucous membrane and in the organs of the chest. The tongue I seldom found coated. though usually red and often ulcerated, especially towards the base, extending into throat and down the œsophagus. The stomach was usually found distended with undigested acid, and sometimes offensive ingesta and flatus. The ileum (small bowel) and colon (large bowel) filled with hard dry feces or with dark liquid, fetid discharges, and distended with gas. The mucous membrane of stomach and intestines, differing with the stage of the disease at which death had occurred, presented the various stages of inflammation and its sequela, from a faint pink blush to a dark red thickened condition. This was the case with the whole surface of the stomach and of the ileum or colon, or more or less extensive portions of each. In some cases the dark thickened membrane could be easily stripped from the sub-mucous coat. Ulcers in the glands of the small intestine and cæcum were frequent. Peyer's glands in two or three cases were very much enlarged and thickened, and covered with hard, dark scabs. In several cases the ileum was so contracted in several places that they looked as if they had been scorched.

The peritoneum was generally more or less inflamed, and in two cases I found in one two and in the other four quarts of straw-colored serum in the abdominal cavity; a portion of which, in the largest, was coagulated, apparently by the great heat of the bowels. (The temperature was 109° F.)

The lungs I found, with two exceptions, in different degrees of inflammation, varying with the period of the disease, which constitute pneumonia. This was the case either in the first stage (that of congestion), the second stage, (no hepatization), or third stage (gray hepatization); though, as is usual in diseases of a low and feeble character, these stages were not always well marked, but often presented more the condition called splenization, caused by the blood not yielding sufficient plastic matter to form the firm, resisting character of hepatization.

The amount of lung involved, of course, varies in each case; in some one lobe, usually the upper if the left lung, and lower if the right; in others, again, all of one lung, and in one case I found the whole of both lungs involved, the left in the third and right in the second stage. In young pigs I found what is known as lobular pneumonia, that is, diseased lobules, of which each lobe is composed, were mixed indiscriminately with healthy lobules, giving this lung a mottled appearance. In one case I found the disease in the upper lobe of the right lung. The inflammation was confined to the air vesicles, and constituted "vesicular pneumonia." In this case I found tubercles scattered through the diseased lung, and in one the upper lobe of the left lung was one mass of tubercles. All of the cases were complicated, to a greater or less extent, either with inflammation of the pleura (the covering of the lung), or of the bronchi (air-passages). In some cases the bronchial tubes were inflamed and filled with a frothy and occasionally a bloody mucus, in others ulcerated and secreting a yellow, offensive pus; the ulceration often extending into the larynx, and even into the nasal passages. In six or eight cases the pleura, especially the right, presented more or less extensive patches of inflammation, with adhesions between the pulmonary and the costal portions, that is, between the portion of the pleura and that lining the chest.

The heart was, in protracted cases, pale and soft, and in one case inflammation of the pericardium (covering of the heart) with effusion into the pericardial sack was observed.

The liver was, in most cases, more or less congested, and in one case very much enlarged and filled with patches of inflammation. The gall-bladder was usually filled, sometimes distended, with dark-green, thick bile.

The pleura was in all cases enlarged, and in one case very dark, almost black, and so friable that it would not sustain its own weight. The kidneys were usually pale and sometimes soft, and in the two cases where there was so much œdema of the lungs and suppression of the urine; the malpighian bodies were of a dark-red color, and the lining of the pelvis (inside of kidney) was very much inflamed and covered with extravasated blood.

With a few exceptions the mesenteric, inguinal, and other lymphatic glands, especially bronchial and cervical, were in various stages of inflammation and enlargement, and in some cases of a peculiar dark-red color.

The brain proper and the spinal cord I found usually in a normal condition. In one case there was effusion into the ventricles. The meninges of the brain and spine were, in protracted cases, congested or inflamed, and in two cases the dura mater (lining of the skull) was thickened and easily separated from the skull.

The cause of the disease has been variously ascribed to feeding, crowding, overdriving, filthy pens, ringing, &c. From information obtained from hog-raisers, from our own observation, and reasoning from analogy, I am satisfied that the real cause of the disease is the present manner of breeding, raising, and feeding the pigs, and as a result of my observation and treatment I found the same remedies as used in remittents in the human subject as the most effectual. I am satisfied that the disease is at least developed by malaria, and relieved, if at all, by the same treatment as malarial diseases in man. Instead of raising pigs from a sow eight or ten months old, and cramming them with slops and dry corn in order to make three-hundred-pound porkers of them in twelve months, select good healthy sows from eighteen to twenty-four months

old; allow them to have but one litter each year; let the pigs grow up
naturally; feed them but little, and give them no dry corn; let them
have plenty of water and clay to drink and bathe in, and give them
a chance to root for a living, and to that end furnish them good
pasturage on soft, moist, and, if possible, shady soil, where various roots
are plenty; in fact, let them "root, hog, or die," and wallow to their
hearts' content. The roots they may get are their natural food, and by
frequent bathing in muddy pools they keep the skin in a lively, healthy
condition, free from dandruff and vermin. A hog looks filthy enough
when he first comes out of his cool bath in a mud-hole; but see him after
he has dried the clay in the sun and rubbed it off on some convenient
stump or fence-corner, and he is a nice, clean, and very presentable animal. After he has attained his natural growth in this manner, say from
eighteen months to two years, he can be fattened on corn, if you will,
without fear of disease. That the disease once started is easily communicated by contagion and infection I have easily found by tracing its ravages in regions of my inquiries. Starting from a diseased hog brought
into the neighborhood, it next showed itself in the herd of the only
neighbor who let his hogs run at large, and whose hogs visited an infected farm. Thence it was conveyed by the hogs of the second party
dying alongside of a large pasture filled with well-fed, well-watered hogs.
Then other neighbors' hogs broke into this pasture and mingled with the
sick hogs, and soon went home to die of the disease and infect others.
Others, again, separated by a large creek, crossed to the infected neighborhood and were soon numbered with the dead. During a dry, south
wind, lasting several days, hogs one mile to the north, separated by the
same creek, developed the disease. Thence it was traced in the same
manner, carried either by straying hogs or dry winds, and in the case of
winds always in the direction of the wind, and then often jumping two
or three farms for favorable material.

Treatment.—Under this head I will necessarily be very brief, for unless
the case is taken early in the disease, *i. e.*, unless the pig-raiser under-
stands the early symptoms of the disease and adopts what might be
called the heroic treatment at once, little, if anything, can be done by
medication.

After fully satisfying myself as to the nature of the disease, I found
by taking the case in its incipiency and giving a good cathartic (calomel
5 to 20 grains, and podophyllin ½ to 2 grains, according to age) in boiled
potatoes at night, to be followed each morning for two or three days by
sulphate cinchoncidia 10 to 40 grains, according to age, in slops, and after
and during this treatment give spirits turpentine (5 to 20 drops), or carbolic acid in slops (1 to 3 drops) every four hours, resulted in a cure in
80 to 90 per cent. of cases treated. In addition I would follow suggestions recommended in prevention of the disease, viz., isolate the sick;
keep them in pastures with free access to water and clay. Clay is one
of our best antiseptics, and the hog knows it, and will when thirsty, if
he can, mix it with the water before he drinks. Give them but little, if
anything. to eat, and, if any, such vegetables as turnips, parsnips,
artichokes. and other food of this class. By no means feed corn, especially dry corn. I really think that if the suggestions as to the manner
of breeding, feeding, and caring for the pig here offered were followed
out there would be but little, if any, need for treatment.

Before closing I wish to acknowledge my indebtedness to Maj. James
S. Mellen, of Saint Louis, for many and valuable suggestions.

I am yours, very respectfully, &c.,

J. N. McNUTT, *M. D.*

PEVELY. MO., *October* 14, 1878.

REPORT OF DR. C. M. HINES.

Hon. Wm. G. LeDuc,
Commissioner of Agriculture :

Sir : Having been honored with an appointment as an inspector of diseases of domesticated animals, under the direction of the Department of Agriculture, I accepted the same on the first day of August, and at once took the necessary steps to find a field for an investigation, which had reference more particularly to the diseases of swine.

After diligent inquiry I found the disease was not sufficiently extensive in the State of Kansas for any extended inquiry into the cause and remedy for "hog-cholera," or, the infectious fever of hogs. Under instructions from the department I therefore proceeded to Cass county, Nebraska, where it was said to prevail as an epidemic in the neighborhood of Eight-Mile Grove.

Upon my arrival at Plattsmouth, the county seat of Cass county, I was informed that no "hog-cholera" had prevailed in that region for nearly three months prior to my arrival.

After a detention of several days in the vain effort of finding a proper conveyance into the country, I at last succeeded, being aided by Mr. James Hall, of Eight-Mile Grove. I was assured that I would find but little of the disease, as I was too early in the season, it being more prevalent in cold weather.

As the time for investigation was limited, I determined to make as much of it as possible. Passing through Cass county I found several small herds under treatment by a veterinary surgeon, and in nearly every case I found they were being "doctored" by the owner, or some one professing to cure the disease. Also, other owners, rejecting all interference, were apathetic, and seemed to consider it something beyond human ken, and as one expressed it, left them to "worry through." Indeed, one farmer said that he intended as soon as he was sure the disease was in the herd, to "ship all those large enough for the market"—an example followed by many others, making widespread havoc. From Cass county I proceeded through Otoe, to the borders of Johnson county, passing over a large portion of both counties, returning again to Plattsmouth when the time for the investigation had nearly expired.

In arriving at the conclusions to be found in this article, I must be permitted the privilege of argument, in order to show my reasons for the same, and, first, I would observe that the disease known as "hog cholera," or "infectious fever of hogs," is not, as I think, so difficult of solution, nor has it a protean character. I consider it *one disease* from *two causes* having *two effects.*

The hog is said to be improved by "crossing," and persons ignorant scientifically of its effects, and how far it may be carried with propriety, write and speak learnedly of the matter. They attempt to improve upon nature, and it has been carried to such an extent as to almost obliterate all traces of original breeds. They attempt also to make a distinct, separate, and, as they suppose, *permanent* stock that will reproduce itself. Although all hogs may belong to the one great family, there is a law in nature that, where a great divergence has taken place from any parent stock, a tendency to revert must prevail, or the creature must suffer from the *lex talionis naturæ.* "So true is it that nature has caprices which art cannot imitate."

Persons, otherwise good farmers, who have improved their stock, as

12 sw

they suppose, by crossing or continual breeding in the same stock, do so until they are really ignorant of how close they are breeding, and of its evil effects, for (as in the human being) the penalty for this violation of the law of nature is loss of vitality, less power of resisting diseases, and scrofulous degeneracy.

I have seen pigs not a month old which were totally blind, with large sores on the jaws, and hogs of eight or ten months with great sloughing sores on the body, and I have been told by reliable gentlemen that some lose the flesh from the jaws, leaving the bone exposed. In the older hog this affection may, perhaps, be brought about by feeding exclusively upon old corn that had been exposed to the elements, but time did not allow for proof of this.

Cholera in Kansas and Nebraska seems to attack preferably the Berkshire, and the Berkshire crossed by the Poland-China, which appear to be the kinds preferred in those States. The "common stock," and those not bred so close, are not so liable to the disease as where they have been continually crossed and called "fine-blooded." I have been told by gentlemen who are largely engaged in hog-raising that the common stock and those of pure breed are less liable to the disease—that they have been in adjoining ranges to those diseased, and have escaped the infection. I have no doubt of this fact.

TREATMENT OF THE HOG, HIS FOOD, QUARTERS, ETC.

Of food.—As Dr. Detmers, of Missouri, in a report upon this same subject says, "Because he is a hog, must he be treated hoggishly?" Poor hog! Man seems to think he "has no *stomach* that he need respect." With what do they not dose him (in lieu of what he would find for himself, were he at liberty?) *Stone coal, charcoal, ashes, concentrated lye!* Give him sour food, and afterward an alkali to correct acidity of stomach! All very good when intelligently administered, no doubt. But, does not the hog need an acid sometimes as well?

The almost universal food for swine in these States is corn, *nothing but corn.* If, perchance, they get any green food it is green corn cut and thrown to them.

Corn is raised in such abundance and the price is so low, in order that there may be a return for the labor of the farmer it must be converted into either beef or pork ; and as, according to general belief and practice, a hog requires less care than other domestic animals, and can *stand anything,* he is the favorite instrument through which to realize gain, and every farmer has his herd of hogs, large or small.

Of quarters.—The laws of the States of Kansas and Nebraska prohibit the running at large of domestic animals. and, as a consequence, the hog is confined in quarters of various kinds and dimensions, dependent upon the ability, inclination, or industry of the farmer. Thus we find that in a prairie country where fencing is expensive they are not apt to have too much range.

In that part of the State of Nebraska to which my observations extended nearly all the farms were located on water-courses of variable size, and for convenience the hog-pens were on the banks of the streams, in many cases at an inclination of from 15° to 25°. The inclosures were full of manure of perhaps years' standing, mixed with earth of the kind known as the loess deposits, into which the hogs rooted, wallowed, and when sick they would eat, in a vain effort to relieve their sufferings. (In many cases scarcely anything else was found in the alimentary canal.)

They had at pleasure the privilege of indulging in a bath of or drinking the semi-fluid matter in the streams passing through their inclosures, composed of old and recent manure, with an admixture of the black soil and material of a like character conveyed to them from sties fifty or a hundred miles above. They might also at their pleasure, after such recreation, bask themselves in the sunshine (with the mercury in the nineties) on the hill-side the livelong day.

Fed with corn that had been exposed to the snows and rains of *one* and sometimes *two* years; heated by the sun in summer, cooled by the snows of winter, washed by the rains of spring, and fanned at pleasure by rude Boreas, is it to be wondered at that animals so treated, and from which so much is expected, should become diseased and die, and that, following the example of the farmer who said that he would " ship his hogs as soon as he was satisfied disease was in his herd," the " hog-cholera" should continue, being spread by rail over a great extent of country, dropping some *here* and some *there?* True, all are not so treated, and where they are treated in a rational manner few are lost.

If the same attention was given the hog that is bestowed on other domestic animals there would be less cause for complaint, and it is useless to attempt to remedy the matter except by a radical change in the treatment of the animal.

Many farmers keep their corn in cribs without covering, and one who was losing hogs every day told me that he had been feeding them on corn that had been exposed to the elements for *two years*. I have found that in proportion to the care taken so was the ratio of health and disease, all other things being equal.

The causes, then, in my opinion, which develop the disease known as " hog-cholera" are of two kinds. First, *continual close breeding*, which has a tendency to lessen vitality, produce a scrofulous condition of body, with less power of resisting disease; second, want of proper treatment, which includes food, quarters, and general management.

SYMPTOMS OF DISEASE AND MODES OF ATTACK.

First mode of attack.—Generally the hog is sick a considerable time before it is noticed, and he is not cut off as suddenly as many suppose.

The hog's external depurating apparatus is said to be fixed in the posterior portion of the fore leg and the nose. When the disease sets in the discharge from these parts ceases, and often (especially in young pigs) a swelling of the fore leg may be noticed, extending to the shoulder. The nose becomes dry, and the hog now has the fever. His bowels become constipated, and when moved by the administration of a cathartic his discharges are of scybala, coated with mucous or epithelium. His appetite fails, and he eats what is unusual for him in a state of health, such as dirt and herbage, that, when well, he would pass by. He lies down, or leans against the side of the inclosure, and when started up moves wearily. Two moist streaks may be seen, one from each eye; holds his head down, and his ears fall; when lying down rises up and falls down; stumbles along as though he had rheumatism; is weak in the fore legs; becomes lousy, and if he does not die by the disease which fixes itself upon the brain and spinal cord, he may recover, but is often left entirely blind. If recovery or death does not take place in this first mode of attack, he passes into the condition of those under the second mode of attack, and the force of the disease is exerted upon the mucous membranes of the alimentary canal. In this first mode of attack the disease is seated in the serous membranes.

Second mode of attack.—Begins with fever, as in the first mode, but, although the brain is affected, the force of the disease is exerted more directly upon the stomach, bowels, and lungs (upon the mucous membranes). The hog loses his appetite, grows rapidly thin, and instead of the discharge from the eyes it is from the bowels. He lurches from side to side as he moves along, is weak in the loins, has diarrhea, often vomits, and worms are sometimes discharged from both stomach and bowels. The discharge from the bowels is of a yellow color, seemingly mixed with pus. In this mode of attack all the parasites that infest the hog, of whatever character, seem roused to unusual activity, and the hog, unable to partake of a sufficient amount of nourishment, these parasites, fixing themselves in many parts of the body, prey upon its vitals until it succumbs.

Cough is a prominent symptom, sometimes from the first; is of a spasmodic character, and apparently due to some extent to nervous irritation. In some cases, at every fit of coughing there would be a discharge from the bowels.

Character of the disease.—As before stated, it attacks first the serous, secondly the mucous membranes, or it may be confined to either.

In the first mode of attack the fever is of a sthenic character, and presents many of the characteristics of *measles* in the human being. There is fever, discharges from the eyes, sometimes a discharge from the nostrils, and discoloration of the skin. Cough, which is an attendant upon measles in man, is generally absent in the hog in the beginning of this disease. I prefer to consider it a fever developed in the same manner as typhus or typhoid fever is in man; that there is "blood poisoning," and that the disease germs are intangible: that it has no symptom in common with cholera in man, save the diarrhea. The action of the infection upon the blood is quite the opposite to that of cholera, for in the disease in question there is a lack of fibrin and of hæmatin; it is pale, deficient in red corpuscles, and does not "cup." I do not believe that it is dependent upon any particular condition of the atmosphere, except that portion immediately surrounding the diseased animals. I think there can be no doubt that it may be communicated to other hogs, and more readily to those of a like breed, and living under like conditions. Being (as I think) not primarily of a *typhoid* character, I cannot see any reason why this term should be applied to the disease. The truth is, I believe, that the hog is sick some time before it is generally noticed, and that a little attention given him at the commencement will stop it. Is this, then, of a typhoid character? In confirmation of this I will state a little circumstance related to me by a gentleman in this neighborhood. A colored barber called upon him at his farm one day, and while looking at a fine hog, which the owner said would eat but little, and appeared to be sick, the barber said : " Your hog has the cholera. I will cure him"; and immediately, to the great amusement of the gentleman, caught the hog, opened his mouth, made two incisions in the papillæ at the root of the tongue, and then began rubbing the fore-legs of the animal with a corn cob. Telling the gentlemen to give the hog a dose of some purgative medicine, he went his way. In a few hours the hog began to eat and recovered in a short time.

Of infection.—Hogs of the same class, and placed under like circumstances, are more liable to convey the infection to each other than to those differently situated. I met with a farmer in Nebraska who was purchasing diseased hogs at a low figure, and taking them on his farm for treatment, without fear of communicating the disease to his own herd. He had some knowledge of the disease, and had treated his own herd.

Professor Law says, as quoted by the commissioner of agriculture of the State of Virginia, " contagion is the main cause of the disease." We are satisfied that we understand the circumstances under which one may contract the chills or intermittent fever, but I presume no medical man will say that he can *touch* the "disease germs," as they are termed. Contagion cannot *cause* it, but may aid in spreading it.

Prevention of the disease.—In my opinion the surest means of prevention are those of a hygienic character. Do not breed close, give the animal a variety of food, keep his range clean, and protect him from extremes of heat and cold. In a prairie country, where domesticated animals are not allowed to run at large, I would recommend that ranges for the hog should be inclosed by portable fences in sections. Posts should be placed at the proper distances (they might be of iron and driven) and the sections wired together or fastenings might be attached to each section so as to unite at once. Constructed in this manner the range may be changed to another location in a few hours. This should be done once or twice a year at least, and preferably in the spring and the beginning of winter. Raise vegetables especially for them. If possible sow oats, and let them have the range of the field. Give them fresh water to drink, which may be raised by a windmill and conveyed through pipes to the range. Instead of having hogs to "follow" the cattle, as a matter of economy, I would feed them separately, and have the corn for the cattle ground in a horse-power mill.

Eradication of the disease.—This might be effected partly through State laws prohibiting the transportation through the States of hogs showing evidence of disease, attention to hygienic laws, and a greater admixture of the breeds known as "common stock," gradually brought about.

Treatment of the disease.—This is very simple if attended to in time, and very few need be lost. Simply a transfer to a new range and a change of food at the beginning of the disease will save a great many. Give the hog a purgative of soft soap, raw linseed-oil, or any simple purgative; afterward warm mashes and comfortable dry quarters. Very often this is all that is necessary to arrest the disease. As soon as his nose becomes moist and the secretion is restored in his fore-legs, you may count upon his recovery. A farmer told me that his herd had the "cholera," and that he fed the living with the carcasses of those that died, and his hogs recovered. Another that, having more fresh beef than he wanted, fed the surplus to his herd, and they recovered. This food, being unusual, acted upon their bowels, hence their recovery.

In investigating this disease I had many obstacles to contend with. There were no herds to be found within a reasonable distance (nor beyond that I was aware of) which had not been dosed with something, and none so isolated as to be entirely free from contact in some way with other herds. As a consequence I made no use of the clinical thermometer, which would have given no perfect data to discourse upon.

The first herd of hogs treated numbered forty-five head, situated on high and dry land; but the range was dirty from the accumulations of old manure, they having been fed on corn from crib exposed to the elements for a considerable period. Were drinking water from a well. All sick. No other hog had been near the range except a boar, and he was said to be well; neither had any been away from the herd. The breed was Berkshire, crossed with Poland China. Owing to the inland situation and want of necessary articles at the place the troughs were not made with circular holes, but were constructed in the following manner:

The trough was divided longitudinally by a board on edge so that the hogs could feed on either side without permitting the admission of the feet, thus: They were graded 1, 2, 3, 4, according to age and condition. The herd was suffering from both modes of attack, as heretofore described. They were moved from their range and placed on new ground. As a general thing the younger hogs suffered the most.

Pen No. 1 contained the oldest hogs, fifteen in number, from one to three years old.

Pen No. 2 contained fourteen head, from eight months to one year old.

Pen No. 3 contained eleven head, from five to eight months old.

Pen No. 4 contained five head, from five to twelve months old, and was the *dead pen*.

No food was allowed for the space of twelve hours. Nos. 1 and 2 were given salt and water, which they were compelled to drink, being without food or water. This had the effect of causing vomiting and purging. In several cases worms were discharged from the stomach and bowels; principally from those suffering from the second mode of attack. Some had to be pressed forward and urged to drink. After the action of the salt the tincture chloride of iron was administered in water in doses of twenty drops every four hours, for the older, and fifteen for the younger hogs. A mash of bran was made (which was always fed while fresh and sweet), and they were allowed to partake moderately of the iron half an hour after the first dose. They were fed at intervals between the doses of iron, and no other food was given until convalescence began, when they were allowed some corn in connection with the mash. In those suffering from the disease in the first mode, there was constipation of the bowels, dry noses, and watery discharges from the eyes. When the bowels were moved (and in some instances they were very torpid), the passages would be sterceraceous, and covered with a white substance (apparently epithelium), were very hard, and upon examination appeared to be composed almost entirely of earth. These began to improve on the third, and were so much improved on the sixth day that they were allowed a more liberal supply of food. They were not considered out of danger until the eighth or tenth day. It was not necessary to give any other purgative, and gradually the discharges from the bowels became of a proper consistency.

No. 3. Most of these had diarrhea. Some had a cough, and whenever a fit of coughing came on there would be a profuse discharge from the bowels, thin and of a yellow color. Occasionally there would be vomiting also, showing the great irritability of the pneumo-gastric nerves. Worms would also, at times, be ejected from the stomach or bowels. To these were administered from one and a half to two fluid ounces of raw linseed-oil, according to the age of the animal. After the action of the oil the discharges were not so frequent, and the animals seemed more lively. Twenty drops of carbolic acid were then administered to the older, and fifteen to the younger hogs every four hours.

The action of this remedy not meeting my expectations, I had recourse on the third day to the tincture chloride of iron, as in the cases of Nos. 1 and 2. Fifteen drops were given to the older and twelve drops to the younger every four hours with marked improvement. The food given was the same in all cases. Convalescence in this class was slower than in Nos. 1 and 2, it beginning a day or two later, and the recovery was more protracted, with the prospect, in some cases, that a month or longer must elapse before they would be of any value.

No. 4.—*The dead-pen.*—In this pen were five hogs of different ages, ranging from five months to a year old. They were selected for this pen, as there was but little hope of saving them. Two were sick after the first mode of attack, and three after the second mode. Linseed-oil was administered in corn-meal and water. They had to be urged and brought up to drink. One utterly refused, and was too far gone to undergo treatment. He died in a few hours in convulsions, as in the first mode. The morning after two more were found dead, and the next day another died. These latter were after the second mode. One after the first mode recovered. The tincture chloride of iron was administered to these also. As they began to improve, which was in from six to ten days from beginning of treatment, they were fed more liberally according to their condition. The pens were kept clean, the manure being removed at once. Chloride of lime was used as a disinfectant.

The loss was four out of forty-five hogs. Together with the foregoing treatment, the following was administered every four hours, between the doses of iron: Powdered alum, ℥iss; sulphur sub., ℥iij; powdered saltpeter, ℥iss; flaxseed-meal, ℥ix. These were mixed, and two pounds of the mixture was added to every barrel of mash in which it was given.

The second herd treated numbered originally 123 head; several had died, reducing it to 114. The breed was Berkshire crossed with Poland-China. They had been bred very close. This was a bad lot to treat, as they had been dosed with "condition powders," "concentrated lye," and several other articles. They had been fed on corn exclusively. Their range was located on a hill-side, and a stream of water passed through it. It was covered to a considerable depth with old and recent manure, exposed to the sun, and without shelter for the hogs. The stream was thick with mud and manure, where the hogs could wallow at pleasure and bask in the sun all day. There were other ranges above and below; the number I have no idea of, but presume that every farm located near this stream had its range on it, as it was common so to do for convenience. No other hogs had been brought there, and none taken away and returned.

The herd was moved to new ground in the shade, and graded according to size and condition. They were divided into five classes.

First class.—This consisted of eighteen hogs, the ages ranging from one to three years. They were suffering with symptoms belonging to the first mode of attack; had no cough. The bowels of some had been moved by remedies, others not. Could partake of some food, but not heartily. They were treated together.

Second class.—This consisted of twenty-one hogs, ranging from one to two years, and were suffering from an attack after the second mode. They had cough and diarrhea.

Third class.—This consisted of thirty-nine hogs, ranging from five months to one year old, suffering from an attack after the first mode.

Fourth class.—This consisted of twenty-six hogs, ranging from five months to one year old, suffering from an attack after the second mode.

Fifth class;—dead-pen.—This consisted of ten hogs of different ages.

Three were after the first mode and seven after the second mode of attack.

Believing that the theory of *blood-poisoning* was correct, I did not see any reason for a change of treatment from that followed in the case of the first herd. Those suffering from the disease by the first mode of attack were first given salt and water and afterward the iron, as in the case of the first herd. Those suffering from an attack in the second mode received a dose of linseed-oil, and afterward the iron and powder as detailed in the case of the first herd. Many had to be urged and forced to drink. Some refused altogether to partake of anything. I sum up the deaths by class:

Of class 1 .... None.
Of class 2 2
Of class 3 3
Of class 4 5
Of class 5 8

Total number of deaths 18

Nine had died before treatment, making twenty-seven in all.

Post-mortem examinations.—In making *post-mortem* examinations, I was afforded opportunities in Nebraska (besides those under my own observation), by Mr. A. J. Rainey, a veterinary surgeon, who had a large number of animals under treatment. Also, by a Mr. Dudly, an enterprising farmer residing in the neighborhood of Syracuse, who gave me permission to examine his herd, in any manner I saw fit, in furtherance of my object.

In my description of appearances after death, I shall confine myself to one or two dying under each mode of attack.

Hog six months old.—*The blood.*—This had the appearance of water colored yellow. Fibrin broken up, and a want of hematin. Excess of serum and salt. Poured upon the ground it was absorbed, leaving scarcely a perceptible stain.

The brain.—Effusion of serum in cavity of skull, and softening of the brain. Effusion in the membrane of the eye.

The lungs.—Effusion of serum in pleural cavity. Base of lungs somewhat congested, apparently of a passive character.

The heart.—Normal condition, but pale.

The stomach.—Normal condition, the spleen enlarged.

The liver.—There was but little bile in the gall-bladder; the organ was darker in color, with petechial spots. Kidneys pale. No ulceration of intestines. This hog died from the first mode of attack.

Hog six months old.—This hog had recovered from an attack in the first mode. Was left blind, and had an ulcer on one of his feet. He was killed. Was apparently free of disease; the blood was of the proper consistency and color, and coagulated. Blindness was the effect of the disease.

Examination of those dying from second mode of attack.—*Hog six months old.*—This hog was very thin, nearly all the fat having been absorbed. Could detect no disease of the brain. In this case there was the usual diarrhœa, cough, &c., belonging to this class. Heart normal in structure, but pale. No effusion in pleural cavity.

Lungs.—These presented the appearance mentioned by writers on this disease as gray hepatization.

Stomach.—This presented evidences of disease. Two ulcerated patches were found, nearly healed, circular in form, and eight or ten inches in diameter. The dead mucous membrane was still adherent, but was easily removed.

The liver was discolored; dark patches were diffused over its surface. One large worm (*Ascaris lumbricoides*) was found in the duodenum. There was a large ulcer, about an inch in diameter, in the ascending colon, plainly seen on the external surface of the intestine. Its edges were very hard, and the inflammation extended some distance beyond. There were other ulcerations in different parts of the intestines, but less extensive.

The spleen was of natural size, but darker in color.

The kidneys presented a grayish appearance, very pale, and having an appearance as though there had been a deposit of black pigment in their substance. They were easily broken up, the internal portion or belly showing evidences of suppuration.

The bladder.—This organ was intensely inflamed, so much so as to diminish its capacity to *one fluid ounce.* All the organs in the course of the alimentary canal had more or less petechial spots on them.

Hog five months old.—Killed him. He was very much emaciated. Was apparently recovering from the disease, but very slow and doubtful. Found three large worms in the stomach and one in the duodenum. The one in the duodenum had his head inserted in the gall duct up to the gall bladder. There was some chronic inflammation at the upper portion of the duodenum where the worm had fixed himself. The stomach of the hog was full of grass. It seemed that this hog would have to die of inanition, the presence of the parasite interfering with the flow of bile into the alimentary canal.

Geology of the district of country where these examinations were made: The soil is of what is termed the "Loess deposits," and by analysis by Samuel Aughey, Ph. D., contains—

Insoluble (siliceous matter)	81.23	Soda	.15
Ferric oxide	3.86	Organic matter	1.07
Alumnia	.75	Moisture	1.09
Lime, carbonate	6.07	Loss in analysis	.59
Lime, phosphate	3.58		
Magnesia, carbonate	1.29		100.00
Potassa	.27		

Parasites.—Of the entozoa that infest the hog I have seen but three kinds. Two of those are familiar to most persons, and are found in man. The third is a smaller parasite, and is often found in the stomach of the hog, and which is said at times to destroy the pyloric orifice of the stomach. I have seen but one of this species; it was white, and from eight lines to an inch in length.

I append a statement by some farmers in Kansas, who are successful hog-raisers, as to their treatment of hogs. Mr. Jacob Allen, of Neosho county, says: "Last year my hogs had the fever, or 'hog-cholera.' They would eat dirt; dirt was found in lumps in their stomachs; but few worms, and those in intestines and kidneys. No trichina under microscope. Were constipated. I lost some; cured the others by the use of senna and jalap."

Rev. John Schoemakers, of Osage Mission: "Has been a resident here for thirty years, and states that he is of opinion that the disease comes of want of proper management, forcing them with corn, and want of a variety of food." He states that they have a large number of hogs on the Mission farm, but that they lose none by cholera. They are let run in a large field that has been under cultivation. Does not confine them to pens.

Mr. David Bloomer, of Neosho county, feeds his older hogs corn in the winter and spring. Sows oats for them in two separate fields, and at different times. When the oats are four inches high he turns them into the field first sowed, and afterwards into the second field, so as to keep them until corn is "out of the milk," when he cuts and feeds them corn. Feeds his pigs on oats and shorts during the winter. Lets the sows wean the pigs. Breeds his sows twice a year; first litter to come about the 20th of February, next litter the 20th September. After the green oats are gone he turns them into a pasture of 120 acres. They have access to clear running water and to shade in summer. Has cover for pigs in winter, but none for old hogs. Does not "shuck" his corn, and keeps it always under cover. Breed, pure Berkshire, not bred close. Loses no hogs by cholera.

In conclusion I have to state that of other diseases affecting animals in the States of Kansas and Nebraska, there were an unusually small number, and only of those familiar to nearly every one.

In giving a name to the disease known as "hog-cholera," I have no hesitation in saying that the disease in the latter stages has all the characteristics of gastro-enteric fever in man.

Very respectfully, your obedient servant,

C. M. HINES, M. D.

Osage Mission, Kans., October 29, 1878.

PREVALENCE OF DISEASES AMONG DOMESTICATED ANIMALS.

By a perusal of the subjoined correspondence of the department, it will be seen that there has been no abatement of diseases among domesticated animals during the current year. Those incident to swine seem to have been quite as prevalent and almost as fatal and destructive to the animals attacked as they were during the year 1877. The per cent. of deaths for the last-named year was given at 58.94, while this year it is given at 52.75. Now that the disease which has been so destructive to this class of farm animals has been shown by recent investigations to be highly infectious and contagious, proper care and vigilance on the part of farmers and stock-growers will lessen the spread of the plague, and confine it to such limits as to greatly reduce the heavy annual losses of the past few years.

Many diseases of a malignant and contagious character have prevailed among other classes of farm animals the past year, which will receive the attention of the department during the coming season.

ALABAMA.

Bibb County.—The losses from cholera among hogs are annually very heavy. At least 40 per cent. of all the hogs in the county suffer from this disease, and 75 per cent. of those attacked die. Cholera is also prevalent among fowls, and large numbers of them die.

Clarke.—A few horses annually die in this county of farcy, a fatal contagious disease, and a few from want of care and proper attention, the latter mostly owned by negroes. There seems to be no disease among stock-cattle. Both hogs and chickens die of cholera.

Cullman.—There is some murrain among cattle, and considerable cholera among the hogs and chickens in this county. There is but little stock raised in the county.

Escambia.—The only class of farm stock affected by contagious diseases in this county is that of swine. These diseases have been a great drawback to hog-raising.

Jefferson.—Horses suffer severely from distemper. Cattle are occasionally affected with black tongue and murrain, but at this time are unusually healthy. Hogs are seriously affected with cholera, quinsy, and other unknown diseases. The losses have been very heavy this season. Cholera and roupe prevail among fowls.

Lauderdale.—We have had no infectious or contagious diseases among horses or cattle in this county. They suffer terribly, however, during the winter for lack of food and proper attention. At least five hundred horses and mules, and a greater number of cattle, are annually lost from this cause. Hog-cholera prevails here every year, and the losses are sometimes enormous. I estimate that between 7,000 and 8,000 head have been lost during the past year. The condition of farm stock generally is low—worse than at any time since the war.

Madison.—No infectious or contagious diseases prevail among farm animals in this county. Hogs frequently die of so-called cholera. Fowls are afflicted with the same malady. The general condition of farm animals as compared with previous years is good.

Monroe.—A few hogs only have been lost by disease in this county this year.

Saint Clair.—Stock in this county is in very good health and condition. I hear of no infectious or contagious diseases.

Walker.—Horses are seriously affected and frequently die of epizootic distemper. A good many cattle are lost by murrain and black tongue, and many hogs die of cholera. Fowls die of cholera and a disease which affects their throats. There are but few sheep raised in this county; but this industry is on the increase.

ARKANSAS.

Baxter County.—The graded calves of this county have this year suffered severly by a disease called black-leg. The first symptom is a lameness, and they usually die within from twenty-four to thirty-six hours. No remedy has been found. About one-fifth of the calves have been attacked, and nine-tenths of those attacked have died.

Boone.—The only diseases of any moment that have prevailed among farm animals the past year are those incident to swine. The losses have not been very heavy.

Bradley.—Horses, cattle, and sheep are free from disease. About 10 per cent. of all the hogs in the county have died during the past year from eating cotton-seeds and lying in the dust. Cotton is the only product that is raised here for the market.

Fulton.—Hogs in a few localities of this county have been fatally affected with cholera.

Grant.—Chicken-cholera is prevailing here to an alarming extent. The hog-cholera has somewhat abated. There are no diseases existing among horses, sheep, or cattle, of a serious nature.

Marion.—Horses and sheep are very healthy, and cattle moderately so. Many of the latter have died this season of black-leg. Many fowls are annually lost by a disease commonly known as cholera. A great many hogs have been lost this season in this county by an unknown disease. It is not cholera, but more resembles yellow fever in man.

Monroe.—Cholera among hogs and fowls prevails here every year, and usually proves very fatal. All other kinds of farm stock are healthy this year.

Montgomery.—Horses, cattle, and sheep are proverbially healthy, at least the exception is so small that it is not worthy of note. Until this summer hogs have been healthy, but cholera is prevailing extensively among them at this time.

Perry.—The health and condition of farm animals is generally good. Diseases among hogs continue to prevail at irregular intervals.

Pope.—Occasionaly a horse dies here from bots and blind staggers, and sometimes from bad treatment. Fine cattle brought here from other States frequently die of murrain. Hogs suffer terribly from what is called cholera. In some localities it kills almost every animal. Fowls also suffer from cholera. Sheep die of rot and bad management.

Sevier.—All classes of farm stock are healthy save that of swine, and a good many of these animals are dying in the northern part of the county of cholera.

Saint Francis.—A new disease has appeared in this neighborhood among cattle; it first appeared among sucking calves, but has lately carried off several grown cattle. The symptoms are a trembling appearance and gradual prostration, which ends in death in from three to seven days.

Stone.—All classes of farm animals have been unusually healthy during the past year in this county.

White.—At least one-third of the hogs in this county have been afflicted with disease during the past year, and of this number eighty-five per cent. have died.

CALIFORNIA.

Calaveras County.—We have never had any infectious or contagious diseases among any class of farm animals. Every year we lose a greater or less number of animals by starvation. Last winter probably ten per cent. of all the cattle and sheep in this county died from this cause alone. In 1862 fully three-fourths of all the cattle died for the want of feed.

Contra Costa.—Hogs here are subject to cholera and pneumonia. These diseases are brought on by lack of proper care and attention.

Lassen.—There are no infectious or contagious diseases prevalent among domesticated animals in this county.

San Bernardino.—This portion of California has always been remarkably healthy for all classes of farm animals. No contagious disease has ever prevailed here except scab among sheep, and this disease never destroys the animals.

San Diego.—The only disease existing among cattle is murrain; this disease is very fatal, especially in dry seasons. Hog-cholera is not known here. A good many sheep are killed by eating poison-weed after our spring grasses are dried up. We lose a good many fowls from a disease known as swelled head.

Shasta.—Horses in this district are annually afflicted with an epizootic distemper; if properly cared for but few of them die. Cattle and hogs are healthy. Sheep are affected with scab, but when washed and properly treated but few are lost. Our State is turning some attention to the Angora goat. I think the raising of these animals will eventually make the best business of the State. California contains a vast extent of country adapted to the roving of this animal which is fit for nothing else. It subsists entirely on brush, and seldom, if ever, grazes.

Tuolumne.—Farm animals of all kinds are in a healthy condition. The weather is mild, and feed is starting.

Yuba.—A kind of epizootic disease is seriously affecting horses in this county. Diseases among cattle are generally caused by want of proper care in winter. Cholera prevails among hogs and fowls. Sheep become diseased for want of proper care, and keeping too many together.

COLORADO.

Bent County.—Neither horses nor cattle are affected with infectious or contagious diseases here. Cattle-raisers estimate their annual losses at about 5 per cent. A few sheep are lost by a disease known as scab. Stock is in extraordinary good condition at this writing.

Gunnison.—This is a new county, and there are not over one thousand horses in it. About half of these have suffered this year with epizootic distemper, but none have died. There are no hogs or sheep in this county. There are a few fowls, but they are entirely free from disease.

San Juan.—There are no domestic animals raised in this county, and none are wintered here. During the early summer there are quite a number of animals poisoned by eating a weed which the Mexicans call "Loco." The botanic name of this weed is unknown to the writer.

DAKOTA TERRITORY.

Brule County.—There is but little stock in this county. No disease of a serious character prevails.

Lake.—A few horses have died of distemper this year, and a few calves and cattle have been lost by the disease known as black-leg. Fowls are affected with rough, scabby legs, and perhaps 10 per cent. of them die from this disease very suddenly and while in good condition.

Pembina.—This is a very new county, and contains but a few farmers and little stock. What little we have is healthy, and free from all infectious or contagious diseases.

Traill.—A contagious but not very fatal distemper exists among horses in this county. The symptoms are a discharge from the nostrils, and swelling of the throat to such an extent as to prevent even the swallowing of water for three or four days at a time. Pants when driven severely, and his tongue hangs out of his mouth.

DELAWARE.

New Castle County.—Chicken cholera prevails this season. The best preventive is a tea made of smartweed, and placed easy of access to the fowls. Condition of farm animals, "good and improving." We believe the best stock and the best care will insure most satisfactory and profitable results.

FLORIDA.

Calhoun County.—Cattle are generally affected with black tongue, hollow horn, and murrain. Horses suffer to a considerable extent with staggers and scurvy; and many hogs are annually lost by cholera and staggers.

Columbia.—Native horses are generally much more healthy than those brought in from Northern States. Cattle have been generally healthy, except in a few localities, and in these the losses have been quite heavy. About one-third of the hogs in the county have died from cholera, and the thumps; the greatest fatality ever known has occurred among fowls this year.

Dural.—There is no disease of any kind existing among farm animals here, except a disease known as salt sickness, which affects cattle only. Diseases affecting fowls are attributable to lice, and are contagious because they are infected by contact. They receive no care. The breed is wild; they are rarely fed, and the only wonder is that they do not all die.

Lafayette.—There are but few horses in this county, perhaps not over fifty head, and but little attention is paid to raising or caring for them. There are about 3,000 cattle in the county. They depend entirely on wood range for subsistence, and are generally in bad condition. Hogs subsist on mast, and do very well. No sheep are raised here. During the summer fowls are afflicted with cholera.

Levy.—Staggers among horses is very fatal, especially among young animals. Epizootic distemper is the most fatal infectious disease among this class of animals. Cattle are affected with what is known as "salts"; called this, perhaps, for want of a better name. Hogs are subject to cholera, sheep to black tongue, and fowls to "sore-head."

Madison.—Distemper and glanders are the only contagious diseases prevailing among horses in this county. Cholera has been very destructive among hogs. A few cases of thumps have been reported among the same class of animals. Many fowls have died of cholera and sore-head.

Polk.—The losses of cattle in this county, from various causes, amount to about 5 per cent. of the whole number. But few horses are raised here, and sheep are just being introduced. Fowls do not do well; the climate seems to be too warm for them.

Saint John's.—No sort of attention is paid to the raising of hogs or sheep in this county. I have not learned of a single person having an improved breed of pigs. All depend on the "razor-back" or "land pike." But little disease prevails among any class of stock.

Santa Rosa.—Very few cattle die from disease here, but a great many die from want of proper care in the winter, and food in the spring. Some few sheep die of rot or grub. Hogs are sometimes afflicted with fatal diseases.

Sumter.—Pink root or foot disease is quite common among white hogs, but does not affect black ones. Salt-sick is a disease common among cattle. We have no remedy, but some recover.

Suwannee.—Out of 60 head of horses recently brought here from Texas, 36 died, with no apparent well marked symptoms of disease. No other horses were so affected. Hogs are afflicted with so-called cholera, and chickens with what is here known as sore-head. The head of the fowl becomes very sore, and so much swollen that the tongue hangs out of the mouth, the eyes swell shut, and they soon die.

Volusia.—Horses and mules are seldom attacked by disease except blind staggers and sand disease. About 60 per cent. of the first and 20 per cent. of the latter, attacked by these diseases, die. Cattle are affected with salt-sick and hollow-horn. The greater loss is from the former. Hogs and chickens are sometimes affected with cholera and other diseases.

Wakulla.—Horses, colts, and mules die of staggers, grubs, and colic; cattle of hollow-horn and hollow-tail, and hogs of thumps and cholera. Chickens also die of cholera.

GEORGIA.

Charlton County.—During the past twelve months hogs have died in greater numbers than was ever known before. We have no improved breeds, our hogs all being "land pikes." We have no remedy for the diseases which carry them off in such numbers.

Coffee.—Horses in this county are seldom attacked by contagious diseases. A few are affected with epizootic distemper, and a good many die of staggers. Occasionally one dies with colic or sand disease. Cattle are only affected with diseases brought on by poverty in the winter season. Cholera among hogs is the most dreadful and fatal disease we have to contend with. Sheep are sometimes affected with staggers and sore-head, but rarely die except from old age or poverty.

De Kalb.—The value of horses lost by disease in this county during the past year will reach $5,000, and that of hogs $8,000 or more. Immense numbers of chickens have also died of cholera.

Fannin.—Stock of all kinds have been remarkably free from infectious and contagious diseases in this county. Stock here is raised only for domestic purposes.

Forsyth.—Horses are affected with bots and staggers, and a good many cattle die of distemper and murrain, and hogs of cholera.

Hart.—The losses in this county from diseases among farm stock are generally very light.

Jones.—The only disease among cattle here is hollow-horn, and that, as a general rule, is produced by neglect in bad weather. Hogs and fowls have suffered severely with cholera the past two seasons.

Laurens.—We have no infectious or contagious disease among horses, cattle, or sheep. Disease prevails more or less among hogs every year. The general condition of farm stock is good.

Lincoln.—Farm animals this year have generally been exempt from infectious and contagious diseases. In a few localities chicken cholera prevails with more or less fatality.

Marion.—A few horses have died during the past year of epizootic and lung fever. Cattle and sheep are healthy. Cholera is quite prevalent and very fatal among both hogs and chickens.

Murray.—About 5 per cent. of the hogs and sheep of this county are annually lost by disease. Perhaps 2 per cent. of the cattle are lost by murrain.

Pulaski.—We have no contagious diseases among horses except distemper, and that rarely kills. Cattle are healthy, but hogs are more or less subject to cholera every year. The only disease affecting sheep is rot. Fowls have more or less cholera every year, which is generally very fatal.

Randolph.—The most fatal disease among horses which has prevailed here during the past year is staggers. Cattle are subject to a good many maladies, some of which are quite fatal. Cholera and big-shoulder sweep off a great many hogs annually. In some localities almost all the fowls have been destroyed by cholera.

Rockdale.—No diseases of a very destructive character have visited our farm stock during the past year.

Schley.—There are no diseases of any character prevailing among farm stock in this county. This section is most prosperous to the farmer, as there is a full crop of all products and good health throughout to both man and beast.

Screven.—The most prevalent disease here, and the most distressing one to the farmers, is colic in mules. It is very fatal, and generally kills within from five to ten hours. At least five out of every seven of those attacked die. It seems to be caused by an accumulation of wind in the body, and not in the intestines. The body swells to the greatest dimensions, and the most excruciating pains follow and continue until death.

Spalding.—The losses among farm stock in this county from the various diseases incident to the same will probably reach as high as $16,000 for the current year.

Tattnall.—Staggers is the most fatal disease among horses in this county, and black-tongue among cattle, although more of each class die of poverty than from the effects of disease. Cholera is very fatal among hogs and fowls.

Towne.—There are no diseases prevailing among farm animals in this county except bots and distemper among horses and milk-sick among cattle; also, cholera among hogs.

Union.—We have no contagious disease among any class of farm stock, but a good many animals are lost every year from common and well-known diseases.

Washington.—Murrain among cattle and cholera among hogs and chickens are diseases that are proving very fatal here. Nearly all the animals and fowls attacked by these diseases die, as we have no remedies. Young animals and those being fattened seem the most liable to attack. Farm animals in this county are now in a better condition than ever before at this season of the year. One reason for this is that a large number of planters do their cotton-ginning by steam and water-power instead of with animals.

Wilcox.—A greater or smaller number of hogs die every year of a disease called cholera. All other classes of farm stock are measurably healthy.

Wilkes.—Hogs die annually in some localities in this county of a disease called cholera. Some years this disease is much more destructive than in others.

IDAHO TERRITORY.

Bear Lake County.—Horses are occasionally subject to a distemper which is regarded as contagious. The symptoms are heavy discharges from the nostrils.

Idaho.—This climate is very favorable to farm stock. All classes subsist on the abundant bunch-grass of the range during the winter, and disease is rarely known among them.

ILLINOIS.

Adams County.—A good many horses have died during the past year of distemper. As usual, the so-called hog-cholera has prevailed extensively, and has carried off stock to the value of $25,000 or $30,000.

Carroll.—All farm animals have been remarkably free from contagious diseases except hogs. Never before has there been so great a mortality among swine in this county. With pigs and shoats the disease has been most fatal. No remedies seem to be of any benefit, and no sanitary condition is a safeguard against attack. They are affected in a great variety of ways and apparently by different diseases.

Clark.—A mild form of epizootic distemper prevails among horses in the southeastern part of the county, and there have been some deaths. Hog-cholera prevails in a very fatal form in the eastern part of the county, and many hogs are dying. Chicken-cholera is also very prevalent and fatal.

Clinton.—A good many horses are every year attacked by an epizootic distemper, and about 5 p'r cent. of those attacked die.

Crawford.—Hogs frequently die of a disease commonly known as cholera. A great many chickens are annually lost by a disease of like character.

De Kalb.—Diseases among swine have prevailed to a most fearful and destructive extent among the hogs in this county during the past season. The losses are estimated at $50,000 and upwards.

Ford.—Hog-cholera is about the only disease of consequence that prevails in this locality. It is sometimes very fatal and destructive.

Grundy.—At present hog-cholera is prevailing in one or two townships of this county, and many hogs are dying. A Mr. Ely has lost 160 out of a herd of 260 head, and the disease is still raging.

Hancock.—The value of hogs lost in this county during the past year will amount to over $35,000.

Hardin.—Horses in this county are more free from disease at this time than they have been for three years past. Distemper in a rather bad form is the worst disease now affecting this animal. Cattle are also unusually healthy. Hogs are less affected with cholera than for many years past. Poke-root slop is the best preventive we have yet found, in addition to the burning of the dead carcasses. A law should be passed for the fine and imprisonment of any person who neglects this latter precaution.

Henderson.—The mortality among hogs for this year is greater than for any previous year. The losses up to this time will exceed $80,000. I feel confident that injudicious feeding, in connection with insufficient shelter, are the predisposing causes of disease among swine.

Iroquois.—During the past few months the number of hogs lost in this county by disease has been immense. Several breeders of fine Berkshires have lost their entire herds.

Jackson.—Hog-cholera is the only disease that has seriously affected any class of farm animals in this county.

Johnson.—Hog-cholera has prevailed to a limited extent this season, therefore the losses have not been very heavy.

Kankakee.—A large number of hogs have died of disease during the present year. Perhaps the aggregate of these losses would amount to $15,000 in this county. About 1 per cent. of all the fowls die every year from diseases incident to them.

Kendall.—Hogs have been seriously affected with cholera. Other classes of farm stock have had the usual affections.

Knox.—With the exception of swine all other classes of farm animals have remained in good health during the past season. The mortality among hogs has been very great.

Lee.—Domesticated animals in this county have been remarkably healthy for the last year with the exception of hogs, which have died in great numbers. My own opinion is that the predisposing cause has been too close in-breeding, and a consequent weakening of the constitution and loss of vitality.

Livingston.—No infections or contagions disease prevails among horses, cattle, or sheep. Diseases incident to swine and poultry are quite prevalent and fatal. I presume the diseases affecting swine are similar to those existing elsewhere.

McDonough.—The loss among hogs in this county during the past year has been very heavy—perhaps $100,000 would not cover it. Other classes of farm stock have remained in their usual health.

McHenry.—Hogs in this county are seriously afflicted with infectious or contagious diseases. Some 2,500 have died from the plague. About nine-tenths of those attacked die, and the aggregate losses thus far will reach $15,000. Other classes of farm stock are healthy, and their general condition above an average.

Macoupin.—Hogs and fowls are annually affected with cholera, and great numbers of each die of this disease.

Madison.—No infections or contagious diseases have recently prevailed among horses and cattle in this county. Among hogs the cholera is quite prevalent and very fatal, reducing the number at a rapid rate. A great many fowls die from the so-called chicken-cholera and from gapes.

Monroe.—We have had a great deal of hog-cholera in this county. I think the disease is mostly caused by malaria, the result of filthy keeping and careless feeding. We also have chicken-cholera, for which we have no remedy.

Ogle.—Hog-cholera, or disease among swine, still prevails to a limited extent in some localities in this county, but is not so severe or so fatal as last year.

Piatt.—There is no infectious or contagious disease prevailing among any class of farm animals except among swine. Recently a malarial fever broke out among some imported stallions in our county, owned by Mr. Harvey E. Benson, and they all died. There were eight or ten of them in number.

Pike.—Hogs are generally seriously affected here with cholera at two periods of their existence, viz., in July, before they are old enough to wean, or between milk and grass, and again just before they are old enough or large enough to fatten. Some die at all stages and every season of the year from the effects of this baneful and destructive disease.

Pope.—The only disease of any moment prevalent among farm stock in this county is cholera among hogs and chickens. The annual losses among both classes are very heavy.

Pulaski.—But few farm animals are raised in this county, and the losses from disease have been light during the past year.

Randolph.—What is known here as hog-cholera has prevailed in several parts of the county, but has generally been most destructive where large numbers were herded together. Cases are reported of several droves, numbering one hundred or more, where but ten or fifteen head, in all, recovered.

Sangamon.—Horses, cattle, and sheep have been healthy during the past year. There has been the usual loss among hogs and fowls, but the aggregate cannot be given in the absence of reliable data.

Schuyler.—Hogs are the only farm animals that have been affected with infectious or contagious diseases in this county during the past season. Great numbers of turkeys and chickens have also died of cholera, but I can give no idea as to numbers that have been lost.

Shelby.—During the past year hogs have died in great numbers in this county of cholera and lung diseases. The aggregate loss will amount to over $60,000. A few horses have died of distemper, and a good many cattle of dry murrain.

Stark.—The hog-cholera has been very severe on some farms this fall, a good many farmers having lost nearly all their stock hogs and some of their fattening stock.

Stephenson.—The losses of swine in this county have been fearful. The class now dying are mostly shoats—last spring's pigs—and they are dying so rapidly in some localities that it is impossible for the farmers to hunt them up and bury or burn them, consequently the air is tainted with their carcasses.

Tazewell.—Immense numbers of hogs have died in this county during the past year of the various diseases which afflict them. No cure has been discovered for these maladies.

Wabash.—Cholera among swine seems to be the only disease affecting any class of our farm animals. About one-half of all the hogs in the county annually die of this disease.

Washington.—The usual diseases have prevailed among farm animals in this county during the past year, and the losses among all classes will reach $8,000 or $10,000 in value.

INDIANA.

Adams County.—The only class of farm animals affected by disease in our county is the hog. The disease seems to be epidemic and contagious, and has occasioned heavy losses among hog-raisers.

Brown.—The only disease of any consequence that has prevailed among farm animals in this county during the past year is cholera among swine.

Carroll.—Cholera has been very destructive among swine during the present year. The losses in this county will amount to $38,000 or $40,000. The symptoms are various and seem to defy anything like successful treatment.

Clay.—I doubt if this county at any time during the past eighteen years has been clear of the hog-cholera. In most herds it has been very fatal.

Clinton.—The losses in this county during the present year from diseases among swine will amount to over $20,000.

Crawford.—The general condition of farm animals in this county at this time will compare favorably with previous years, and is fully up to an average, if not above.

Dearborn.—There has been but very little hog-cholera in this county during the present year.

Decatur.—All classes of farm animals in this county are healthy except that of hogs. These animals are affected with the usual maladies, and the losses have been very heavy during the last year.

Greene.—With the exception of hogs and fowls all classes of farm stock have been measurably healthy during the past season. Perhaps five thousand hogs have been lost during the year by the usual diseases.

Hancock.—The value of the hogs lost in this county during this past year from the various diseases affecting them will amount to over $60,000.

Hendricks.—Our horses, cattle, and sheep are comparatively exempt from disease, but hogs and poultry are seriously affected. The losses among hogs particularly are very heavy.

Jay.—The only farm animals affected with disease in this county are hogs, and they die by the thousands. The disease affecting them is known as hog-cholera. About one-half of those attacked die. I think the disease is contagious.

Kosciusko.—Nearly 50 per cent. of the hogs in this county have died this season.

Some farmers have lost as high as 80 per cent. Some die of a disease resembling lung fever, some of cholera, while others are literally eaten up with worms. The flesh of the hogs fairly swarms with these worms.

Marion.—Hog-cholera is the only disease of any consequence prevailing in this county. The losses have been heavy.

Miami.—The losses to the farmers of this county from diseases among swine will amount to over $20,000 for the present year. Cholera has been very destructive among fowls.

Ohio.—But few losses have been sustained during the past year from diseases among horses and cattle. Cholera prevails among hogs, and is frequently very fatal. Those fed on soap-suds and kept out of the dust seem to be exempt from disease. Plenty of lime, sand, and pure water will prevent cholera among fowls.

Shelby.—Horses, cattle, and sheep in this county are measurably clear of disease. Hogs and fowls, however, are reported as largely afflicted with cholera, from which many of them die. No remedy seems to prove effectual.

Stark.—The disease among hogs in this county is commonly known as cholera, although the symptoms are varied. The disease has not been very destructive this season.

Switzerland.—Some distemper exists among horses, but the losses have been comparatively small. Cattle are healthy and free from all contagious diseases. Hog cholera is prevalent, but not sufficiently general to discourage hog-raising. The losses from this disease will perhaps amount to one per cent. There is some chicken-cholera prevalent, but not sufficient to impede the business.

Tippecanoe.—For this and for several years past it would be safe to say that 50 per cent. of all the swine pigged in this county have died of what is usually termed hog-cholera. This year nearly all the farmers in this region have been afraid to feed their hogs, and have shipped them as soon as shippers could handle them at the summer packing-houses. The fine heavy hogs that the Wabash Valley used to produce are things of the past. All other kinds of farm stock are healthy.

Tipton.—Hog-cholera has prevailed to an alarming extent during the past year and has been very destructive. The disease is of varied symptoms. Some die very suddenly, while others linger for a few days or weeks.

INDIAN TERRITORY.

Cherokee Nation.—It will be several years, under the most favorable circumstances, before our people can hope to be as abundantly supplied with farm stock as they were before the late war; but it is encouraging to know that our people, by their vigilance and industry, have increased the number of their cattle, horses, hogs, &c., and now have not only sufficient for home supply but a small surplus to ship each year to distant markets. Among cattle the most serious and fatal disease we have to contend with is murrain. Hogs are afflicted with various diseases which are classed under the general name of cholera. The principal disease among horses is distemper, though they are occasionally afflicted with blindness and big-head.

IOWA.

Adair County.—Diseases are prevailing among horses and swine in this county. The losses in hogs have been heavy.

Buchanan.—The only epidemic that has prevailed among any class of farm animals during the past year has prevailed among hogs. The mortality among this class of animals has been very heavy.

Crawford.—Hogs in this county have been largely affected by cholera, and but few attacked by the disease recover. The greatest destruction has occurred among pigs. The losses are estimated at $40,000 for the year.

Des Moines.—A few horses and sheep and a great many hogs have been lost in this county during the past season by disease.

Emmett.—A few colts have died here with a disease known as distemper. No disease among other classes of farm animals.

Fayette.—Perhaps $3,000 would cover all the losses of farm animals in this county for the past year from purely contagious diseases, but the losses from all other causes would no doubt swell the aggregate loss to twice or three times this amount.

Franklin.—Hogs have remained healthy until within a few weeks past. Recently a number of fat hogs and shoats have been lost in this locality.

Guthrie.—Distemper is the most common disease among horses, and black-leg among cattle. The latter is more prevalent and fatal among calves than among grown cattle. Cholera and quinsy prevail among hogs, and these diseases are quite destructive. Chickens have cholera, and I never knew one attacked by the disease to get well. Sheep are healthy.

13 SW

Harrison.—Hogs are annually attacked with a disease known here as cholera, and a great many of them die. The past year has proven as disastrous as former seasons.

Humboldt.—This county has been remarkably free from all infectious and contagious diseases among farm animals. There has been some cholera and roup among chickens. The largest loss of hogs that I have heard of was five out of a herd of over one hundred head.

Ida.—Hogs are dying in this locality this year of inflammation of the lungs. About one-half the herds affected die. The animals die in about one week after the first symptoms are noticed. Those that recover from the disease do not amount to much. This is the first year that hogs have died of any disease in this county.

Iowa.—Among horses the only contagious disease prevailing seems to be a very serious distemper. It affects young horses to a greater extent than old animals. Quinsy and cholera have prevailed among hogs this year, but to a less extent than usual. The losses will amount to $18,000 or $20,000.

Jackson.—Cattle have been remarkably healthy, and so have hogs until within three or four months past. From information recently received I am inclined to believe that the losses will be heavy—heavier, perhaps, than ever before.

Jefferson.—Horses and cattle are healthy in this locality. Hog-cholera prevails in some sections of the county, but not in as malignant a form as usual. Still the losses have been quite heavy. Fatal diseases prevail among fowls, for which we have no remedy. The general condition of farm stock is above the average.

Johnson.—No disease has prevailed this year among either horses, cattle, or sheep in this county. The losses, therefore, are merely nominal. The loss of hogs is not so great as last year. The largest number of those that have died were young hogs, and therefore were of less market value. The disease, in all cases, was supposed to be cholera.

Lyon.—Until the past summer all classes of domestic animals have been extremely healthy in this county. During the past summer some herds on the Little Rock were affected with a disease claimed to be black-leg, which I doubt, but of which quite a number died. I notice that all animals well cared for through last winter have escaped. We have never had a case of hog-cholera in the county. I hear of no diseases among fowls.

Marion.—Hogs and sheep are less affected by disease than usual at this season of the year. No infectious or contagious disease exists among horses and cattle. Some seasons a great many fowls die of disease.

Marshall.—Hogs have suffered to a greater extent from disease the past season than ever before. The losses have been heaviest among pigs and shoats. The losses are estimated at from $85,000 to $90,000.

Monona.—Lung fever has caused some heavy losses among horses in this county during the past year. There have been some losses among cattle from black-leg and other diseases. Hog-cholera prevails, and the losses, as usual, have been very heavy. Almost all those attacked die. The few that recover are worthless.

O'Brien.—Cholera or influenza kill a good many hogs in this county every year, although the disease has never appeared as an epidemic.

Palo Alto.—The only contagious disease known among horses here is glanders or nasal gleet. Our young cattle are sometimes attacked with black-leg. I have never known farm animals to be in a more thrifty and healthy condition than they are this year.

Poweshiek.—Horses are afflicted with an epizootic distemper, which has caused many deaths. The mortality among hogs, from a disease supposed to be some kind of fever, has been terrible. The losses the present year, in swine alone, will aggregate from $36,000 to $40,000.

Sioux.—This is a new county, and but few farm animals are raised. The few we have are in good health and condition.

Story.—Hog-cholera has prevailed extensively, and has been most virulent and destructive during the past season. About one-half the hogs in the county have been attacked, and 99 per cent. of those attacked have died. The losses will amount to over $30,000.

Woodbury.—The assessors' returns showed 9,982 hogs in this county this year. Ten per cent. of these were attacked by a disease known as cholera, and about all those affected died. I believe the disease to be an affection of the lungs. Horses are troubled to some extent with lung diseases, but other classes of farm stock are healthy and in good condition.

Washington.—No diseases of consequence have recently prevailed among farm animals in this county, aside from those incident to swine. Diseases among these animals seem to be most destructive where corn is the only diet. The losses during the year will reach $40,000.

Wright.—During the last two years we have been greatly troubled with hog-cholera in this county. It is about the only disease of consequence among our farm animals that we have to contend against. It has been very destructive to swine. We lose a few young cattle every year with a disease called black-leg.

KANSAS.

Allen County.—Cholera and congestion of the lungs carry off a good many hogs in this county every year. Scab in sheep and cholera among fowls also prevail to some extent.

Brown.—Hog-cholera is the only disease that has prevailed to any considerable extent in this county during the present year. The losses have been quite heavy.

Chautauqua.—A few horses, perhaps 200 head, have been lost in this county by an unknown fever. Grown cattle are affected with murrain and young ones with black-leg. Cholera and pneumonic fever are prevailing among hogs, and these diseases are proving quite fatal. A good many fowls are dying of a disease called cholera.

Clay.—Our stock is annually visited by one kind of disease and another, and sometimes our losses are very heavy. This year our losses among all classes of animals will aggregate from $12,000 to $15,000.

Cloud.—The only disease of an infectious or contagious character is that prevailing among swine, and generally known as hog-cholera. It is not so prevalent this year as formerly. The condition of all kinds of farm animals is from 25 to 50 per cent. better than in any previous years.

Crawford.—The principal disease among horses is lung fever, brought on for want of shelter and proper attention. Cholera prevails very extensively among fowls.

Davis.—Cattle are occasionally fatally affected with black-leg, and cholera prevails to a limited extent among hogs where many are kept together.

Elk.—Horses and cattle suffer from various diseases, and the losses will this year perhaps amount to $7,000 or $8,000. No epidemic has prevailed among hogs during the year, but a great many fowls have been lost by the usual disease.

Ford.—A disease called Texas fever prevails here among cattle. It hardly ever proves fatal to cattle brought from Texas, but when it attacks native cattle it is very severe, and generally fatal. A disease like cholera affects chickens, and seems to be contagious.

Franklin.—There are no diseases of any kind prevailing among farm stock in this county. Stock-raising of every kind is greatly on the increase here.

Jackson.—The prevailing disease among cattle is black-leg, and that is confined principally to calves and yearlings. Hogs are afflicted with cholera, but the disease is not so prevalent this season as usual. Chicken-cholera annually destroys a great many fowls. The condition of farm animals is fully an average.

Kingman.—Cattle are afflicted with wolf-tail and hollow-horn. Four out of every ten horses that are brought here from the East die before they become acclimated. Hogs and sheep are healthy.

Labette.—Spanish fever has prevailed among some cattle infected by stock brought in from Texas and the Indian Nation. Cholera has also prevailed to a limited extent among hogs.

Leavenworth.—Horses are rarely sick, but when they are attacked by disease they usually die. Hogs are afflicted with various diseases, and they nearly all die that are taken sick, as nobody tries to doctor them. Fowls also die rapidly.

Lincoln.—All classes of farm animals in this county have been exceedingly healthy during the past year. Stock is in very good condition.

Miami.—The only disease reported among farm animals is that existing among swine. This disease was very destructive last year.

Mitchell.—Murrain, black-leg, and lung-fever have prevailed to some extent among cattle during the past season, and cholera, quinsy, thumps, and fever among swine. A lamentable ignorance seems to prevail in regard to the nature and cause of disease among swine.

Nemaha.—No disease of consequence has prevailed among our stock this season. There has been some hog disease, and a greater loss than usual from diseases among fowls. In some instances parties have lost all they had.

Reno.—A number of horses in this locality have been sick with blind-staggers and glanders, and some have had a mild form of the epizootic. The first two diseases prove quite fatal, one of my neighbors having lost five animals, another three, and so on. Cattle are usually healthy. A few cases of black-leg, or something like it, have occurred. A disease is prevalent among hogs, which causes them to lose the use of their hind parts, and from which they die in about six weeks. Sheep are very healthy. Fowls are frequently seriously affected in the fall with cholera.

KENTUCKY.

Breathitt County.—A good many hogs have been lost this year by cholera, and many sheep with foot-rot. Horses are suffering with distemper, and cattle are frequently attacked with murrain.

Bullitt.—Cholera prevails to a considerable extent among hogs in this county. It seems to be more fatal among pigs and shoats than among older hogs.

Calloway.—Distemper prevails among horses, cholera among hogs and fowls, and rot among sheep. Cattle are healthy.

Carroll.—The disease known as hog-cholera is not so prevalent this year as usual. Many of our swine, however, have a delicate and unhealthy look, and do not improve fast even with the best treatment. This we regard as an evidence that the disease is hereditary.

Clay.—The so-called cholera among hogs has proved very disastrous to the farmers of this county during the past year.

Cumberland.—A few horses have died of distemper in this county. There are no infectious or contagious diseases prevalent among cattle. Hogs are afflicted with cholera, and a great many have died.

Estill.—The only diseases of consequence prevailing among cattle are hollow-horn and murrain. Horses are afflicted with distemper, and occasionally die of some kind of lung-fever. Hogs are badly afflicted with cholera. I should say from 80 to 90 per cent. of those attacked with the disease die. A good many hogs also suffer from thumps, and about two-thirds of them die. Cholera also prevails among fowls.

Fleming.—Horses are afflicted with distemper and lung diseases, from which about one in twenty die. The most destructive disease we have to contend with is cholera among hogs. At least one-fifth of all the hogs in this county annually die from this disease. Two years ago I lost one hundred and seven head, and this fall I have already lost sixty head more.

Hart.—Hogs, in some portions of the county, are more or less affected every year with cholera, but the loss this season is small compared with other years. Horses and cattle are free from all contagious diseases. Chickens suffer from various diseases.

Kenton.—There is but little stock raised in this county, and the only disease that has caused material loss to farmers is that among hogs.

Knox.—Hogs in this county have been seriously afflicted with cholera and blind-staggers. Murrain has also prevailed extensively among cattle, and distemper among horses.

Lewis.—This is one of the largest poultry-raising districts in the State. The loss by disease runs into the thousands. The shipments from this post-are about one thousand chickens per week.

Martin.—The most prevalent and destructive disease among any class of farm animals is that of cholera among hogs. This disease is very fatal, and makes its appearance semi-annually. We have no remedy. Fowls also suffer with a disease generally known as cholera.

Ohio.—Hogs, as well as fowls, are still afflicted with cholera. The mortality among the former has been very large.

Oldham.—Distemper is the only disease afflicting cattle in this locality. Hog cholera prevails more or less all the time. Sheep are affected with various diseases, among others those of rot and scab. Fowls are suffering from cholera and roup. We have been unusually free from diseases of all kinds this year.

Pendleton.—Hog-cholera has not been so destructive this season as in previous years. The losses this year will, perhaps, not amount to over $16,000 or $18,000.

Rowan.—Hogs die in great numbers from cholera. There is no other infectious or contagious disease prevalent among the farm animals in this county. Fowls also die in great numbers from a disease generally called cholera.

Russell.—The disease commonly called hog-cholera has prevailed to a fearful extent in some portions of this county. The losses have been at least 75 per cent. of those attacked. Out of a herd of 75 head I lost 55. I hardly think the disease is contagious. So fearful has been the ravages of the disease that there will not be enough pork raised in the county to supply the home demand. Other classes of farm stock are in good health.

Shelby.—But few horses are raised in the county. The assessors report 9,588 head of cattle in the county. Owing to the ravages of hog-cholera there has been a falling off in the number of swine. Sheep husbandry is largely on the increase, and aggregates nearly double that of any previous year. At least 45,000 head have been placed on the farms of the county this fall for breeding purposes. No diseases of consequence, except hog-cholera, are prevalent.

Warren.—The losses have been quite heavy from diseases among hogs. Other classes of farm stock are healthy.

Whitley.—Distemper is quite prevalent among horses, and occasionally we have a case of murrain among cattle. Hog-cholera frequently prevails, and is often very fatal.

LOUISIANA.

Bienville County.—Horses here are subject to bots, colic, distemper, and blind-staggers. Perhaps 50 per cent. of the losses are occasioned by bots. The most common disease among cattle is known as screw-worm or "wolf" (in the back), hollow-horn, and, occasionally, murrain. Hogs are subject to cholera and mange. The former is much the more fatal.

Claiborne.—Cholera among hogs is the most destructive disease now prevailing in this county. Domestic fowls are also dying rapidly from the effects of the same disease. We have recently lost some fine cattle, hogs, and mules by hydrophobia. They were bitten by mad dogs.

De Soto.—The only destructive disease among farm animals that we have to contend with here is a disease among swine, which kills about one-half of those attacked.

Jackson.—Horses frequently die here of blind-staggers and bots, and cattle of hollow-horn or head disease. A good many hogs are annually lost by cholera and thumps, and sheep with scab.

West Feliciana.—Charbon, which has prevailed in a mild form among horses and mules, and distemper among sheep, are the only affections among any class of farm animals. A few deaths have occurred among horses and mules, and many sheep have died of distemper.

MAINE.

Piscataquis County.—No infectious or contagious diseases prevail among farm stock in this county. About 10 per cent. of the fowls are annually carried off by disease.

Waldo.—The only contagious disease we have to contend with here is an epizootic distemper among horses, and this is fatal in but few cases.

York.—The usual number of diseases have prevailed among farm animals in this county during the past year, and the losses will amount to from $10,000 to $12,000.

MARYLAND.

Alleghany County.—Hogs have what we call cholera, and but few of those attacked recover. Fowls also have what we term cholera, and nearly all that are affected die.

Baltimore.—Lung fever has prevailed among cattle in the vicinity of Baltimore for the past twelve or fifteen years, and the losses have been considerable. Hog-cholera prevails in a few localities in the county, and a number of animals have died. The losses in fowls seem to be less than in former years.

Dorchester.—Hog-cholera prevails to a limited extent in this county.

Howard.—Some seasons the losses from hog-cholera are very heavy, and perhaps amount to as high as $5,000. The annual losses from chicken-cholera will amount to that sum.

MICHIGAN.

Alpena County.—As this is a lumbering county a large number of horses and cattle (oxen) are used, but very few of them are raised here. A few milch cows and a few stock bulls, however, have been raised in the county. No disease has prevailed since the epizootic in horses.

Cass.—Distemper has prevailed among horses, milch fever among cattle, and so-called cholera among swine and fowls.

Chippewa.—This is a new county and we have but little stock as yet, and it is entirely healthy. Grass is grand for dairy cattle. It is always green and nutritious. All animals that run at large in the summer are rolling fat in the fall.

Clinton.—Farm animals in this locality are free from all infectious or contagious diseases.

Delta.—This is comparatively a new county, and what little stock it contains is in a healthy and thriving condition.

Emmett.—A few horses have been affected with colds and a discharge from the nose, but none have died.

Houghton.—Diseases among hogs have prevailed here for three years. Some have died suddenly when in apparent health and in good condition. A number of cattle are affected with cancer or worm in the tail.

Huron.—Distemper prevails among horses, but the disease seldom proves fatal. All other farm animals are free from infectious and contagious diseases.

Kalamazoo.—No disease has prevailed this year among farm animals except cholera among swine. This disease has prevailed to a limited extent this fall.

Kent.—There have been no infectious or contagious diseases prevalent among farm animals during the past year.

Manistee.—The general condition of farm animals in this county is good, and rather above the average.

Muskegon.—The proportion of farm animals that are attacked and die with infectious and contagious diseases in this county is very small. Of horses perhaps 1 per cent. are lost; of sheep, one-half of 1 per cent. I hear of no losses among cattle and hogs. Of fowls perhaps 5 per cent. die annually of disease.

Oakland.—One year ago the disease known as hog-cholera created a good deal of uneasiness, but the low price of pork has caused the "thinning out" to such an extent that we now hear but little complaint. The losses this year will perhaps amount to $9,000 or $10,000.

Otsego.—Several hogs have been lost by the farmers in this county during the past year from some disease thought to be contagious. All other classes of stock are healthy.

Presque Isle.—Horses, hogs, and fowls in this locality are measurably healthy, but calves seem to be affected with a contagious disease.

Saginaw.—About 1 per cent. of the cattle and hogs raised in this county annually die of disease. As a rule all our stock is housed in the winter and comfortably cared for.

Saint Clair.—Several horses have died in this county during the present year of contagious diseases. Cholera prevails among swine in one locality, but has not appeared in a very malignant form.

MINNESOTA.

Beltrami County.—The only disease prevalent among any class of farm animals is distemper among horses. This is an Indian agency, and there are but few animals in the county.

Faribault.—A few cattle die annually in this county of a disease known as black-leg, and a few sheep with the scab. Other classes of farm stock are healthy.

Houston.—The only disease among our stock that we have had to contend with the past season has been the so-called cholera among hogs.

Isanti.—Cattle in this county are frequently attacked with black murrain, and a disease that causes a rising and running sore on the head. There is no remedy known for the latter disease, and when an animal is attacked by it, it is generally killed. Hogs are subject to cholera and a disease called staggers, both of which are fatal.

Lac-qui-parle.—Black-leg is quite prevalent among cattle, but is principally confined to young animals. All die that are attacked with the disease. The general condition of farm animals is above the average.

Martin.—A few horses die annually in this county from epizootic, and perhaps 5 per cent. of the cattle from black-leg. Stock generally is in good health and condition.

Nicollet.—Glanders prevails among horses in this county, and is the only contagious disease with which these animals are afflicted. Quite a number of cattle died of black-leg during the past spring, and about one hundred more from the effects of eating smutted corn.

Olmsted.—No diseases prevail among any classes of farm animals in this county that I am aware of. In some localities cholera exists among chickens.

Pope.—Epizootic has prevailed among horses, of late, with some fatal cases. Cattle have been suffering more or less with black-leg, which is fatal, with but few exceptions. Hogs and sheep have been healthy, so far as I can learn.

Rice.—All classes of domesticated animals in this county are in good health. I have not heard of the prevalence of any infectious or contagious diseases during the year.

Rock.—There has never been a marked case of any infectious or contagious disease among farm stock in this county. There yet lingers some traces of epizootic distemper in horses, but few, if any, deaths have occurred from that cause this year.

Saint Louis.—Our farm animals are remarkably free from all infectious or contagious diseases. The diseases peculiar to fowls are roup, &c., much of which is due to in-breeding.

Scott.—Recently there has appeared among horses here an epidemic or endemic disease somewhat akin to the epizootic of some years ago. The first symptom is a mucus discharge from the nose, culminating in ten or twelve days in an affection of the kidneys. After having reached this stage the disease generally proves fatal. If taken in time the patient can be cured.

Swift.—The only disease of a serious character that has visited any of the farm animals in this county, during the past year, is black-leg among cattle.

Yellow Medicine.—A few horses in this county have been afflicted with distemper, but none have died. No other contagious disease is prevalent, and all classes of farm stock are healthy.

MISSISSIPPI.

Calhoun County.—In this county, horses and cattle are rarely if ever affected with infectious and contagious diseases. Hogs are frequently afflicted with cholera, and the estimate given ($7,500) is hardly high enough during a year of its general prevalence. Sheep are hardly ever afflicted with any disease save rot. Fowls of every breed occasionally have cholera, and when it attacks a flock it generally kills them all.

Choctaw.—Cattle suffer with charbon, horn-ail, and murrain, and hogs frequently die of cholera or swine-pox. Sometimes a farmer loses nearly all his hogs by these maladies.

Covington.—Owing to the extremely hot weather during the summer we lost at least 20 per cent. of our farm horses by staggers. All ages were affected alike. At this time all classes of farm animals are in fine condition.

Franklin.—The number of hogs affected with diseases during the past summer was greater than usual, and at least 50 per cent. of those affected died. Other animals have remained healthy.

Holmes.—A good many colts die in this county every year from distemper. Hogs die in great numbers of cholera, lung fever, and quinsy. Fowls are subject to cholera and roup, and frequently one-half of them are lost by these diseases.

Leake.—From the most reliable information I am able to obtain, I am led to believe that about 8,000 hogs were lost in this county during the past season, a large majority of which died of cholera.

Lee.—A very destructive disease prevails among fowls in this locality. It made its appearance here four or five years ago, and has continued with more or less virulence ever since. It frequently sweeps off whole flocks. I myself have this year lost 300 game fowls. It is not cholera, but a disease more resembling paralysis. They are taken very suddenly, lose the use of their limbs, fall down and flutter until they die, which is generally within from twelve to forty-eight hours. If they linger beyond that length of time they are apt to recover. The disease is singularly sudden and fatal, and causes a heavy loss to the people of this locality.

Lowndes.—Since the prevalence of the epizootic some years ago no contagious disease has prevailed among horses in this locality. Murrain is the most fatal disease we have among cattle, and it annually proves very destructive. The losses among hogs from a disease called cholera are very heavy. Fowls also die oftener of cholera than of any other disease.

Marshall.—The usual diseases prevail among all classes of farm animals, and the aggregate losses this year will perhaps amount to from $8,000 to $10,000.

Noxubee.—All classes of farm animals, with the exception of hogs, have been free from disease this year. The losses among swine have been very heavy, and will perhaps aggregate $10,000. Pastures were quite good throughout the summer, but very little cattle feed has been housed, and a spring report will no doubt tell a tale of starvation, &c.

Prentiss.—A few cases of hog-cholera have been reported in the county, but the disease has not been very destructive.

Rankin.—Charbon and blind-staggers occasionally prevail among horses, and various fatal diseases among hogs.

Scott.—The only diseases of any consequence that have occurred among farm animals in this county during the past year have been among swine. Between one and two thousand head have died.

Tippah.—Last year a number of hogs died here from a swelling of the head. The head would swell until the skin would break, and the hog would bleed to death in a few hours. I cured a cow recently of murrain by giving her kerosene oil, lard oil, and epsom salts, in doses a few hours apart.

Tishomingo.—Diseases of a mild type have prevailed among all classes of farm animals during the past year. The losses have been light.

Wilkinson.—Hogs in this county frequently suffer and die of pneumonia and congested liver, as do also fowls.

Yazoo.—A great many horses die annually in this county from a disease called bighead or big-jaw—an enlargement and softening of the bones. It is caused by feeding corn exclusively. Hogs annually suffer severely with cholera.

MISSOURI.

Andrew County.—This year has been remarkably favorable to all kinds of farm stock. I have heard of no infections or contagious diseases among any class except hogs.

Barton.—The losses among horses, cattle, and hogs from disease will probably amount to $10,000 or $12,000 for the present year. Diseases have not been so prevalent among farm animals during the past season as usual.

Benton.—All classes of farm animals, with the exception of hogs, have remained healthy during the past year.

Buchanan.—Horses, cattle, and sheep are free from serious diseases, but cholera exists among both hogs and fowls. With my own hogs I noticed that all those that had diarrhea recovered. The most of those afflicted were costive and had high fever. Cattle are frequently attacked with hoven, caused by eating white clover.

Clay.—Heavy losses have been sustained by the farmers of this county during the past year in the loss of hogs, sheep, and fowls by various contagious and malignant diseases. Horses and cattle have remained healthy.

Franklin.—The so-called hog-cholera has not been so prevalent and wide-spread in this county the past season as during previous years.

Henry.—No epidemic has prevailed among horses here for several years past. Texas fever among cattle has prevailed to a limited extent, but only when parties here violated quarantine laws regarding it. Hog-cholera has prevailed extensively, and the estimated loss is put at lowest figures $20,000 annually.

Hickory.—Hog-cholera has prevailed to some extent in this county during the past year, but the losses have not been as heavy as usual.

Jasper.—Hogs and fowls are afflicted with cholera, and cattle with a disease generally known as murrain.

Lawrence.—Black-leg prevails to a considerable extent, and is very fatal among calves and yearlings. There is also some murrain among older cattle. Cholera (so called) is quite prevalent among hogs, but the greatest fatality seems to be among pigs and shoats. I do not think one hog out of a thousand, however, dies of cholera. The disease is more like lung fever or congestion of the lungs, and has been very destructive the past year, especially among young stock.

Lewis.—I have heard of no infectious or contagious diseases among domesticated animals in this county, except among hogs. The diseases which affect hogs are manifested by various symptoms. The annual losses are very heavy. We have no remedy, but generally separate the sick from the well hogs immediately on discovering that they are sick.

Marion.—The disease prevailing among swine and poultry in this locality is commonly called cholera, and that among horses and sheep is designated as distemper.

Miller.—Hogs and fowls in this county are dying at a rapid rate of a disease commonly known as cholera. All other kinds of farm stock are healthy.

Mississippi.—A few cases of blind-staggers among horses and murrain among cattle have occurred. Cholera prevails among swine, but it is impossible to give the amount of annual losses.

New Madrid.—The diseases most prevalent here among farm stock are cholera among hogs and fowls, distemper among horses and mules, murrain and hollow-horn among cattle, and rot among sheep.

Nodaway.—A contagious distemper prevails among horses, but it is not of a very fatal character. Black-leg and Texas fever have been very destructive to cattle. Hog-cholera also prevails and seems to be much more fatal to pigs than to older animals. Sheep are to a limited extent afflicted with scab and grub in the head.

Pettis.—Cholera and lung diseases have prevailed among hogs during the past year, and have been very fatal. Fowls have also suffered considerably with what we term cholera.

Phelps.—A few horses have died of distemper, and some cattle of Texas fever and murrain. A heavy loss has been sustained by the farmers of the county from diseases among swine.

Pike.—Hog-cholera is the only disease that has prevailed among any class of farm animals in this county during the past year. The losses will amount to from $12,000 to $15,000.

Platte.—There is but little demand for horses here, hence stock-raisers have turned their attention to raising cattle. They find them more profitable and less liable to disease, and ready for market at a much earlier age. When cattle are well cared for we lose but few by disease. The most skillful farmer, with the assistance of our best physicians, have completely failed to find a remedy for diseases of hogs. All die that are attacked, and the same can be said of fowls that are attacked by disease. But few sheep are raised in this county.

Polk.—Cattle are affected to a limited extent with Texas fever and black-leg. Other classes of farm stock are healthy, with the exception of hogs, and a good many of these have been lost by the various diseases incident to them.

Putnam.—The class of animals mostly affected with disease in this county is hogs, a great many of which die of a disease generally known as cholera. The remedies used, as a rule, I do not think amount to much. The general condition of farm animals is better than last year.

Shelby.—Horses, cattle, and sheep are very healthy, but our hogs die at a fearful rate with a disease commonly called cholera. It prevails at almost all seasons of the year, but with more virulence during some months than in others. Sometimes it will kill nine-tenths of all the hogs in a herd, at others perhaps one-half, and at still others but a few will die. We do not know what causes the disease, nor have we a remedy for it. Chicken-cholera also prevails to a fearful extent, and sometimes carries off as high as nine-tenths of the crop. The general condition of farm stock, aside from hogs, is good.

Stoddard.—Our principal losses are from cholera among swine and fowls. Horses, cattle, and sheep are moderately free from diseases.

Stone.—The so-called hog-cholera is more fatal this season than usual. The losses up to this time are estimated at from $40,000 to $50,000.

Worth.—All farm animals except hogs are free from disease. These animals are afflicted with cholera. Chickens also occasionally suffer from cholera.

Wright.—No disease of any moment exists among farm stock in this county. Last year about one-third the hogs in this county died of cholera.

MONTANA TERRITORY.

Lewis and Clarke County.—Stock of all kinds in this county are comparatively healthy. A few sheep introduced from Oregon show the scab to some extent.

NEBRASKA.

Cass County.—The only diseases among farm animals in this county are confined to swine. The losses are not very heavy.

Cedar.—Many cattle die in this county during the month of November from a disease contracted by feeding on corn-stalks, but the disease does not seem to be infectious or contagious.

Clay.—Deaths frequently occur among horses from colic, fevers, and inflammatory affections. Cattle are affected with murrain and black-leg. There is no cholera at present among hogs, but there has been a loss of about 600 head of sheep in the county during the past year from diseases incident to this class of stock. Many fowls, especially young chicks, die of roupe and gapes.

Cuming.—There are a few cases of epizootic and distemper among horses, but the diseases are of a mild type, and but few animals have been lost. Black-leg is about the only malady among cattle. Hog-cholera prevails to a greater or less extent every year.

Dakota.—From ten to fifty head of horses annually die in this county, supposed to be from the effects of alkali. About one hundred yearling calves die annually of a disease called black-leg. We occasionally lose hogs by cholera, but the disease is not prevalent this year. Fowls also frequently die of cholera.

Furnas.—We have a new disease among cattle here that has killed a great many in the past two weeks. They are attacked by a twitching and jerking of the nerves of the whole body, bloat a little, are in great pain and agony, and die within from six to fifteen hours. I examined four animals to-day, and found two with the galls bursted, another very large, and a fourth blood-shotten. The cattle thus attacked have been running in corn-fields after the corn had been harvested. The disease is new to us, and we do not understand it.

Greeley.—A good many horses die annually in this county, but in almost every case the loss can be traced to exposure and ill treatment.

Knox.—Black-leg prevails extensively among calves and is very fatal. There are no diseases prevalent among other classes of farm animals.

Merrick.—But little disease prevails among domesticated animals in this county. Perhaps $3,000 or $4,000 will cover the annual losses for all classes.

Nuckolls.—A very bad distemper prevails among horses in this county. There is also prevalent a mild form of epizootic which few horses escape. Cholera also prevails among hogs.

Pawnee.—Cholera has prevailed to a fearful extent among hogs in this county during the past season, and the losses, in value, will exceed $10,000.

Platte.—Young cattle frequently die within a few hours after being attacked with a disease supposed to be caused by eating smut found on the stalks of corn. A good many hogs have died from that pest of the farmer, the cholera, but at present the disease seems to be confined to one locality in the county.

Red Willow.—While this is a remarkably healthy climate for farm stock, a good many cattle and sheep annually die from diseases incident to these animals.

Richardson.—The disease commonly known as cholera has carried off a great many hogs in this locality during the past season.

Saline.—This is a new county, and what few farm animals we have are in a very healthy condition.

Sarpy.—Black-leg is quite prevalent and destructive among calves. There seems to be more cases this fall than usual. Hogs are affected with a lung disease, which made its appearance here last year. The disease is chiefly confined to pigs from three to six months old. At least 10 per cent. of those attacked die.

Saunders.—The only diseases prevalent among horses are glanders and distemper, or quinsy. Some six head of horses have died of these diseases. The increase in the production of hogs is 7,543. Of this number 33 per cent. were affected with cholera, and one-third of them died. The only disease among sheep is foot-rot.

Valley.—This is a new county, and what little stock we have is healthy, and free from all infections and contagious diseases.

Wayne.—No disease of any consequence has prevailed among farm animals in this county since the epizootic some years ago.

Webster.—The farm animals in this county are entirely free from all diseases of an infectious and miasmatic character.

York.—Some hog-cholera prevails in this county, also cholera among chickens. Horses and cattle are healthy.

NEVADA.

Nye County.—All kinds of stock range our mountains and plains, and are only gathered in once a year to corral. No diseases of an infectious or contagious character prevail among them at present. All losses occur from starvation and exposure in winter.

NEW JERSEY.

Burlington County.—The prevailing disease among cattle in this county is pleuro-pneumonia. It is very fatal, and the losses in this class of animals have been very heavy. Hog-cholera is prevailing extensively and in a very fatal form. The same might be said of diseases among fowls. The losses among all classes of farm animals will annually amount to over $100,000.

Camden.—There are no diseases prevalent among horses, except those peculiar to colts and young horses. These are generally of a mild character. Hog-cholera prevails to some extent; so does cholera among fowls.

Cape May.—During the past year horses have suffered severely from a disease called blind staggers. A good many animals have been lost. Hog-cholera prevails to a limited extent.

Middlesex.—A contagious lung fever prevails among cattle in this county, but it has not as yet appeared in a very malignant form.

NEW MEXICO TERRITORY.

Colfax County.—Scarcely any disease is prevalent among domesticated animals in this county save scab among sheep. The losses among this class of stock perhaps amount to 1 per cent. of the whole number raised. There is scarcely any infectious or contagious disease among our larger farm animals.

Dona Ana.—A few flocks of sheep, of improved variety, have scab in a mild form. Hogs and fowls are free from cholera. Occasionally a cow is lost by hoven, brought on by eating green alfalfa; and I lost two merino rams the past summer from the same cause. I also lost two Angora bucks from an unknown disease.

San Miguel.—There are about 100,000 sheep annually raised in this county, and about 1 per cent. of them die of an affection of the milt.

Taos.—Horses are generally affected with epizootic distemper and cattle with Texas fever. A great many die in the spring of the year from the effects of eating poison-weed. Hogs, sheep, and fowls are generally healthy.

NEW YORK.

Allegany County.—Our horses are frequently attacked with strangles or distemper, a disease which is believed to be contagious. A few cows are annually lost from milk fever, and a limited number of calves die of murrain. Sheep are affected with foot-rot, and many fowls die of cholera.

Chenango.—I have heard of a few calves dying of worm in the lungs, which seems to be a new disease in this county. A farmer who lost three examined the lungs and found large quantities of worms about an inch long and the size of an ordinary thread. I lost a few calves by the disease known as black-leg, and on opening the lungs found a few worms.

Fulton.—Owing to abundant feed, stock of all kinds is in fine condition this fall.

Genesee.—The value of farm stock lost by various diseases during the past year in this county will amount to from $20,000 to $25,000. The heaviest losses have occurred among horses and swine.

Montgomery.—A great many cows suffer annually from abortion; and the loss by accident and the various inflammatory and congestive diseases will average one cow for every dairy of thirty-five cows. No contagious disease prevails in any of our flocks or herds.

Niagara.—Light diseases, with but few fatal results, have prevailed among swine during the past season. Other classes of stock have remained healthy.

Seneca.—All classes of farm animals in this county have been unusually healthy during the past year.

NORTH CAROLINA.

Alleghany County.—We have no diseases among either horses, cattle, or hogs. Sheep occasionally die from distemper and a disease called rot. Young chicks frequently die of gapes.

Brunswick.—On consultation with the best informed persons in the county I do not find that any diseases have prevailed among farm animals during the past year.

Large numbers of fowls have died, but it is impossible to estimate the number or give any name to the disease.

Cherokee.—There are no diseases of a contagious character prevalent among the farm animals of this county. There have been some losses of young chickens and turkeys by gapes, but I do not think this disease is contagious.

Cumberland.—The loss of hogs from the various diseases to which they are incident, but all of which are called cholera, has been very great. A great many fowls have also died from a disease generally known as cholera.

Currituck.—The only disease of any consequence that we have had to contend with among the farm stock in this county has been that commonly known as cholera among swine. The loss so far has been quite heavy.

Halifax.—All classes of farm stock have been more free from disease this year than any year during the past ten.

Haywood.—Hog-cholera prevails to a greater or less extent every year and kills a great many animals of all sizes, but is more fatal among pigs. Chicken-cholera is also quite prevalent and fatal. Stock generally is in better condition.

Henderson.—A large number of both cattle and hogs have been lost during the past year; perhaps the aggregate for these two classes alone will amount to $18,000 or $20,000.

Hertford.—Horses and mules are affected with but one contagious disease—that of glanders or farcy. All that are attacked die. Thousands of hogs die annually of disease, but whether it is contagious or not we have not determined. Cholera is generally prevalent and very destructive among fowls.

Jackson.—Infectious and contagious diseases among horses, cattle, and sheep are almost unknown in this county. Last year nearly all the hogs in the county died of disease, but during the previous five years but few were attacked.

Madison.—The only class of animals affected by disease during the past year has been that of swine. Cholera has been quite prevalent and fatal among fowls in some localities.

Mitchell.—Large numbers of hogs and fowls are annually lost in this county by a disease commonly called cholera.

Orange.—Some four or five thousand hogs have been lost by disease in this county during the current year. A few horses and cattle have also died from diseases peculiar to these classes of farm animals.

Pamlico.—The most prevalent and fatal disease we have to contend with is that of cholera among hogs and fowls. The disease annually carries off numbers of both hogs and domestic fowls. The condition and quality of farm animals is better than for years past and is gradually and surely improving.

Perquimans.—Hogs are much diseased in this county and are very cheap. Young pigs attacked with cholera seldom recover.

Person.—The prevailing disease among farm animals here is that of cholera among hogs, which is very destructive. Trichinæ destroy many of the pigs and shoats. Sheep are healthy, but a great many fowls die of cholera. The goose and peafowl are the only species of domestic fowls that do not suffer with it.

Robeson.—Hogs in this county are more affected by disease than any other class of animals. Cholera is the prevailing disease among them, and for which we have no remedy. The general condition of farm animals is 50 per cent. better than for previous years.

Sampson.—No epidemic has visited horses, mules, cattle, or sheep so far as I have been able to learn. At least one-third of the hogs of the county die every year from a disease known as cholera. If any recover they are of no value, as the disease either leaves them deaf, blind, or afflicted in some other way. Fowls die in about the same proportion from a disease of like character.

Transylvania.—We have no contagious diseases among cattle. Our losses are occasioned by exposure and want of feed during winter. No unusual disease is prevalent among any class of farm animals.

Wake.—Horses, cattle, and sheep are free from infectious and contagious diseases. Hogs suffer a good deal from cholera and lung diseases. When these diseases appear in a herd there seems to be no cessation until the last animal is destroyed. Fowls are subject to all sorts of diseases, and frequently the mortality among them is very great.

Wilkes.—We have some distemper among cattle, but are at a loss to know what causes it. It seems to prevail mostly where the people have the typhoid fever. Cholera is the prevailing disease among hogs and chickens. It has been very destructive during the past summer.

Yadkin.—Hog and chicken cholera has prevailed here for several years past. When the disease gets among a class of fowls it kills nearly all of them.

Yancey.—Distemper prevails among horses and sheep, and murrain and hollow-horn among cattle. Hogs have been seriously affected with cholera and some kind of fever; a good many fowls are also lost by cholera. The condition of all kinds of farm stock is better than usual at this season of the year.

OHIO.

Ashland County.—We have no infectious or contagious diseases among domesticated animals in this county. There are a few sporadic cases of disease and death, but the aggregate loss is very small. About one farmer out of every twenty-five loses his chickens every year by cholera.

Athens.—The so-called cholera prevails among hogs and fowls in this county. I believe the greatest number of deaths among cattle have occurred among cows, which have died of milk fever. The disease shows itself from the first to the third day after calving, and generally attacks the animal after the fourth calving. Select breeds and good milkers, and those in good condition. are generally the ones that suffer. The symptoms are loss of appetite, staggering gait, wild look, and cessation of rumination; they fall down and cannot rise; the brain seems to be affected; the animal will dash about, striking her head and horns against the ground, when she soon dies. We have no remedy.

Anglaize.—Hogs seem to be the only farm animals seriously affected with disease. They suffer with the disease generally known as cholera. The losses so far this year will amount in value to from $25,000 to $30,000.

Brown.—A few colts are lost in this county by distemper, and a good many hogs and fowls annually die of a disease commonly called cholera. Sheep die of grub in the head and of neglect while young.

Erie.—There is no special disease prevailing among the farm animals in this county.

Fairfield.—With the exception of swine the live stock of this county has been comparatively free from disease during the past year. Swine have suffered with cholera, though not so extensively as in former years.

Franklin.—Large numbers of hogs and fowls have died in this county during the past year of the various diseases common to them.

Gallia.—Hog-cholera prevails to some extent in this county, but in a rather mild form this season. Chicken-cholera is quite prevalent and malignant, and the losses are heavy.

Geauga.—No disease of consequence exists among any class of farm animals in this county. The general condition of farm stock is good.

Guernsey.—The prevalent diseases among horses are those affecting the lungs, principally lung-fevers. Hogs are affected with cholera and cattle with murrain.

Hardin.—Hog-cholera has prevailed to a limited extent in the northern part of the county, but in this locality we have not suffered from the disease this year.

Jefferson.—No diseases of a malignant character have prevailed among farm animals in this county during the past year. Owing to abundant pasturage farm stock is in very high condition.

Knox.—There is no disease among farm animals here. Chicken-cholera prevails every year and carries off a great many fowls.

Meigs.—There have been few, if any, deaths among farm animals in this county during the present year, except from natural causes. During the past eight or ten years chicken-cholera has prevailed from time to time, and is prevalent in some localities at this writing.

Mercer.—The so-called cholera still prevails among hogs in some localities in this county. Cholera among fowls is also prevalent, but the disease is not so fatal as formerly.

Miami.—But little disease exists among farm animals in this county aside from the so-called cholera among hogs. The loss among this class of animals is, in some years, very heavy.

Monroe.—No infectious or contagious disease has prevailed among farm animals in in this county during the past year.

Montgomery—Chicken cholera has prevailed as an epidemic during the past season, and many fowls have been lost. Cholera among hogs has also been very destructive.

Ottawa.—The only disease prevailing among any class of farm animals is a disease among hogs, and this is confined to two townships of the county. The animals have a diarrhea, vomit, and wheeze as one afflicted with asthma. They die very suddenly.

Paulding.—The mortality among horses has been unusually large in this county during the past year. The same can be said of cattle. During the two years hog-cholera has been very extensive and fatal. Fowls are also subject to cholera.

Richland.—It is estimated that the product of chickens in this county will aggregate 150,000 per annum, and that 25 per cent. of these die of cholera. Several diseases annually prevail among farm animals, and frequently the losses are very heavy.

Summit.—Several fatal diseases are prevalent among horses, among others inflammation of the lungs and bowels, and distemper or epizootic. Cattle are afflicted with hollow-horn and murrain. The prevalent diseases among hogs are cholera and blind staggers. But few of these animals recover. Consumption carries off a good many sheep, and cholera is very destructive among fowls.

Trumbull.—The production of farm animals in this county has decreased in the past four or five years, but stock of all kinds has improved in quality. No infectious or contagious diseases are prevailing.

Tuscarawas.—No epidemic disease has recently prevailed among farm animals in this county, but a good many domesticated animals have been lost during the year by the various maladies incident to this class of property.

Wood.—Horses and cattle have been free from infections and contagious diseases during the past year. Hogs and chickens have suffered severely from a disease commonly known as hog and chicken cholera. The losses among hogs have heavy been very, as some farmers have lost entire herds. Sheep have healthy.

Wyandot.—Cholera has prevailed among hogs to a limited extent in this county the past season. Cholera has seriously affected the fowls, in some cases sweeping off whole flocks.

OREGON.

Clackamas County.—No disease among horses. A good many cattle die annually for want of proper attention. A few hogs die every year from liver-disease.

Linn.—Horses here are suffering to a limited extent with contagious distemper; cattle are healthy, but sheep are subject to scab and other diseases.

Polk.—Cattle, hogs, sheep, and fowls are afflicted with the usual diseases, though the losses are never very heavy.

Tillamook.—There are no diseases of a contagious nature prevailing among the farm animals in this county.

PENNSYLVANIA.

Armstrong County.—Cholera prevails to some extent among hogs in the eastern part of the county. One man recently lost twenty head by this disease. Chicken-cholera also prevails, and is fatal in most cases. A good many sheep annually die of foot-rot and grub in the head.

Blair.—Distemper and lung-fever prevail among horses, and cholera among hogs and chickens. Foot-rot also seriously affects sheep where not properly treated and cared for.

Erie.—No special diseases have prevailed among farm animals in this locality for some years, and hence the losses have been comparatively light.

Lycoming.—There have been no infectious or contagious diseases among farm animals in this county the past year.

McKean.—The condition of farm animals is good compared with previous years. Horses are overworked in the oil regions, and many die from abuse and lack of proper attention.

Northampton.—There has been no contagious diseases among farm animals in this county so far as I have been able to ascertain.

Perry.—Losses among horses and cattle from various diseases will perhaps reach $3,000 annually in this county. Losses among hogs, when no epidemic disease prevails, will probably amount to $900 or $1,000 per annum. Some years cholera is very destructive among chickens, so much so as to kill about all in some localities.

Wayne.—But few horses are lost here by contagious diseases. A good many young pigs and chickens die of cholera.

SOUTH CAROLINA.

Barnwell County.—The only disease known among horses in this county is staggers or blind staggers. We know nothing about the pathology of the disease and have no remedy. Ninety-nine animals out of a hundred that are afflicted with the disease die. Occasionally our hogs are afflicted with cholera. Sometimes one farmer will lose two-thirds of his entire stock of hogs while his next-door neighbor will lose none. Fowls die by the hundreds when closely confined in coops that have remained on the same ground for a number of years.

Colleton.—Many hogs are annually lost in this county by a disease generally known as hog-cholera. Great numbers of fowls also die of a disease called cholera.

Lexington.—Hogs have suffered less this than last year from cholera. The losses last year were frightful. Fowls have this year suffered beyond all precedent from so-called cholera. I had a fine lot of 100 Brahmas, from which I had 150 dozen eggs during the early spring. As warm weather came on they were attacked and nearly all died—only one of those attacked survived. Unless something can be done to prevent the annual recurrence of this fatal epidemic, we will have to stop trying to raise fowls.

Oconee.—We have no infectious or contagious diseases among either horses, cattle, or sheep. In a few localities of the county hogs have suffered from cholera. There are a few localities along the Blue Ridge range of mountains where the cattle greatly suffer from milk sickness. An appropriation by Congress for the discovery of the cause of this disease would be eminently proper.

Orangeburgh.—Hog-cholera has prevailed this year in some localities in this county.
Pickens.—The diseases common to cattle are distemper, murrain, and milk-sick. Distemper is regarded as contagious, and a similar disease prevails among horses. The prevailing diseases among hogs is commonly called cholera, and it nearly always proves fatal.

TENNESSEE.

Bedford County.—There is no disease here among cattle except murrain, which was brought in from other States and seems to be contagious. The most fatal disease among hogs is cholera, for which we have no remedy.

Benton.—Horses, cattle, sheep, and fowls are affected with the usual diseases. We have had no hog-cholera this year. This disease usually kills nearly all the hogs in this section about once in every three or four years.

Bradley.—All kinds of domestic animals are exceedingly healthy in this county.

Blount.—Horses and mules suffer from distemper, epizootic, and glanders: cattle from murrain and sore tongue; hogs from cholera and quinsy; sheep from rot; and fowls from cholera. These diseases prove fatal in many cases.

Dyer.—Horses, cattle, and sheep in this county suffer very little from disease of any kind. Hogs and chickens frequently suffer terribly from the ravages of cholera. The disease seems to be infectious or contagious with both classes, and is very fatal, as but few of either class recover. The malady is not at all understood, and no remedy that amounts to much has as yet been discovered.

Fentress.—There are but few horses in this mountainous county, but cattle are plentiful. Hogs could be raised here in great abundance were it not for the ravages of the disease known as cholera. Fowls frequently die of gapes.

Hamblen.—Several horses have died during the past season with blind staggers or brain fever. The condition of farm animals is better than usual.

Hardeman.—There were some losses of horses and cattle last spring from starvation and bad treatment. We have suffered greater losses, however, from diseases among hogs than of any other class of farm animals. The disease is called cholera by some, and by others red mange, and by still others measles. The hog at first presents a mangy appearance; afterwards it breaks out in pimples or sores, and soon dies. A black hog of mine which recovered from the disease is now gray.

Hardin.—Milch-cows and oxen have suffered severely during the past season from murrain. Cholera prevails among swine of all ages.

Henderson.—Blind staggers is about the only disease that proves destructive among horses. Every disease incident to the hog is called cholera, and diseases are more prevalent among swine than among any other class of animals. Rot prevails among sheep, and cholera among fowls.

Jackson.—The great bulk of the annual losses of hogs in this county occurs from a disease known as cholera. Fowls die of a similar disease.

Macon.—Cholera is the only disease that affects hogs in this county. The disease has been quite prevalent and fatal during the year. Chickens also die of cholera.

Marion.—Horses are subject to distemper and blind staggers, from which many of them die. Cholera prevails among our hogs, and has proved very fatal. During some years almost all the fowls die of cholera. All kinds of stock suffer for want of proper attention.

Monroe.—There are from two to three thousand horses annually raised in this county. There is but little disease among this class of animals—nothing worse than common distemper, and an occasional case of bots or colic. About five thousand cattle are annually raised, and they are seldom affected with disease. Formerly hog-cholera prevailed extensively, and the fatality was very great, but of late years the disease has been very mild and has not prevailed as an epidemic. But little interest is taken in sheep. Fowls are raised by almost every family, and have become an important matter of trade among the ladies of the county in buying little items in stores.

Morgan.—Diseases in various forms have prevailed extensively among our hogs and fowls for years past. They have not been so prevalent during the past year.

Obion.—Horses, cattle, and sheep are remarkably healthy. Cholera exists among hogs, and a good many animals have been lost, but the disease is not very extensive this season. Cholera also prevails among fowls in some localities.

Overton.—Our cattle do not often suffer from contagious diseases, but many of them die for want of proper care and attention. Hogs and fowls suffer from cholera, and sheep from rot.

Perry.—The loss of hogs from cholera in this county during the past year will amount to not less than $12,000. Sheep have been affected with rot, and a good many fowls have died with cholera.

Sequatchie.—Swine are affected with what seems to be diseases of a local character. Many of these diseases are no doubt brought on by careless treatment.

Sevier.—But little disease has prevailed among farm animals in this county during the past year. A good many hogs have been lost, but the diseases among them have not been so widespread as in former years.

Van Buren.—The disease among hogs in this county is generally called cholera, although it manifests varied symptoms. Chickens are also affected by a disease designated as cholera.

Weakley.—With the exception of slight affections among hogs and chickens, all classes of farm animals have been unusually healthy during the past year.

TEXAS.

Austin County.—The losses of horses by infectious and contagious diseases varies greatly, but for the last two years they have been unusually large. The losses have been heaviest among stock horses on the prairie, and the disease affecting them seems to be a distemper or kind of croup. A strange disease has been prevailing among cattle in the northern part of the county, and every animal attacked has died. The diseases among hogs are cholera, lung affections, measles, inflammations of the throat, &c. Most of the animals attacked die. Cholera prevails among chickens, and losses have been very heavy.

Bandera.—All classes of domesticated animals have been unusually free from disease during the past year. Fowls are afflicted with various diseases, some of which are very fatal.

Bexar.—A few diseases annually prevail among domesticated animals in this county, and the annual losses among all classes will probably aggregate from $8,000 to $10,000.

Camp.—Stock generally is in good condition in this county. I have heard of no diseases prevailing among any class of farm animals.

Comal.—The only disease among horses consists in a swelling of the glands of the throat, frequently ending in ulceration. The disease prevails more extensively in spring when the weather is cold and wet. The majority of the animals that die are colts. The affection seems to be an epidemic, produced by scanty pasturage and rough weather. There are no contagious diseases prevalent among cattle, hogs, or sheep.

De Witt.—Horses, cattle, hogs, and sheep are generally healthy and in good condition in this county. The losses are so small as to attract but little or no attention. Fowls frequently die of a disease known as cholera.

Eastland.—A good many horses die in this county of blind staggers and big head, caused principally by feeding unsound corn. Diseases among cattle are not so fatal this season as they were last year. We have no special diseases among hogs, but a great many of them have died this year for lack of feed. Foot-rot and scab prevail among sheep. Fowls die of various maladies.

Harrison.—A fewer number of horses have died from disease in this county during the past twelve months than for several years past. Our hogs die in considerable numbers from a wheezing disease caused, no doubt, by eating cotton-seed, picking them up from about our gin-houses, or where they have been dropped by cattle. Our chickens and turkeys have died by the thousands with a disease we call the cholera.

Hays.—Our farm animals are in remarkably good health and condition. We have been free from all contagious diseases for three years past.

Hill.—A disease heretofore unknown has been quite troublesome to horses in this neighborhood. Our stockmen generally designate it as "loin distemper." Cholera among hogs and fowls is frequently quite prevalent and fatal.

Hopkins.—Horses in this locality are affected with glanders, and cattle with bloody murrain. Hogs are affected with cholera, and a disease which causes wheezing and choking, as of a hard lump in the throat. These diseases generally follow an acorn crop. Sheep die with scab, and a great many fowls are lost by cholera.

Jasper.—Farm animals are in much better condition than for several years past. No contagious disease prevails except among hogs, and the losses are quite small, as we raise but few hogs in this county.

Kerr.—I have never known an infectious or contagious disease to prevail among horses and mules in this county. Fifty-four cattle have died during the present year of dry murrain. A large number of goats and sheep have died of foot-rot and scab. A great many hogs have died of a disease termed sore eyes, and many fowls have died of cholera. So destructive has been the latter disease that many farmers are entirely without chickens. The condition of all farm animals, however, is a little better than the average.

Lavaca.—Ticks kill a good many colts in this county every spring. We have some distemper among horses, but it has rarely been fatal. Until this year hogs have always been healthy, but for several months past cholera has prevailed among them, and in some neighborhoods all have died. The disease seems to be contagious, and I think was introduced by the importation of fine breeds.

Llano.—Owing to depredations by Indians, but few horses are raised in this county.

Cattle are moderately healthy, and hogs entirely so. Sheep have scab occasionally, but the disease is cured by dipping. Fowls are subject to cholera, and I never knew one to recover from an attack of this disease.

Marion.—We have had, and still have, hog and chicken cholera in several localities in this county. It is very destructive during some seasons.

Matagorda.—Cattle in this locality have been healthy this year with the exception of an epidemic of opthalmia, which seemed to be atmospheric in its origin; or, in other words, it was caused by excessive heat and moisture. A good many horses have died from the effects of bites and stings of insects, which were never so bad before. Ticks, screw-worm, and the large horse or cow fly have destroyed many animals. Measles prevail among young pigs and shoats, and a good many of those attacked die. This disease is both contagious and infectious.

Maverick.—There are no hogs raised in this county. Horses and cattle are healthy. There is some scab among sheep, but much less than in former years.

Menard.—The disease called scab prevails among sheep, yet I believe a greater number die from careless management than from this or any other disease. Other classes of stock are healthy and in good condition.

Montague.—Cholera among hogs prevails to some extent this season, but the disease is not so general as in former years. The general condition of farm stock is good.

Navarro.—The losses among hogs in this county from cholera and other diseases will aggregate for the past year not less than $15,000 in value. Chickens also die of cholera, and sheep from liver-rot and scab. Horses and cattle are healthy.

Rusk.—Hogs and chickens are suffering with a disease called cholera, which seems to visit some portions of the county annually. Various preventives are used, but no specific has as yet been found for it.

San Patricio.—There are no diseases affecting any class of farm animals in this county. I have resided here twenty-seven years, and this is the first year within that time that any important disease has prevailed among fowls. About nine-tenths of them have died in this town and surrounding localities of cholera, at least the disease was so pronounced by those who know the symptoms.

Somerville.—Stock of all kinds in this county have been unusually healthy this year.

Titus.—Infectious and contagious diseases affecting horses are not so fatal as heretofore, though glanders and distemper kill a great many. A large number, however, are lost by staggers and bots. Cattle are affected with murrain and black tongue, and nearly all die that are attacked by these diseases. Many also die from feeding on acorns. A great many hogs are annually lost by cholera and red mange or measles, thumps, and staggers. Scab, rot, glanders, and black tongue produce fearful ravages among sheep. We have no remedy for these diseases.

Upshur.—A good many cattle and hogs have died in this county during the past year of diseases peculiar to these classes of farm animals.

Uvalde.—Horses die of blind staggers and a kind of lung fever, and cattle of lung fever and spinal diseases. The principal disease among hogs is cholera. Sheep die of scab and lung fever, and chickens of cholera. We have no successful remedies for any of these maladies.

Victoria.—This has been a very disastrous year for sheep, owing to the great amount of wet weather. There is no disease among native cattle, but about one-third of those imported die of Spanish fever. Horses, and especially colts, die of distemper.

Williamson.—The principal disease among horses is distemper. The disease prevails to a greater or less extent every fall, and principally among colts. The losses among cattle are generally occasioned by the same disease. We occasionally have hog and chicken cholera, but the losses are not very heavy from this disease. A few sheep are annually lost by scab, but not so many as in former years.

Young.—All kinds of stock and poultry have been unexceptionably free from disease during the past year.

UTAH TERRITORY.

San Pete County.—Contagious diseases have prevailed among cattle in this county during the present year. A few cases of bloody murrain have also occurred.

Wasatch.—There is no malignant disease prevailing among farm animals in this county. Fowls are subject to croup, of which a good many die.

VERMONT.

Addison County.—Horses are afflicted with a distemper which is regarded as contagious. Cattle have what is called murrain or black-leg, a disease which seems to be epidemic and contagious, but is confined mostly to calves. Murrain never attacks lean cattle. Many deaths occur among cows in early summer from milk fever. Cattle generally are in fine condition.

Caledonia.—No infectious or contagious diseases of a serious nature have prevailed among farm animals in this county during the past year. The season has been favorable, and stock generally is in good condition.

Chittenden.—The number of cows have increased in this county since the last census, nearly one-sixth. Horses and swine remain about the same. Sheep have fallen off materially, say three-fourths. At present there is a lively interest in fowls, especially of pure bloods. The number has increased at least one-third. Our animals are all in a healthy condition.

Grand Isle.—Since the prevalence of the epizootic some years since, there has been no infectious or contagious diseases among horses in this county. Cattle, sheep, and hogs are almost entirely free from disease. I ascribe such exemptions to the better care they receive in good feed, protection from storms and cold weather, and better care generally.

Rutland.—Foot-rot has prevailed extensively and fatally among sheep in this county during the year, but the disease is now diminishing.

VIRGINIA.

Accomac County.—The annual losses from disease among all classes of farm animals in this county will amount to from $12,000 to $15,000.

Alexandria.—Upwards of 100 head of cattle have died in this county during the past year, principally of pleuro-pneumonia. The origin of the disease has been traced to Georgetown. It occurred first among cattle there about two years ago, and has since gradually traveled down the river to a distance of about 25 miles. It has not, as yet, extended over 2 miles from the river toward the interior.

Botetourt.—There has been no infectious disease among horses in this county for several years past. The disease affecting cattle seems to be confined to those under one year old, and is known as bloody murrain. Those that are best kept are more liable to the disease than those poorly cared for. Flour of sulphur given with salt once a day for three days is the best remedy known here. Hog-cholera, the only disease affecting swine, is less malignant than in former years. Cholera in fowls has been successfully treated by placing iron in the water which is given them to drink.

Brunswick.—Cattle sometimes die in large numbers from a disease called murrain. A good many hogs are also annually lost by cholera, a disease which appears under many different forms or symptoms.

Campbell.—We have had no infectious or contagious diseases either among our horses, cattle, hogs, or sheep, with the exception of an unknown disease that has prevailed among cattle in the vicinity of Lynchburg. Cholera among fowls sometimes depopulates an entire henery in a few days.

Dinwiddie.—A good many horses, cattle, hogs, and fowls have died this season from the various diseases incident to them. Cholera prevails more or less every year among hogs in this county.

Essex.—I have heard of no infectious or contagious diseases prevailing among any class of farm animals, except swine.

Floyd.—The only contagious disease among horses here is a distemper which affects very seriously the head and throat of the animals. The most fatal disease among cattle is black-leg. Its attacks are more frequent among young animals of from one to two years of age. The only disease affecting swine is known as hog-cholera. All other animals are healthy.

Gloucester.—A large number of horses, hogs, and sheep have been lost by disease in this locality during the past season. The disease affecting hogs is the so-called cholera.

Greene.—The infectious and contagious diseases prevailing here are distemper among horses, and cholera among hogs and chickens.

Halifax.—An infectious distemper is prevailing among horses in this county, and murrain and distemper among cattle. The latter seems to be a contagious fever, and kills nearly all attacked. The prevailing diseases among hogs are measles and quinsy. There are no diseases among sheep. The condition of all farm-stock is good, better than last year, as pastures have been abundant. The cattle distemper, which is a high grade of fever, and generally considered contagious, is the worst disease farmers have to contend with. No effectual remedy has been found, but the following has proved generally quite a successful preventive: 1 gallon common salt; ¼ pound flour of sulphur; 2 ounces saltpeter and 2 ounces copperas. Dissolve these ingredients in three gallons of water and mix with red clay to the consistency of plastering mortar, and put in troughs for the cattle to lick. The troughs should be kept supplied from the first of July to the first of November. It rarely fails as a preventive.

Henrico.—Hog-cholera has been very severe here during the past season. About all the hogs affected have died.

Highland.—The prevailing diseases among horses are lung fever, distemper, diarrhea and mad stagger or fits. Cattle have horn-ail and hogs are afflicted with cholera. Domestic fowls also have cholera.

James City.—Horses are frequently affected with a distemper, the results, no doubt of former attacks of epizootic. Hogs are afflicted with cholera and mange. In cases of cholera a change of both pasture and food is recommended. Raw turnips fed on alternate days are thought to be a preventive.

Page.—Perhaps $5,000 or $6,000 will cover the loss among all classes of domestic animals by disease in this county during the past year. Farm stock is in much better condition than usual.

Patrick.—There have been some losses of horses in this country by distemper and cattle by murrain. Hogs, as usual, have been seriously afflicted with cholera.

Pittsylvania.—The principal disease among horses in this county, during the past year, has been pneumonia or lung-fever, which is not considered contagious. Cattle are afflicted with murrain, and hogs with various diseases.

Rappahannock.—The number of horses and cattle affected with contagious diseases in this locality is small. Some few hogs die of cholera. Sheep are healthy.

Roanoke.—No diseases have prevailed among farm animals in this county during the past year. There has been some cholera among chickens, but the disease has not prevailed as an epidemic.

Rockbridge.—Black-leg is the only disease that affects cattle fatally in this county, I do not think it contagious, though all the animals attacked by it die. Hogs and fowls are sometimes fatally affected by cholera.

Smyth.—Distemper has prevailed to some extent among the horses in this county, but it has proved fatal in but few instances.

Spottsylvania.—No diseases of a malignant character have prevailed among farm stock in this county during the past year.

Sussex.—Cholera has, and still is prevailing to a considerable extent among hogs and fowls in this locality. The losses have been quite heavy.

Washington.—Distemper and murrain prevail among horses and cattle, and cholera among hogs. A disease similar to murrain also prevails among sheep. Fowls are annually lost in great numbers by a disease called cholera.

Wise.—Diseases have prevailed to a considerable extent among hogs in this county during the past season. All other classes of farm animals are healthy.

WASHINGTON TERRITORY.

King County.—There are no infectious or contagious diseases among horses, cattle, or hogs. Sheep, however, are afflicted with scab, and fowls more or less troubled with vermin.

San Juan.—No infectious or contagious diseases exist among either horses, cattle, hogs, or sheep in this locality. There are frequently losses among all classes of farm stock from accident or lack of proper attention.

WEST VIRGINIA.

Boone County.—Epizootic distemper prevails among horses, murrain among cattle, and cholera among hogs. Fowls are subject to both cholera and gapes.

Doddridge.—Distemper is the only contagious disease prevalent among horses and that of foot-rot among sheep. Murrain also exists among cattle. All of these diseases are destructive, but none of them so much so as cholera among hogs. Many persons regard this disease as contagious.

Gilmer.—No diseases of any consequence are prevailing among farm stock in this county. Occasionally a horse is affected with distemper, but I have never heard of a case of lung-fever. Cattle are sometimes affected with foot-evil, which is the only disease I ever hear of as affecting this class of animals. I have heard of a few cases of cholera among hogs.

Grant.—I have made diligent inquiry, but can hear of no infectious or contagious diseases existing among the farm stock in this county.

Greenbrier.—There are no infectious or contagious diseases prevailing among farm animals in this county. About 13,000 chickens and other domestic fowls are annually lost by the people of this county from disease.

Harrison.—The only disease we have among horses is an infectious disease, which, for the want of a better name, we call "distemper." Horses that have it, generally recover. Among young cattle we have a disease called "black leg," which annually kills a number of cattle, and generally those in high condition. Our losses are sometimes very heavy from this cause. We also have a disease among cattle which affects their feet, and which we call "foot-evil." This disease does not kill cattle, but hinders them in their growth and deteriorates them in value. If a simple and sure remedy could be found for this disease it would save our people a considerable annual loss. We occasionally have a case of hog-cholera, and those attacked usually die, but, as a general thing, hogs are very healthy in this county. Chickens, in some localities, have suffered with cholera, and the flocks attacked generally all die.

Lewis.—No diseases of any consequence have affected the farm animals of this county during the past year. Pasturage is very abundant, and cattle are generally in good condition. Hogs are in good health, which we attribute to their consumption of bituminous coal.

Logan.—Horses here are much given to distemper. Sometimes the disease proves fatal; but the greatest loss occurs from colic and bots. Cattle sometimes have a distemper much like that of horses, but this seldom occurs. Diseases are prevalent among hogs, and they often die quite suddenly. Sheep and fowls are in pretty good condition. Sometimes the former are afflicted with rot.

Morgan.—The number of hogs and chickens that have died in this county during the past year of cholera has been very large. Other classes of farm stock have been healthy.

WISCONSIN.

Door County.—No infectious or contagious diseases prevail among any class of farm animals in this county.

Dunn.—Last winter horses in this county were seriously affected with distemper; but it disappeared in April and has not since made its appearance. Several animals were lost. Cattle and hogs are healthy, and are in very good condition.

Green.—The so-called cholera has prevailed among the hogs in this county for the first time during the present year. A great many have died. Cholera prevails extensively among fowls also, and many thousands have died.

Iowa.—Diseases among swine have been very prevalent during the past year, and the losses to the farmers of this county have, consequently, been very heavy, as the maladies have been of a fatal character. Other classes of stock have been healthy.

Jackson.—There are no diseases whatever of a destructive nature prevalent among farm animals in this county.

Juneau.—Distemper and inflammation of the lungs, or lung fever, are the only diseases of a serious character prevailing among horses. Other classes of stock are healthy.

Kewaunee.—Aside from a few horses afflicted with glanders, all classes of farm animals are in good health.

Monroe.—All classes of farm animals are in good condition and measurably healthy.

Ozaukee.—With the exception of a few cases of hog-cholera, I have heard of no other disease among farm animals in this county.

Portage.—Domestic animals in this county are exempt from contagious diseases to a remarkable degree. Indeed, I hear of but few farm animals dying of any disease, except sheep and hogs—sheep from grub, and hogs from black vomit, or something like it.

WYOMING TERRITORY.

Laramie County.—There are no diseases whatever among farm animals in this county. The losses among cattle, caused by eating the poisonous loco weed, will perhaps not exceed 1 per cent. About 300,000 head of cattle come into the Territory annually from Texas, Oregon, Montana, Idaho, and Nevada.

CORRESPONDENCE RELATING TO THE MORE COMMON DISEASES OF DOMESTICATED ANIMALS.

ALABAMA.

Mr. Robert Wardland, Tuscumbia, Colbert county, Alabama, says:

Not having had much experience with farm animals, I will confine my remarks to fowls and the ailments to which they are subject. I grow them for my own table, and not for market or fancy purposes. Long years ago I devoted considerable attention to fowls, and soon became satisfied that the majority of the diseases incident to them were induced by carelessness and inattention to their sanitary condition. I have found that prevention is much better than cure, and now, if I desire a sick chicken to experiment, I am compelled to go to some of my neighbors for the subject. In cases of what is known as cholera, the liver of the chicken is found very pale, much enlarged, and literally rotten. The whole internal viscera is more or less deranged. With such cases it is the veriest quackery to attempt a cure. A careful examination of the disease known as gapes has convinced me that it has its origin in parasites. These, and that other great pest, lice, produce many of the diseases which result so fatally to fowls.

My treatment of fowls, which has proved very successful, is very simple. I give them a well ventilated yet cheap house, provided with plenty of roosts, nests, &c.

Next, I have none but healthy birds to breed from, and am very particular to keep their quarters perfectly clean. I have my hen-house cleaned once a week during summer, and once in every ten days during the winter season. I remove the contents and have them stored under cover for use as a fertilizer for my crops. I use quick-lime and wood-ashes as disinfectants, and charcoal as an absorbent. The result is clean houses and healthy fowls.

I pay close attention to their food. Too much corn makes them fat and indolent. Once or twice a day is as often as they should have grain. They should be provided with grass lots for grazing, as the amount of this kind of food they will consume would astonish any one who has not given the subject attention. Pure water and plenty of it is indispensable. Sick birds should at once be separated from the well ones, but the best plan is to cut off their heads and bury them.

I am partial to dark-colored fowls, as I am of the opinion that they are more hardy than the light-colored ones. I am careful not to overstock my flock, and breed only from those that are peaceable, and as a result have no games or ill-natured fowls.

Mr. V. C. Lavmore, Valley Head, De Kalb county, says:

My observations and experience with farm stock extends to a period of near forty years. In the care of horses I am particular to give them good grazing and sound feed. In winter I give them good shelter and feed both hay and grain. I also give them salt and ashes, slaked lime, and copperas or saltpeter. During the summer months I keep the nits cleanly scraped off from their limbs and bodies. I practice about the same treatment with cattle, and in addition use sulphur, rosin, and turpentine in the summer and fall to keep off the ticks. I use the same preparation to remove lice from my hogs. When disease is in the neighborhood I give them salt and ashes, and sometimes turpentine. My hogs have been visited but once with cholera, and then they had it very bad. I tried everything I could hear of, but to no purpose until I separated them into three different lots. I put the well ones into a field by themselves, those that looked feeble into another, and the sick ones I turned into a meadow through which a stream passed. I drove them through this creek once or twice a day. I burned all the dead carcasses, old beds, and even the woods where they had been running in the mast. I had about two hundred head, and many of them died, but they commenced to improve soon after I commenced this treatment, and soon the disease disappeared.

Mr. R. Tucker, Marion, Perry county, says:

Hog-cholera seems to prevail throughout the United States, and perhaps more hogs die from the effects of this disease than from all other causes combined. I have been using preventives for years, and when I attend strictly to this duty I hardly ever lose a hog by cholera or any other disease. I use copperas, lime, ashes, charcoal, sulphur, and tar. The most of these articles are good for worms and keep the hog in a healthy condition. Cholera makes its appearance in various forms, and in many cases, I think, what we call cholera is caused principally by worms.

In this latitude we have a disease called murrain among cattle, which, perhaps, is more destructive than any other to this class of domestic animals. It usually makes its appearance in the spring of the year. We have what is known as both the dry and bloody murrain. As preventives we use salt and sulphur freely, and keep tar in the feeding-troughs. When a severe case makes its appearance it is hard to cure, though soap and oil have been used in cases of dry murrain with some success.

Blind staggers seem to be the prevailing disease among horses and mules. A horse properly fed on sound corn and hay, with lime, wood-ashes, tar, and sulphur constantly in their troughs, will never have the blind staggers. Bots and colic also kill a good many horses. Oil and chloroform will generally effect a speedy cure in such cases.

Dr. George T. McWhorter, Chickasaw, Colbert county, says:

In connection with my report of the hog-disease, which prevailed so fatally here during the past season, I desire to call your attention to reports from Van Wert and Preble counties, Ohio, Iroquois county, Ohio, and Lauderdale county, Alabama, found in the Report of the Commissioner of Agriculture for 1876, page 108. I am convinced that the "new disease" mentioned by correspondents from these counties is the same that prevailed here, and that it is caused by the worm, specimens of which I sent you. You will observe that all call attention to the lung trouble, some stating that the lungs were the only parts affected. By careful examination I found, as stated in my report, the lungs, liver, stomach, and bowels infested by these worms, but in every case the lung-tissue had suffered most, in some cases being entirely broken down.

I suspect also that much of the pneumonia (page 109, same report) reported from Kentucky, Illinois, Ohio, Indiana, and Kansas is due to the same cause. The trouble is so much more patent in the lungs than elsewhere that it might reasonably be overlooked in other situations. You will remember that the worms taken from the lungs were much larger than those from the bowels. I attribute this to the inferred fact that the lungs afford better conditions for their development than the other organs. The fact that their presence in the lungs is so much more deleterious to the health of the animal, and manifests itself by such decided symptoms, is perhaps the reason that some have supposed that they alone were affected. I am still of the opinion that the alimentary canal is the *nidus* in which the egg is hatched, and from which the young worm starts, producing violent and noticeable symptoms when the lungs are reached and perforated.

ARKANSAS.

Mr. William B. Turman, Waldron, Scott county, says:

As hogs are the only class of farm animals affected by disease in this locality, I will confine my remarks to the malady generally known as hog-cholera. The symptoms are a cough, followed by constrained breathing, producing, in many cases, a movement similar to thumps in horses. The animal refuses food. After a while great thirst prevails, and scarlet red spots, from the size of a pin's head to that of a man's hand, appear on the surface of the body. At this stage of the disease they refuse to leave their beds. In some cases death ensues within a few hours, while in others the animal may linger for several days. Perhaps one hog in ten survives a mild attack. An examination after death reveals the lungs, to all appearances, greatly affected, and in many cases much decomposed. In some cases the blood is also found coagulated in and around the kidneys, and the entire flesh in a more or less putrid condition.

I am informed by Mr. W. M. Johnson that for the last twenty years he has kept his hogs healthy by giving them, with their food, common pine tar, occasionally smearing some on the hair of his hogs. He has not lost a single hog by this very common disease.

Mr. Dearman keeps his hogs healthy by giving them soap, pine tar, and sulphur. Mr. A. J. Gentz keeps his in good condition by mixing boiled garget or poke root with their feed. Mr. A. H. Hooper gives sulphate of iron and salt, which has proven an excellent preventive with him.

Mr. W. W. Hughey, Warren, Bradley county, says:

There has been no disease in this immediate vicinity that has seriously affected swine since 1873. During that year fully three-fourths of the hogs in the county died of what is commonly known as hog-cholera. The first symptom of the disease was a refusal to eat, followed by a dull, stupid appearance. Frequently eruptions about the size of a pea would appear on the body, and death would then ensue in from five to twelve hours. In a few hours after death the carcass would swell to such an extent as to break the skin in many places, from which a yellowish water would run.

About the 20th of December last, a similar disease made its appearance in the western part of the county, which is proving quite fatal to grown and fatted hogs. Not more than one in five of those attacked recover. We expect it to spread throughout the county by the first of May, as it did on its former visit. Hogs are not raised here for market, yet most farmers endeavor to raise a sufficient number to provide themselves with their own meat.

Mr. J. N. Deaderick, Wittsburg, Cross county, says:

The most fatal diseases we have among horses here are staggers, Spanish fever, and charbon. In sleepy staggers a disposition is shown to move around in a circle. The general treatment is blistering over the brain and profuse bleeding from the nose. The disease lasts from one to two days, and the fatality among those attacked is about 90 per cent.

In Spanish fever the symptoms are extreme languor, stupor, and high fever. The duration of this disease is from five to fifteen days, and the per cent. of deaths about the same as in staggers.

Charbon first makes its appearance by a small hard lump, somewhat resembling that caused by the sting of a wasp. This lump grows and spreads very rapidly, and frequently chokes the animal to death in a few hours. The remedy generally used is to paint with iodine.

Cattle are affected with murrain, Spanish fever, and charbon, and occasionally a disease resembling dropsy in the human system. When attacked with the latter disease they generally drop dead without a struggle, and on tapping them, very often as much as a barrel of water will exude from the incision. The fatality in murrain is about 95 per cent., and in dropsy all die.

Hogs are affected with cholera, quinsy, and mange. The symptoms of cholera are varied. In the most violent cases there are discharges from the bowels, bladder, and lungs. In other cases a loss of appetite is occasioned, and there is a disposition to bed up during the night; and during the hottest weather, if driven from their beds, they will shiver as though suffering with a hard chill. The loss is about 75 per cent. of all attacked. A great many remedies are used, but with little success. I value soft soap more than anything else. Pine tar is a good remedy for quinsy. Mange or scab is very fatal to young pigs. It appears as ulcers in the mouth, throat, and in the body. Carbolic acid, sulphur, and turpentine are used with considerable success. The fatality in this disease is about 50 per cent. of those attacked.

Sheep are sometimes affected with rot, a disease somewhat resembling Spanish fever or dry murrain in cattle. The fatality is about 50 per cent.

Chickens are liable to cholera, and often drop dead from their roosts without warning. Others have copious discharges of filthy, green matter, their combs and gills become very pale, and after lingering a week or two die in a very emaciated condition. It frequently happens that some farmers will lose their entire flocks by this disease, while others living near by will not lose any. There seems to be no remedy, and about all die that are attacked by the disease.

FLORIDA.

Mr. T. K. Collins, Mikesville, Columbia county, says:

A disease commonly called "thumps" is perhaps the most fatal disease that affects hogs in this part of Florida—more fatal from the fact that no remedy has ever been found for it, at least to my knowledge. I have resided here seventeen years, and during that time have not known a single case cured, notwithstanding 8 per cent. of our hogs die of it annually. The first symptoms of the disease are a cough, shortness of breath, thumping or bellows-like motion of the sides, with loss of appetite, and ultimately, like in cases of consumption in man, waste away and die a mere skeleton. The duration of the disease is from one to three months. I can offer no remedy for this disease, or even suggest its cause. Some old stock-raisers say that this disease is always worse after a heavy pine mast, which my own experience confirms.

Staggers is also a common disease among hogs here, but it is seldom fatal. Cutting the ears or scarifying the head generally gives relief, but cold applications or sunstroke treatment, when applicable, is considered better.

Cholera made its appearance among swine here this season, and cut our meat crop short. Most of those attacked died suddenly, many of them even before they were known to be sick. This disease is new to us, and as yet we have found no remedy for it. These are about the only diseases that attack swine in this locality.

Mr. Chester S. Coe, Coe's Mills, Liberty county, says:

With the exception of cholera among hogs we have but few other diseases among any class of farm stock. As regards this disease we have never been satisfied as to its origin, as hogs take it at any time and under all circumstances, those running at large in the range as well as those kept in inclosed pastures. During many years' experience I have noticed that those which we term yard hogs—i. e., that are fed on dishwater and kitchen slops—seldom or never take the cholera, and that if those that take it in the range are confined in pens and fed on kitchen slops, with the addition of a little copperas and sulphur, they generally get well. As for a preventive, we have never found a positive one, though I am of the opinion that if hogs are frequently fed on slops seasoned as above stated they will seldom take the cholera.

In an every-day experience of over sixty years in the use of horses and mules I have never lost but one, and that one I lost by blind staggers. Good care in feeding, watering, and driving, with an occasional handful of salt mixed with a little lime or strong ashes, has always kept my stock in health and good order.

I have had no disease among my fowls for thirty years. We keep a supply of nux vomica on hand, and twice or thrice a week mix it with their feed, giving from a fourth to a teaspoon level full, according to the number to be fed. This has kept them free from all disease; and more than that, if a hawk ever takes one he will never come back for another. There is no perceptible difference caused in the taste of the meat. The drug may be used by bruising two or three buttons and steeping them in hot water. Then add a few spoonfuls in mixing up their feed.

INDIANA.

Mr. D. C. Smith, Vincennes, Knox county, says:

The disease known as hog-cholera is caused by worms. There are two kinds of worms. One works upon the kidneys, liver, heart, and lungs, and is more dangerous when it is in the region of the heart. It looks like a kidney-worm, but is somewhat smaller. It penetrates to all parts of the body. I have found it between the leaf-lard and the intestines, and between the shoulder and the ribs. The other worm works upon the stomach and small intestines, and causes the diarrhea. When they are in the liver they cause a dry, hacking cough; when in the lungs the cough is more severe, and the hog will bleed at the nose, and a bloody foam will run from its mouth. The symptoms of the disease may be seen in the hair of the hog standing up straight, and the discoloration of the skin behind the ears, which sometimes turns yellow and at others assumes a bluish cast. The hog will walk very slowly, and when it stops will drop its head and look as though it were standing on its nose. Some will become lame in their fore legs. When the worm is in the stomach the hog will purge and vomit.

I have taken as high as ten worms out of the liver of one hog.

I have a remedy for these diseases, which I have used with great success for ten years. Since using it I have never lost any hogs by cholera. The remedy is as follows: Mix two tablespoonfuls of spirits of turpentine in a half-barrel of slop, stir well, and feed three times a week every other week. Give this amount to fifteen or twenty head. While they are eating pour a tablespoonful of coal-oil across the back and shoulders of each hog. This will penetrate the skin and drive the worms inwardly, when the turpentine will kill and expel them.

The following are extracts from a letter from Mr. Lewis Bollman, of Bloomington, dated August 26, 1878:

I see that you have appointed a commission to investigate the hog-cholera. I hope that it may result in some greater practical utility than prior commissions have effected. Allow me to add to this communication the little I know about it.

I have always understood that the disease originated at Aurora, in this State, a town on the Ohio River, in Dearborn county. A large distillery is there, and years ago it fed about 4,000 hogs on the distillery slops. This excessive crowding and unnatural feeding generated the disease, and from there it slowly but steadily spread over the West.

While a farmer here years ago, I raised from 50 to 75 hogs annually, and for three years my neighbors lost many hogs with this disease. One year my adjoining neighbor lost about 70 head; there being between our hogs the common rail-fence only. I never had a hog sick from the cholera, and I attribute this exemption to my practice of salting my hogs with a mixture of salt and pulverized brimstone and copperas. Of three parts, two salt; of the remaining part, two parts brimstone and one part copperas. I adopted this salting to destroy intestinal worms and lice. I strictly adhered to this practice twice a week in summer and about every ten days in winter.

A farmer here told me the other day that he lost hogs one year only from the disease, but having adopted this feeding with sulphur and copperas he never since had any of his hogs sick with it.

Whether this salting is really a preventive I cannot certainly say. I but state my experience. In its modes of infection the hog-cholera is much like the rinderpest when in Great Britain. If well animals crossed the track of diseased ones they caught the disease, as with cholera. If I remember rightly British authorities were forced to confine their cattle to the farms of their owners and to prohibit the sales of unfattened cattle at fairs where such are generally purchased by those purposing to fatten.

So far as my observation extends I believe this moving of our hogs, and allowing them to run outside of their owner's inclosure, is the cause of the continued existence of the disease.

A farmer here recently rented an out-field to hog down, located about a mile from his home. The first thing he knew was that that field emitted a stench from his dead hogs. About $300 worth died in a few days, nearly all that he had. So I learn that many have died around this place in consequence of their running at large. The greatest fatality exists on our river bottoms where the hogs are collected by purchase and driven on the extensive corn-field to hog down the corn.

I suggest to your consideration a careful examination into the consequences of this mode of moving stock hogs in order to fatten them, and if found to be a common and fruitful source of the spread and continuation of the disease that the exclusion resorted to in Great Britain be enforced here. It is a quarantine regulation such as is now sought to be enforced in our Western cities to stay the spread of the yellow fever.

Believing that he had made an error in attributing the cause of the disease in this herd to the fact of its removal to another farm, Mr. Bollman writes as follows under date of September 2, 1878:

A few days ago I wrote you a letter, chiefly on the topic of hog-cholera, mentioning a recent case of a farmer here who had lost about $300 worth of hogs by that disease. I attributed the loss to moving the hogs to another farm. I saw him since writing, and learn that it is probable the hogs were liable to the disease before their removal. He has raised hogs very extensively, and heretofore has lost heavily from the cholera. His recent loss, I am now satisfied, was the result of overcrowding his farm with hogs—an error so certain to this result that I now again write that the attention of your commission may be directed to it.

One of the greatest difficulties a farmer has to encounter arises from having a large number of certain kinds of stock, which cannot safely be crowded, no matter how complete may be his arrangements to grow them, or any one of them.

Hogs, sheep, fowls, the silk-worm, &c., cannot be raised in large numbers together without soon exhibiting a liability to epidemical and other diseases. The diseased condition of the sheep at the close of the war obliged farmers to sell their flocks at the lowest prices. All attempts to raise chickens in large numbers, or the silk-worm, have failed from large losses by epidemical diseases. And so with the human race. An army generates camp-fever, measles, and other diseases, no matter how strictly every sanitary regulation may be enforced.

As a farmer, I found it was easy to raise twelve or fifteen hogs, but difficult when the number was increased to fifty or eighty. I mentioned my exemption from hog-cholera, as I suppose from the regular salting with copperas, but I am satisfied that as long as any farmer, from year to year, grows many hogs together, the hog-cholera cannot be eradicated. Few farmers understand this tendency to fatal diseases from too great numbers, and I hope the commission may give it a thorough examination.

ILLINOIS.

Dr. Joseph Sybertz, V. S., Bellville, Saint Clair county, Ill., contributes the following paper on the disease commonly known as "hog-cholera":

We must regard this affection of pigs as a disease peculiar to this species of the family *suida*, having close affinities with the scarlet fever of man, yet essentially distinct. Few diseases are designated with a greater number of names than this one. For instance, it is called enteric fever, typhus, pig distemper, epizootic influenza of swine, measles, scarlatina, gastro-enteritis, anthrax, &c. Some authorities advocate the theory that the affection known as hog-cholera is in reality typhoid fever (abdominal typhus). Veterinary authorities agree that it is a form of anthrax or carbuncular fever. But there is an essential difference between anthrax and typhoid fever.

In the first-mentioned disease, the presence of bacteria in the blood is invariable, these parasites, indeed, being considered the cause of the affection. In typhoid fever, bacteria have never been discovered, either in the blood of the patient or in the characteristic lesions of the disease, the determining symptoms in this affection being ulceration of the glands of Peyer, as shown in *post-mortem* examinations. Now, in all forms of anthrax this ulceration is never seen, although mycosis (fungus) of the intestines is frequently noticed.

The line of demarkation between these affections is, then, sufficiently broad; but to which of them does hog-cholera belong?

Hog-cholera is a disease peculiar to pigs of this part of the country; the virus is not communicable to other domestic animals, so far as is ascertained up to this time by the veterinary surgeons of this country.

For the sake of brevity, I will, in dealing with the disease, call it by the conventional or rather common name of hog-cholera.

Hog-cholera is a contagious, febrile, and exanthematous disease, and embraces scarlatina in degrees of virulence in all stages.

Course of the disease.—*a*, stage of incubation; *b*, stage of florescence; *c*, stage of desquamation (scaling off).

The contagion poisons the blood, and produces local inflammation and ulceration in various parts of the system, though more frequently in some portions than in others. The action of this contagion possesses the peculiarity that it affects chiefly the skin and the throat, and originates in both a diffuse inflammation.

Symptoms in general.—*First stage:* Fever with a full and frequent pulse; the pharynx presents an exanthematous flush, but there is no effusion; general debility; appetite smaller than in health; thirst increased; skin hot and dry; sometimes a profuse diarrhea, and in single subjects delirium or spasm. The urine remains of its natural color.

Second stage: More intense fever; elevation of the temperature of the rectum to 35° 40° Celsius; tremulous motions of the cervical muscles; pharynx inflamed; deglutition difficult; the amygdalæ swollen; the mucous membrane presents a vivid red appearance. There is occasionally vomiting or diarrhea, usually constipation. A dry, hard cough is one of the symptoms in early stages, and continues to the last; quick and vibrating pulse, and occasionally epistaxis (the state of bleeding from the nose). Increased heat and redness of the skin; the eruption is not so generally distributed as in the former affection; it disappears often suddenly, and returns after an uncertain period of time. By the effusion of the red points, the disease passes on to the—

Third stage: The symptoms are of a graver type, even in the first accession. In fatal cases the patient is, in fact, by an elevation of the temperature to 43° Celsius, stricken dead by the poison in a few hours, before any eruption or local symptoms come on. The eruption does not present scarlet appearance, but is more of a livid hue, and frequently interspersed with petechiæ. In young animals convulsions and coma are frequent concomitants: in grown, delirium and deafness, sometimes great restlessness, running round towards one side, until at length the patient breaks down and lies helpless and insensible, or in a muttering delirium, till at length death approaches silently, and life ends without a struggle. The temperature is high until death approaches and bloody urine flows, when it very perceptibly diminishes.

The sequelæ of the disease are: Anasarca (effusion of serum in the cellular substance); ophthalmia (inflammation of the membranes or coats of the eye or eyeball); otitis (inflammation of the ear); enteritis (inflammation of the intestines); and cynanche parotidia (inflammation of the salivary glands), causing difficulty of breathing and swallowing; in grown hogs, affection of the sub-maxillary (mandibular) and inguinal gland, the last mentioned causing the staggering gait in young animals.

A secondary stage frequently follows, mostly caused by catching cold or by a disturbed crisis; then metastasis (a sudden and complete removal of the disease from one part to another) often occurs. This would seem to account for the fact that medical experts found so many different lesions by *post-mortem* examinations.

The next cause of the disease is an atmospherical contagion, which is always transferable. The infection is therefore double, atmospherical and individual.

Only constant lesion (and it is questionable whether it can be considered entirely characteristic) is the want of coagulability of the blood and the petechial eruption; all other lesions may be considered incidental; sometimes scarcely one organ of the body is found that is not the seat of some anatomical lesion.

If we consider the hog-cholera as an independent disease, and the malignant throat disease as a partial symptom of it, or the latter disease as an independent typhus disease, an infection of the blood, and the first as a partial symptom of it, has been up to this time, so far as I know, not ascertained, and the process of this epidemic is still a mystery, as in other epidemics.

In single cases and in epidemics it has sometimes the character of a local affection (malignant throat disease); in others, more the character of a general illness (infection); or it may be distinguished by this, that it occurs in all forms intermixed.

I have seen in different hygienic conditions swine affected with the disease, but, by perfect cleanliness, which necessitates the separation of the sound from diseased, and the free use of disinfectants, the poison, even generated or introduced, will be virtually starved out. In neglected hygienic conditions, I saw patients without care and treatment recovering, and on the contrary, the best rules and remedies designated for the prevention and most careful treatment could not prevent them from dying. These are sporadical cases. If the epidemic has existed for a length of time, the disease will seem to become more mild, and a much larger proportion will recover, while the first cases that occur will be very severe and will nearly all prove fatal.

In my practice as a veterinary surgeon I have tried many recommended remedies, but without much success.

I have adopted the following rules: As a preventive, disinfection of the atmosphere and the surrounding objects, and disinfectants for the free use of the animal. Protect them from the hot-bed of manure and close sleeping-places, where they are huddled together in great numbers; supply them with sufficient fresh straw for bedding in different places, as far as possible from each other. Supply them with fresh water and a succulent diet.

When the disease exists the sick should be placed by themselves, and the healthy ones taken to a fresh and disinfected place. Very sick hogs, without any hope of recovering, should be instantaneously taken from the herd, killed, the carcasses interred very deep, and with quick-lime and sulphate of iron overstrewed, so that no noxious emanation takes place.

For disinfection of fecal matters of stables, pens, or other places giving rise to noxious emanations, fill up a bucket with a strong milk of lime, add about one-half pound of sulphate of iron before separately dissolved in water, and sprinkle it upon the places which you intend to disinfect.

For disinfection of surrounding objects, as stable-walls, troughs, pen-rails, &c., take a strong solution of chloride of lime (1 pound to 12 pounds water), and whitewash the objects. This operation develops much chlorine, which destroys the contagion and purifies the surrounding air.

A specific remedy in general never will be found; disinfectant, diaphoretic, sedative, refrigerant, astringent, saline, cathartic, antiseptic, and antizymotic agents, one or more of them, according to the demand of each form and stage of the disease, are beneficial.

Of greater importance, and more useful than the medical treatment, is the prevention of it. From the peculiar construction of the larynx in hogs it is sometimes not possible to give medicine in form of a drench without their vomiting a part of it, or dying from suffocation; beside, this is not practicable with a great number of animals, and would hardly compensate for the trouble and expense necessary to secure the life of diseased hogs. For this reason the best way is to select such remedies as the animals are apt to use willingly. The medicine should be given in a form suitable to their small appetite, and in a way that they may get an approximately full dose of it, according to their age.

IOWA.

Mr. George T. Gibbs, College Springs, Page county, says:

As I have been broken up by the so-called hog-cholera, I have come to the conclusion to give you my theory in regard to the disease. I believe the whole difficulty lies in the manner of breeding which has been practiced for the last fifteen or twenty years. We hold to the maxim that like produces like, and pay high prices for short-horns to improve our cattle and large sums for fine hogs to improve our swine, and then give the lie to our theory by our practice. The practice by most hog-raisers, and especially by those that have been supplying the country with fine stock, has been to breed their sows at the age of from six to eight months, then fatten them and breed from the pigs at the same age. I claim that this has been kept up until the constitution of the hog has been ruined, and any little thing will bring on disease, which sometimes becomes epidemic and appears to be contagious. If you breed from animals whose bodies are immature and constitutions already weakened, if like produces like, you are getting an animal weakened from infancy. The old way of breeding was to allow stock hogs to make a little bone and muscle as well as fat, to mature their bodies before allowing them to breed, and when you once got a good breeder to keep her as long as she would bear pigs. In those days we never heard of hog-cholera, and we could raise eight, ten, and twelve pigs from one sow. My father kept one sow for several years, which raised ten pigs every litter. He sold the pigs all over the county for breeders. They were not hazel-splitters either. I have helped to butcher some of this breed that dressed 250 pounds at six to eight months old, and some that were kept until four years of age weighed 800 pounds. Now hog-raisers get two or three pigs from a sow, sometimes only one. A great many object to fine stock on this account; but we can raise eight or ten pigs at a litter from thoroughbred Poland-China, Berkshire, or Chester Whites, if we treat them properly.

I expect to be laughed at by the wise and scientific, but I have watched this matter closely for the last five years, and I am satisfied I have found the true solution or the difficulty.

SOUTH CAROLINA.

Dr. C. J. Faust, Graham's, S. C., writes under recent date as follows:

I see much written in regard to hog-cholera, as it is termed in the Northwest and in our own Southern country. So far as my own observation goes I am inclined to think that Dr. J. M. Johnson, of Locksburg, Ark., is correct in regard to its symptoms, cause, treatment, and pathology. Last winter I lost ten or twelve head myself out of a herd of twenty-four. They were all in fine order.

We also had an epidemic of staggers among horses and mules in our neighborhood, which proved fatal to a great many animals. The disease generally lasts from six to forty-eight hours. An animal attacked with it rarely recovers. I lost seven head of horses myself last winter, which cost me $1,200, and many of my neighbors lost a greater or less number. The disease known as staggers, however, was not the cause of the death of all of them. The animal, when first attacked, seems to be stiff in his fore legs, is very dull in riding, and when touched with the whip springs off very suddenly for the moment; but this is soon over. The nervous sensation seems to be very acute, and when allowed to run on an hour or so the animal does not seem to have power to lift his feet high enough to keep him from hitting them against the smallest rise on the surface of the earth or any small object in his way. He soon commences to go around in a circle, say 80 or 100 feet in diameter, and when once broken off from this circle he will go over anything in his course, and will even plunge into a dwelling. He becomes dangerous to those around him, and will go on until he is thrown down by running over some large object, when he soon dies in great agony. Our treatment has been full blood-letting, even to fainting, and copious drenching with a free purgative, composed of 300 grains of aloes, 150 grains jalap, and 80 grains of calomel, made into a bolus. This is placed upon a long paddle, two and one-half inches in width, and the paddle put down the horse's throat as far as it will go. The bolus rolls off without trouble and the animal swallows it. It soon acts thoroughly on the bowels. If this treatment should have the desired effect the horse should not be allowed to eat anything for two days, and then only bran mashes and a little green food. This should be continued for several days, when the horse will begin to slowly and gradually recover.

Mr. Charles M. Keyser, Cedar Point, Page county, says:

Having had some experience with the disease commonly called hog-cholera, I will try and relate the result of my investigations made recently. The disease was close to me and there were some cases in my immediate vicinity. About October 10th last I penned my hogs to fatten in their usual health, as I thought. About the 1st of December I found that they began to refuse some of their food, so I butchered them, and, upon examination, I found their lungs and livers in a very bad condition. The lungs were very much darkened and decayed, and the pores or small tubes were filled with worms about the size of a hair; they varied from one to three inches in length, and seemed to completely choke the hog. In color they resembled that of the kidney-worm, though they were not so large. I had no microscope and could not make a close examination. The liver was full of boils, and seemed to be in a perfectly torpid condition. The bowels seemed to be in a healthy condition.

My former experience concerning the disease is that the lungs and liver are the points most affected. The symptoms of the disease were manifested in a dull and drooping condition of the animal, coughing, and a heaving of the flanks—a beating and working like a bellows. In some instances the animals would turn quite a complete somersault and fall over dead. In other cases they would die quite easy.

I do not think there is any cure for the disease after it gets a fair hold on the animal. It seems that hogs that run at large—roam through the woods and fields—are more liable to the disease than those that are kept in clean, comfortable pens and are well cared for. The use of tar in the troughs and wood-ashes (hickory preferable) spread on the ground where they are fed, in a dry time or in a dry place, is a very good preventive, if not a cure, in some cases. They will eat some and inhale a little, which has a good effect on the animal.

PLEURO-PNEUMONIA OR LUNG FEVER OF CATTLE.

The following letter, addressed by the Commissioner of Agriculture to Hon. A. S. Paddock, chairman of the Senate Committee on Agriculture, on the 14th day of February last, gives all the facts in regard to the prevalence of pleuro-pneumonia among cattle in this country, so far as they were then known to this department:

SIR: I have the honor to acknowledge the receipt of your letter of recent date, asking for such information as may be in my possession relating to the subject of pleuro-pneumonia among cattle. The subject is one that is attracting great attention in this country at present; hence information is rapidly accumulating in this department, the more important portion of which I herewith transmit for the information of your committee. I shall first give a brief statement of the action of the department in the matter, and then submit such letters, telegrams, and other information of an important character bearing upon the subject as have recently come into my possession.

In August, 1877, within one month after my accession to the position of Commissioner of Agriculture, I instituted a preliminary examination of diseases of domesticated animals. For years I have been cognizant of the loss of immense numbers of swine and other farm animals by disease, supposed to be of an infectious and contagious character; and, with the very limited means at my disposal, I opened a correspondence with leading farmers and stock-growers in almost every county in the United States for the purpose of eliciting definite information in regard to these maladies, and the probable annual losses occasioned thereby. The result of this correspondence was the accumulation of a vast amount of important information on the subject under consideration, which, by request of the Senate, was communicated to that body on the 27th day of February, 1878, and was afterward published as Senate Ex. Doc. No. 35.

In order that a thorough examination might be made into some of the more destructive diseases affecting farm animals, and such remedial and sanitary measures instituted as would prevent the spread of such maladies as were well known to be both infectious and contagious, an appropriation of $30,000 was asked, and the sum of $10,000 was granted. In my letter of transmissal to the Senate in February, 1878, the following language is used:

Our wide extent of country and its great diversity of temperature and variation of climate, the severity of frosts in some sections, and the intensity of heat in other localites, render farm stock liable to the attacks and ravages of almost every disease known in the history of domestic animals. So general and fatal have many of these maladies grown that stock breeding and rearing has, to some extent, become a precarious calling instead of the profitable business of former years. This would seem especially true as it relates to swine. Year by year new diseases, heretofore unknown in our country, make their appearance among this class of farm animals, while older ones become permanently localized and much more fatal in their results. Farmers, as a rule, are neglectful of their stock, and pay but little attention to sporadic cases of sickness among their flocks and herds. It is only when diseases become general, and consequently of an epidemic and contagious character, that active measures are taken for the relief of the afflicted animals. It is then generally too late, as remedies have ceased to have their usual beneficial effects, and the disease is only stayed when it has no more victims to prey upon.

This interest is too great to be longer neglected by the general government. Not only the health of its citizens, but one of the greatest sources of our wealth, demands that it should furnish the means for a most searching and thorough investigation into the causes of all diseases affecting live stock.

At the time this communication was made it was not known that the destructive disease known as contagious or malignant pleuro-pneumonia among cattle was prevalent to any considerable extent in any section of the country. There may have been, and no doubt were, isolated cases of the disease, but they were not sufficient in number to attract attention or cause alarm. During the past summer and fall my attention was called to the prevalence of the disease in several localities widely separated from each other. Among other letters addressed to me on the subject, I cite the following.

J. Elwood Hancock, of Burlington County, New Jersey, writes:

The prevailing disease among cattle in this county is pleuro-pneumonia. The disease is very fatal, and the losses among this class of animals from this malady have been very heavy.

Mr. J. E. Hancock, of Columbus, Burlington County, New Jersey, states that the disease has been prevalent in that county for some years. He says:

I have had some experience with pleuro-pneumonia among cattle, having lost one-third of my herd from its ravages in 1861, when I succeeded in eradicating the disease after a duration of about six months. I had a second visitation of the malady in my herd in the early part of 1866, when I lost 6 head from a herd of 23. Of the animals affected I am satisfied that not more than one-third will recover.

N. W. Pierson, Alexandria, Va., writes as follows, under date of October 12, 1878:

The principal disease among cattle in this locality is pleuro-pneumonia. The disease started from Georgetown, D. C., two years ago, and has gradually spread down the Potomac for a distance of about 25 miles, extending back from the river not more than 2 miles.

B. A. Murrill, Campbell County, Virginia, writes, about the same date:

An unknown disease has prevailed this fall among cattle in the immediate vicinity of Lynchburg, but has not spread elsewhere. [This disease was pronounced pleuro-pneumonia by competent authority.]

R. L. Ragland, Halifax County, Virginia, writes that the cattle in that county are affected with a contagious distemper which is supposed to be pleuro-pneumonia.

C. Gingrich, Reistertown, Baltimore County, Maryland, says:

Lung fever (pleuro-pneumonia) has prevailed among cattle in the vicinity of Baltimore for the past twelve or fifteen years, and the losses from the same have been quite heavy.

A report from William S. Vansant, veterinary surgeon, contained in the report of the New Jersey State board of agriculture for 1876, shows that nineteen different herds of cattle suffered from this disease in Burlington County of that State during the year above named. It would seem that while the disease has been almost constantly present in New Jersey for many years past, no organized effort on the part of the State has been made for its suppression and extirpation.

With no means at my command for the suppression of the malady, in November last I caused an examination to be made of some of the afflicted cattle in the vicinity of Alexandria, Va. The investigation was conducted by Dr. Alban S. Payne, of Fauquier County, Virginia, who, as will be seen by his report below, pronounced the disease a contagious type of pleuro-pneumonia. The results of his investigation are thus given in the following brief extract from his report:

I visited Mr. Roberts's mill, one mile south of the city of Alexandria, Va., with as little delay, under existing circumstances, as possible. I found Mr. Roberts, in connection with his other business operations, carrying on a dairy. On his farm were sixty-two milch cows, and of these forty have had pleuro-pneumonia. Twenty-two have not as yet taken the disease. I also found almost in the heart of Alexandria City two cows sick with the disease. One of these cows belonged to Mr. Townsend Baggott and the other to Colonel Suttle. I also examined about the suburbs of Washington City some sick cows. All the cases I saw were, without doubt, cases of pleuro-pneumonia of the non-malignant variety.

Knowing the insidious and destructive character of this disease, and that it was liable to assume a contagious form and cause the destruction of millions of dollars' worth of property, and interrupt and perhaps destroy one of our greatest commercial interests and sources of income, I called the attention of Congress to the existence of this fatal malady in my preliminary report, bearing date of November last, and asked the immediate intervention of the government by the enactment of measures for its suppression and extirpation. The following is a brief extract from this report:

One of the most dreaded contagious diseases known among cattle is that of pleuro-pneumonia, or lung fever. It was brought to this country as early as the year 1843, and has since prevailed to a greater or less extent in several of the Eastern and a few of the Southern States. It made its appearance about a century ago in Central Europe, and has since spread to most European countries. With the exception of rinderpest, it is the most dreaded and destructive disease known among cattle. Unlike Texas cattle fever, which is controlled in our northern latitudes by the appearance of frost, this disease " knows no limitation by winter or summer, cold or heat, rain or drought, high or low latitude." It is the most insidious of all plagues, for the poison may be retained in the system for a period of one or two months, and even for a longer period, in a latent form, and the infected animal in the mean time may be transported from one end of the continent to the other in apparent good health, yet all the while carrying and scattering the seeds of this dreaded pestilence.

Since the appearance of this affection on our shores it has prevailed at different times in the States of Massachusetts, Connecticut, New York, New Jersey, Maryland, Delaware, Virginia, and in the District of Columbia. It has recently shown itself at two points in Virginia (Alexandria and Lynchburg), where it was recently prevailing in a virulent form.

At present the disease seems to be circumscribed by narrow limits, and could be extirpated with but little cost in comparison with the sum that would be required should the plague be communicated to the countless herds west of the Alleghany Mountains.

This disease is of such a destructive nature as to have called forth for its immediate extirpation the assistance of every European government in which it has appeared, many of them having found it necessary to expend millions of dollars in its suppression. The interests involved in this case are of so vast a character and of such overshadowing importance, both to the farming and commercial interests of the country, as to require the active intervention of the Federal Government for their protection, and for this reason the considerate attention of Congress is respectfully asked to this important matter.

Prof. F. S. Billings, V. S., temporarily residing in Germany, writes under recent date as follows:

BERLIN, *January* 16, 1879,
14 *Louisen Street.*

MY DEAR SIR: I intended in my last to have mentioned some ideas for your consideration upon the so-called contagious pleuro-pneumonia of cattle in the United States. I have given the subject a long-continued consideration, and it seems to me the views which now appear conformable to our case will find their approval with you. The disease is one which is rather a new thing to us, and while we find cases coming to pass in many sections, still we cannot say it has acquired any devastating extension. I truly believe that by using what means we have at command, and by fixing two or at the most three points by which cattle can be imported from Canada, and by furthermore exacting that such cattle be accompanied by attested health certificates of competent men, and furthermore that all such cattle, except when destined for immediate slaughter, be compelled to undergo twenty days of quarantine at point of entry when unaccompanied by such certificates, like rules applied to sea-ports—if we can make and enforce such regulations, then in one year at the most we can stamp the disease out of the United States and keep it out. For us the inoculation should be absolutely forbidden and severely punished. It is only of value in localities where the disease has become almost domesticated, and where of the two evils the lesser must be chosen, and that is, as is being attempted in Saxony, to inoculate every animal, and produce as soon as possible the artificial disease; all newly-introduced animals to be by law at once inoculated.

This renders the losses less severe to such a community, probably not over 25 to 30 per cent., if as much; statistics as yet are unreliable. But it is self-evident this is also the way by which the disease is rendered a constancy—it becomes domiciled, a thing we do not desire. Hence I recommend to your consideration the absolute killing of every infected and exposed animal, or, perhaps, utter quarantining—isolation of the latter under rigid inspection. The slaughtered animals to be paid for at full market price, real, not fancy, by the respective State governments, or, better, by the general government; for, if we are to have a general law, then the general government must take care of it. I earnestly recommend your bringing this to the attention of Congress, and you yourself must see the recommendation is logical and true to the country's interest. The first cost might be a little startling, but the final results equally fortunate. The rinderpest was at last reports limited and decreasing.

Your obedient servant,

F. S. BILLINGS.

To Hon. WM. G. LE DUC,
Commissioner of Agriculture, Washington, D. C.

Professor Gadsden, of Philadelphia, who recently made an examination of infected and diseased cattle on Long Island, writes as follows:

134 NORTH TENTH STREET,
Philadelphia, January 29, 1879.

SIR: I consider it my duty to report to you that the contagious disease known as "pleuro-pneumonia" exists to a frightful extent among the cows near Brooklyn, Long Island. On the return of Professor McEachran, the cattle-inspector of Canada, from Washington, he asked me to accompany him to New York State, and find out for ourselves if the report was true that a contagious disease existed. We found it too true, as at a distillery at Williamsburg we found a large byre, or cow-house, containing about eight hundred cows, with very many of them in the last stages of "contagious pleuro-pneumonia." Others had this disease in a milder form. The place was very dirty, the cows very much crowded, ceiling low, and everything favorable for the rapid spread of this disease.

The cows belong to a number of milkmen, who keep them there very cheap on hot swill (from the distillery) and hay, which increases the milk very much. This place is a regular pest-house for the disease. We were informed, on good authority, that just before the cows die they are killed and dressed, then sent into the New York market as beef, where we are told that they bring a good price because they are tender and not too fat. Others are sold, when the milk dries up, to farmers on Long Island.

This disease is very prevalent within a few miles of Brooklyn, and has been for some time. Cannot you, sir, try and stamp it out? as I am afraid if it spreads from there the English Government will not receive any cattle from our ports, as they have a law ready to put in force as soon as they are satisfied this or any contagious disease exists in cattle. I have made inquiry from several veterinary surgeons in this State; they all answer there are no contagious diseases in cattle in their district. I have no reason to believe there is any in Pennsylvania or in the Western States; so I do hope this disease on Long Island will not interfere with the sending of live cattle from Philadelphia to England, as I know they are making great preparations for this spring's trade.

Respectfully, &c.,

J. W. GADSDEN, V. S.

Hon. WM. G. LE DUC,
    Commissioner of Agriculture.

On the morning of the 30th of January, 1879, the following telegrams appeared in the metropolitan journals:

TORONTO, ONTARIO, *January* 30.

Intelligence of the slaughtering of cattle lately shipped to Liverpool on a steamship creates an anxious feeling among dealers here. On or about the 14th instant the steamship Ontario sailed from Portland for England with a cargo of cattle, the shippers being Messrs. T. Crawford & Co., of this city. The cattle numbered 265 head, and were, according to Mr. Crawford's statement, in sound condition, having been examined by competent men at both Montreal and Portland. The Ontario reached Liverpool on Sunday last, and on the following day Messrs. Crawford & Co. received a cable dispatch from their agent there that the cattle had been detained for inspection by order of the British Government. This inspection was evidently attended with unsatisfactory results, for on Tuesday the agent cabled that the cattle had been condemned on account of disease and were to be slaughtered. The disease was said to be pleuro-pneumonia. The Toronto Exportation Company and Messrs. Crawford & Co., the two firms that do the largest shipping business in their line in the city, were instructed by their agents to ship no more. The first named have a cargo of 170 head on the steamship State of Alabama, which it is anticipated will arrive at Liverpool on Friday next. What will become of these remains to be seen. The general feeling is that it is not at all likely that a trade which was rapidly becoming a necessity for England will be allowed to suffer interruption for any great length of time without a good cause for the embargo being adduced.

OTTAWA, ONTARIO, *January* 30.

Information having been received that the British Government has totally prohibited the importation of cattle from the United States, the cabinet met last evening to consider the situation. The result of the meeting was the adoption of a resolution that steps would be taken to prevent any injury being done to Canada.

MONTREAL, QUEBEC, *January* 30.

Considerable anxiety exists in regard to the order from the imperial government prohibiting the importation of Canadian cattle into England. It is said if the order is continued cattle will be slaughtered here and the meat will be taken across in refrigerators.

The following letter from the president of the American Veterinary College will explain itself:

AMERICAN VETERINARY COLLEGE,
    New York, February 1, 1879.

SIR: In returning from Washington, where he had the honor of seeing you, Professor McEachran, of Canada, asked me if pleuro-pneumonia was to be found in New York State. I took him to Long Island, and there had the opportunity to show him a barn where a large number of cows (some 600) are kept, and where we found ourselves in the difficult task, not to detect diseased animals, but to discover healthy cows. *Post mortems* confirmed our diagnosis, so that no doubt can be had of its correctness.

The milk and the carcasses of these diseased subjects find their way to our market in New York City. Our boards of health have no veterinarian to detect the disease and enforce the laws! Our market meat-inspectors are deficient in detecting diseased from healthy meat! Our cattle are exposed to the spreading of that fearful disease! Our exportation is now impeded to such extent that to-day I am told animals exported to France even must have a clean bill of health, and England is threatening closing her ports to our stock!

May I respectfully be allowed to call your attention to this state of affairs, and to place myself at your orders for whatever professional assistance I may be able to give your department in overcoming this great danger to our European cattle trade and to our own live stock.

I am, sir, your obedient servant,

A. LIAUTARD.

Hon. W. G. Le Duc,
    Commissioner of Agriculture.

On the 4th instant I received the following telegram from Mr. J. B. Sherman, superintendent of the Chicago Union Stock-Yards:

UNION STOCK-YARDS, Chicago, Ill., February 4, 1879.

The COMMISSIONER OF AGRICULTURE:

The most important blow struck at the interest of this city, State, and Northwest is the report in circulation in reference to the prevalence of cattle disease in the West, and these reports are absolutely false. I have sent a telegram to the Secretary of State, on whom I wish you would at once call.

This business of the export of live cattle to England has developed immense proportions in the last year, and we must not, cannot, remain quiet and see it destroyed. It is worth millions to the country, and affects directly every farmer in the Northwest, while the whole country feels the effect of this large increase in its exports. The action of the British and Canadian Governments is based on a misconception of the facts, and we need such final investigation as will put the matter at rest.

J. B. SHERMAN,
            Superintendent.

To which the annexed reply was at once forwarded:

DEPARTMENT OF AGRICULTURE,
        Washington, D. C., February 4, 1879.

J. B. SHERMAN,
    Superintendent Union Stock-Yards, Chicago, Ill.:

The disease to which your telegram refers appeared in this country as early as 1843, and there is no more reason for the present action of the British Government in this matter than has existed for years past. Pleuro-pneumonia has never troubled the cattle-breeders of the West, from whence alone cattle for exportation are derived, but the existence of the disease on our eastern coast at all is a constant threat to the cattle-raising country beyond the Alleghany Mountains, for the extermination of which I have asked authority of Congress. I hope and expect that action will be taken that will speedily remove all excuse for the objectionable orders of the British Government.

WM. G. Le DUC,
        Commissioner of Agriculture.

On the recommendation of gentlemen largely interested in the live-stock trade, I at once made the following appointment of an examiner for the port of New York:

DEPARTMENT OF AGRICULTURE,
        Washington, D. C., February 5, 1879.

SIR: You are hereby appointed an examiner, and directed to make as thorough inquiry and examination as the owners and shippers of stock will permit into the condition of the live stock sent, or about to be sent, from your port, and certify daily to this department the health of each particular shipment, so far as possible, examining particularly as to pleuro-pneumonia in cattle, and noting the presence or absence of this disease in each case. You are authorized to give a copy of your certificate for the department to the shippers, if desired.

WM. G. Le DUC,
        Commissioner of Agriculture.

Dr. JOHN J. CRAVEN,
    Jersey City, N. J.

I also forwarded a like appointment by telegraph to H. J. Detmers, V. S., Chicago, Ill., and received prompt replies from both accepting the positions tendered.

These examiners were also directed to furnish a certificate of health to such shippers of live stock as might desire it, a copy of which is herewith appended:

NSPECTION OF CATTLE FROM THE PORT OF ——, AUTHORIZED BY THE UNITED STATES GOVERNMENT, AND UNDER THE IMMEDIATE DIRECTION OF THE COMMISSIONER OF AGRICULTURE.

This is to certify that I have this day inspected —— beef cattle, owned by Messrs. —— ——, to be shipped by them upon —— —— sailing February —, for the port of Liverpool, England, and found the animals sound.

Dated February —, 1879.

(Signed)

——— ———,
Inspector.

These letters were promptly followed by the following, addressed to the Secretary of the Treasury, informing him of the action taken by this department:

DEPARTMENT OF AGRICULTURE,
Washington, February 5, 1879.

SIR: I have the honor to inclose for your information a copy of a letter this day addressed to Dr. John J. Craven, of Jersey City, N. J. I have also telegraphed to Dr. Detmers, of Chicago, substantially the same instructions as are noted in Dr. Craven's letter.

So far as the limited funds at the command of the department will permit, the proposed examinations will be continued, with the view of furnishing shippers information relative to the health of the stock, and thus prevent the shipment of any that are diseased; and the certificate of the veterinary surgeon of this department making the examination will be in the nature of a "bill of health," and should go far towards allaying any apprehensions, real or fancied, which may be entertained by persons who receive the stock.

This department is ready to second any efforts made by the Treasury Department to quiet the unnecessary excitement now apparent in Europe and in our own country on this subject.

Respectfully, your obedient servant,

WM. G. LE DUC,
Commissioner.

Hon. JOHN SHERMAN,
Secretary of the Treasury.

To which the following reply has been received:

TREASURY DEPARTMENT,
OFFICE OF THE SECRETARY,
Washington, D. C., February 7, 1879.

SIR: I am in receipt of your letter of the 5th instant, inclosing a copy of one addressed by you to Dr. John J. Craven, of Jersey City, N. J., authorizing him to make inquiries into the condition of live stock about to be sent from that port to foreign countries, and to certify daily to your department the health of each particular shipment as far as possible.

I inclose herewith for your information twelve copies of a circular issued by this department, under date of the 1st instant, requiring as a condition precedent to the shipment of live cattle abroad an examination thereof by the customs-officers with reference to their freedom from disease, and the issuance of a certificate by the collector that they are free from such disease, if the facts shall be found to warrant it.

Doubtless Dr. Craven, and any other person appointed by your department for the purpose named, could give valuable aid to the collectors of the ports from which such shipments are made, and this department would be pleased if you would instruct the experts selected by you to afford aid to the customs-officers in this respect as far as possible.

You will see that the circular requires that the officers of the customs shall also furnish this department from time to time such information upon the subject as they may be able to procure, and I would be pleased if you will also forward such information as you receive it.

This department has furnished the State Department with copies of the circulars before mentioned, and the Secretary of State has doubtless furnished them to the proper representative of the British Government.

This department perceives the importance of protecting its export trade in live animals as far as possible, and will do all in its power to attain the desired object.

Very respectfully,

JOHN SHERMAN,
Secretary of the Treasury.

Hon. WM. G. LE DUC,
Commissioner of Agriculture.

15 SW

The following is a copy of the circular inclosed by the Secretary of the Treasury:

[Circular.]

INFORMATION IN REGARD TO CATTLE DISEASE.

TREASURY DEPARTMENT,
Washington, D. C., February 1, 1879.

To collectors of customs and others:

By department's circular of December 13, 1878, it was directed that live cattle shipped from the various ports of the United States might be examined with reference to the question whether they were free from contagious diseases, and that, if found to be free from such diseases, a certificate to that effect should be given.

By that circular such inspection was not made compulsory, but the certificate was to be issued only upon the application of parties interested.

As the export trade in live cattle from the United States is of vital importance to large interests, every precaution should be taken to guard against the shipment of diseased animals abroad, and such a guarantee given as will satisfy foreign countries, especially Great Britain, that no risk will ensue from such shipments of communicating contagious or infectious diseases to the animals in foreign countries by shipments from the United States.

Collectors of customs are, therefore, instructed that in no case will live animals be permitted to be shipped from their respective ports until after an inspection of the animals with reference to their freedom from disease, and the issuance of a certificate showing that they are free from the class of diseases mentioned.

Notice of rejected cattle should be promptly given to this department.

In order that this department may be fully informed in regard to such diseases in any part of the United States, collectors of customs are requested to promptly forward to this department any information which they may be able to obtain of the presence of contagious or infectious diseases prevailing among live animals in their vicinity.

It is probable that if the disease prevails to any considerable extent it will be noticed in the local press, and collectors are requested to send copies of any such notices to this department for its information.

JOHN SHERMAN,
Secretary.

The following letter has been received from Prof. James Law, who, it will be seen, has been ordered to the port of New York by the governor of that State:

ASTOR HOUSE, *New York, February* 8, 1879.

DEAR SIR: I came down here last night in accordance with instructions from the governor of New York to ascertain and report as to the existence of the lung fever in cattle. From what I have seen to-day I have no doubt of its existence in Kings and Queens Counties, but I hope very soon to be able to report on the *post-mortem* lesions as well as the *ante-mortem* symptoms.

I hear that the malady exists in Watertown, Conn., perhaps at Ratonah, Westchester County, New York, and around Newark, N. J. The two first places I expect to visit in the interest of New York, and I shall find out what I can about the vicinity of the shipping yards for the stock exported to Great Britain. Would it be well for me to visit Newark also before returning?

I strongly commend the position you have taken in this matter, as the only just and tenable one. If we should ever suffer from a temporary suspension of the foreign trade in cattle, it will be well expended if it should lead to a thorough extinction of the lung plague in the United States.

Yours, very truly,

JAMES LAW.

Hon. WM. G. LE DUC,
Commissioner of Agriculture.

The following late telegrams, showing the action of the British Government, are appended:

THE AMERICAN CATTLE TRADE—NO FURTHER INTERFERENCE EXPECTED.

LONDON, *February* 8.

A committee of the Cattle Trade Association at Liverpool, in order to avoid interruption to the trade, have offered to erect the necessary lairage and abattoirs to comply with the requirements of the Privy Council. It is believed, however, that, in consequence of the growing importance of the trade to Liverpool, either the authorities or the corporation or the dock board will undertake the work. All arrivals of cattle from America since the steamer Ontario's cargo have been found entirely free from disease. The severity of the weather, therefore, it is believed caused the outbreak in that instance. The British Government is, under the circumstances, not inclined to interfere with the importation of cattle from America, provided there is adequate inspection before shipment and provision of the required lairage at Liverpool to put them in position to meet such cases as the Ontario's. It is not believed that slaughter on the quays will be enforced where no disease exists. Persons in the trade say that under these conditions American shippers need not fear any interference with the business.

LONDON, *February 9.*

In regard to the importation of cattle from America, no action of the Privy Council has been made known since the notice read in the Liverpool town council on February 5, that cattle cannot be landed at the Liverpool docks after March 1, unless provision is made for slaughter on the quay.

THE CATTLE EXPORT TRADE—EFFECT OF THE BRITISH ORDER IN COUNCIL.

LIVERPOOL, *February 11.*

The order of the Privy Council adopted yesterday revoking after March 3, 1879, article 13 of the foreign animals order so far as it relates to the United States was a great surprise to the trade here. All cattle from the United States after March 3 will have to be slaughtered in abattoirs now being prepared on the dock estates of Birkenhead and Liverpool within ten days after landing.

I also forward you articles on the subject of pleuro-pneumonia, clipped from the National Live Stock Journal of March, 1878, and November, 1878, from the pen of Dr. James Law. They were inclosed to me and my attention directed to them by Mr. J. H. Sanders, the able editor of that Journal.

[From the National Live Stock Journal of March, 1878.]

THE GREATEST DANGER TO OUR STOCK—THE LUNG FEVER—CONTAGIOUS PLEURO-PNEUMONIA.

The Journal has frequently called attention to the great dangers that beset our live stock from imported plagues of foreign origin. During the past year the sudden invasion of Western Europe and England by the rinderpest roused the agricultural community from their dream of safety, and called forth from the Treasury an order remarkable alike for its promptitude and good intentions, and for the fatal blunders which rendered it worse than a dead letter. Once more there seems a prospect of a renewal of these apprehensions, the Russo-Turkish war having led to an extension of this cattle plague into Hungary, from which the Atlantic coast and Great Britain may be any day infected, owing to the activity of the stock trade. Should this unfortunately take place, it will find us no better prepared than we were a year ago, and our Treasury order, now in force, will freely invite the disease to enter, provided it makes its advent respectably—in the systems of *blooded stock,* and not in poor cross-bred animals, which it would be ruinous to import, even if sound. A similar welcome is extended, by implication, to all those ruminants which are devoted more particularly to luxury, and have not been degraded to such vulgar utilitarian objects as the production of meat or wool. Yet all ruminants are subject to rinderpest, and this malady was carried to France, in 1866, by two gazelles, as other plagues have often been carried to new countries by the privileged *blooded stock.*

But we started out to notice a danger which is no longer separated from us by the broad barrier of the Atlantic, and whose malign presence is not to be dismissed by any one of ten thousand contingencies, as is the case with the possible advent of the rinderpest. This danger stands in our midst, and is steadily gaining in force as it encroaches farther and further, showing how certain it is, if unchecked, to lay the whole country under contribution, and inflict most disastrous and permanent losses. The lung fever of cattle, imported into Brooklyn, L. I., for the first time, in 1843, in a Dutch cow, has never since been at any time entirely absent from our soil. From this center it has slowly and irregularly extended over a portion of New York, New Jersey, Pennsylvania, Maryland, Delaware, and Virginia, besides having repeatedly invaded Connecticut. The slowness of its extension has begotten a false sense of security, and no real apprehensions of serious consequences remain from an animal poison which has been for over a third of a century hidden away in the near vicinity of the Atlantic coast.

To disturb this comfortable and restful condition of the public mind is an unpleasan

task, which nothing but the imperative sense of duty would compel us to undertake
But this disease has a history, which we can only ignore at our peril; and as its records
can now be drawn from all quarters of the globe, we can have before us an unequivocal testimony as to what will inevitably happen under given conditions of climate,
surroundings, and treatment.

England imported the lung fever of cattle in 1842, just one year before we did, was
soon very generally infected, and has continued so to the present time. Up to 1869, it
is estimated that England had lost, almost exclusively from this disease, 5,549,780 head
of cattle, worth £83,616,854 (say $400,000,000). For the succeeding nine years, up to
1878, the losses have been, in the main, as extensive, so that we may set them down as
now reaching at least $500,000,000 in deaths alone, without counting all the contin-
gent expenses, of deteriorated health, loss of markets, progeny, crops, manure, &c.,
disinfection, quarantine, &c. With us no attempts have been made to estimate the
losses, but they cannot exceed an inconsiderable fraction of those above named; and
thus we have slept on in a pleasant dream of immunity.

It is even alleged that the disease has, in a great measure, been shorn of its virulent
power, by being transplanted to the shores of the New World, and that we may comfort ourselves with this, and continue to ignore its presence. If, on the other hand, it
can be shown that the difference is in no material respect affected by climate, but
altogether determined by the surroundings, it will be well for us to attend to the facts
of the case, and face the real danger. The lung fever, which had really entered England, by a special importation, some time before the free trade act of 1842, was, by
virtue of this act, thrown upon her in constantly accumulating accessions. The ports
at which the continental cattle were landed, and the markets in which they were sold
—London (Smithfield Market), Southampton, Dover, Harwich, Hull, Newcastle, Edin-
burgh, &c.—insured the mingling of the imported stock, week by week, with the
native store cattle. Then, if they failed to find a profitable sale, they were sent by
cars to other and inland markets, where they were again and again brought into contact with numerous herds of store cattle, by which the germs of the disease were taken
in, and carried all over the country.

With us, on the other hand, the disease was long confined to the dairies of Brooklyn
and New York, where the cows were kept until they died, or were fattened for the
butcher. A few doubtless found their way to the country, and by these the disease
was carried to different farms, which were thus constituted centers of contagion from
which the adjacent country became infected. But any such movement from the city
dairies was necessarily of the most restricted kind, and it never took place to any great
distance. It would have been folly to move a common milch cow, worth $40 to $70,
to the West, where she could be bought for one-half or one-third of that sum. The
same deterrent condition existed in the case of the farms on which the diseased city
cows had been brought. Sales were no doubt occasionally made from infected herds,
to secure the apparent value of an animal which the owner had good reason to believe
to be doomed, and as such animals would, for obvious reasons, be sent as far from
home as possible, this became a principal means of the formation of more distant centers of contagion and the wider diffusion of the malady. But with us the disease has
hitherto had to fight against the heaviest obstacles—the current of cattle traffic having been almost without exception from the cheaply-raised herds of the West to the
profitable markets of the East. The exceptions have only been in the case of thoroughbred stock, and hitherto our Western stock has escaped contamination by this means.

The wonder is not so much that the plague has failed to reach the West, but that in
the face of such tremendous obstacles it has succeeded in invading all of the six or seven
States that are now infected. In Great Britain, where some would have us believe
that the disease is more virulent, we can point to a more satisfactory record. There
the great body of the country has been infected for thirty-five years, but the greater
part of the highlands, exclusively devoted to the raising of cattle and sheep, has enjoyed the most perfect immunity. Here, under nearly all possible predisposing causes
of lung disease—altitude, exposure, cold, chilling rains, and fogs, the piercing blasts
of the Atlantic and German Oceans—this contagious lung disease has never penetrated,
though severely ravaging the lowlands immediately adjacent. The explanation is,
that these hills support none but the native black cattle, and other breeds are never
introduced. In spite of the alleged virulence of the disease in England, it has proved
powerless to enter this magic circle from which all but the native stock is excluded.
The same holds true concerning some parts of Normandy, Brittany, the Channel
Islands, Spain, Portugal, Norway, Sweden, &c.

The fact that the disease has maintained a foothold among us for thirty-four years,
and in spite of all obstacles has made a slow but constant extension, is sufficient
ground for the gravest apprehensions. A disease-poison which shows such an obstinate vitality and such persistent aggressiveness cannot be allowed to exist among us
without the certainty of future losses which will eclipse those of Great Britain by as
much as our herds of cattle exceed those of that nation. A recent outbreak in Clinton, N. J., caused by a cow brought from Ohio, suggests the possibility of the disease

having already reached the latter State; an occurrence which was inevitable sooner or later, but the actual existence of which must enormously increase our dangers. With every such step westward there is the introduction of more diseased and infected cattle into the natural current of the traffic, and the earlier probability of the general infection of all parts to the east of such ultimate centers of disease. There is, further, the infection of more cattle cars, which, carried West, may be the means of securing a rapid extension of the plague to our most distant States and Territories.

RELATIVE DANGERS OF THE POISONS OF LUNG FEVER AND OTHER PLAGUES.

The persistent vitality of the *lung-fever poison*, in comparison with that of any other animal plagues, is noteworthy. It has held a tenacious grasp on the United States for over a third of a century, though forbidden by circumstances to make a wide extension. *Aphthous fever* (foot and mouth disease), on the other hand, though twice imported into Canada within the last ten years, and on one occasion widely spread in New York and New England, was on each occasion easily and early extinguished, and with little or no effort on the part of the States. It might indeed almost be said to have died out of itself. Even the dreaded *rinderpest* has its poison early destroyed by free exposure to the air, in thin layers, at the ordinary summer temperature. Numerous experiments on hides hung up and freely exposed in warm weather have shown that the infecting power is lost as soon as they are quite dried. But the poison of lung fever maintains its virulence for months in the dry state in buildings, and we have known parks, with sheds, that proved regularly infecting year after year to all cattle turned into them. In other cases we have known the virus carried for miles on the clothes of attendants, and thus introduced into new herds.

A far greater danger lies in the lengthened period during which the poison of lung fever remains dormant in the system. This averages about three weeks or a month, but may extend, in exceptional cases, to not less than two months. An ox or a cow which has been exposed to the contagion may, therefore, be carried from one extremity of the continent to the other, may be exposed in a succession of markets, and may change hands an indefinite number of times, and be all the while in the best apparent health, though infallibly approaching the manifestation of the disease, and for the latter portion of the time spreading the germs of the malady to others. There is here an opportunity for the unscrupulous to sell off exposed and infected animals without the purchaser having the least suspicion of foul play. There is also the strong probability of animals that have contracted the disease by accident, in cars or otherwise, in passing to a new home, mingling with the herd of the new owner and infecting them extensively before there is a suspicion that anything is amiss. This long period of incubation after the animal is infected, and the equally long period of latency of the malady in animals he has infected, one or two of which only will be attacked at intervals of a month, lull suspicion as to the presence of contagion, and it is too often only after great damage has been done that the truth dawns on the mind.

In aphthous fever and rinderpest, on the other hand, the disease shows itself in from one to four days after infection, and the surrounding animals are so rapidly attacked after the coming of the infected stranger, that there is no room for hesitancy as to the existence of contagion. Nor can the victims of these diseases be carried far from the point where they have been infected and disposed of as sound animals; so that in the very vigor and promptitude of their action we have an excellent basis for their restriction and control.

DANGER OF INFECTION IN OUR UNFENCED STOCK RANGES.

It is needful to note the above-named insidious progress and stealthy invasions of the lung fever, and to contrast them with the more prompt and open manifestations of the other animal plagues, in order to show the great peril to which we are subjected by the presence in our midst of a *pestilence* which literally *walketh in darkness.* Let us now consider the prospective infection of our great stock ranges. That this is inevitable, though slow, at the present rate of progress of the plague, has been sufficiently shown. That it might occur any day by an animal infected in an Eastern farm or stock-yard, or in a railroad car in which it was sent for the improvement of the Western herds, must be abundantly evident to every one who has read this article. If we now add the fact that more than one *thoroughbred Ayrshire* and *Jersey* herd has been infected with this disease during the past year, we are at once confronted with a strong probability of an early Western infection. Let us remember that thoroughbreds alone are carried West for improvement of native herds, and that a bull of the Ayrshire, Jersey, Holstein, or short-horn breed, taken from a herd now or recently infected, may be carried to any of our Western Territories and mingle for a month with the native herds before his own infection is so much as suspected, and we can conceive how imminent is the danger when the infection has reached our *Eastern thoroughbred cattle.*

To illustrate the result of the infection of our unfenced stock ranges, I must quote another page from the history of this disease in other countries. The instance of Australia is the most recent as well as the most striking. The lung fever was introduced into Melbourne in 1858. by a short-horn English cow, which died soon after landing. Having been confined to an inclosed place, there is every reason to believe that with her the disease would have ended, had not a teamster turned his yoke of oxen into the infected park under cover of the night. These oxen working on the streets infected others, the disease soon spread to the open country, and the mortality increased at an alarming rate. Vigorous measures for its suppression were adopted, thousands of infected and diseased cattle were slaughtered, but all proved of no avail. Not only were the free, roaming herds infected, but so many places were contaminated that it was soon perceived that help from this source was not to be expected. Destroy a whole infected herd, and you still left the infection in the station from which, in its unfenced state, other herds could not be excluded, and where they were certain to take in the germs of the malady. After enormous losses had been sustained by the combined operations of the pest and the pole-ax, it was concluded that the remedy was worse than the disease, and the colonists reluctantly fell back on the expedient of inoculation. This is based on the fact that the disease is rarely contracted a second time by the same animal, and it can be practiced on all calves with losses at the rate of from two to five per cent. only, so that the mortality is insignificant as compared with the thirty to fifty per cent. which perish where the affection is contracted in the ordinary way. The great objection to inoculation is, that it can only be practiced at the expense of a universal diffusion of the poison, and of its maintenance in a state of constant activity and growth. With such a universal diffusion of the virus, the stock owners are virtually debarred from introducing any new stock for improving the native breeds, or infusing new vigor or stamina, inasmuch as such new arrivals would almost certainly fall early victims to the plague. Australia, therefore, now suffers from the permanent incubus of the lung plague, and can only import high-class cattle at great risk.

This is an occurrence of yesterday, but it is only a repetition of the immemorial experience of the steppes of Russia. There we find the same conditions of great herds roaming free over immense uninclosed tracts, and all the facilities for an easy and wide diffusion of animal poisons. There, accordingly, we find the home, in all ages. of the animal plagues of the Old World. To these endless steppes Europe and European colonists owe their frequent invasions of *lung fever, rinderpest, aphthous fever*, and *sheep-pox*. To these are to be charged the losses, to be estimated only by many thousands of millions, which have repeatedly fallen on the other civilized countries of the world. From these steppes the disease has spread over the continent on the occasion of every great European war, dating from the expulsion of the Goths from Hungary by Attila and his Huns. in A. D. 376, down to the present Turkish war, which has secured the extension of the rinderpest to Hungary at least. On these steppes, too, the Russian veterinarians believe the rinderpest, at least. to be an imported disease derived from Eastern and Central Asia, yet all their efforts to crush out this or the lung fever, though receiving the freest support from the Russian Government, have failed. The same conditions exist, to a large extent, at the Cape of Good Hope ; and there, too, the lung fever, imported in 1854, has acquired a permanent residence.

PREVENTIVE MEASURES DEMANDED.

Such is the history. Now comes the question pregnant with weal or woe to our future stock, agricultural, and national interests: Shall we learn from the disastrous experience of others and extirpate the lung plague from the United States while it is still possible, or shall we sit quietly by with folded hands and await the inevitable, early or late, infection of our open Western stock ranges, and then repeat, for the benefit of other nations, the already twice-told tale of a desperate and extravagant but fruitless attempt to suppress a plague which we have criminally allowed to pass beyond our control? With or without a prodigal but vain effort to crush out the poison, the results may be thus summed up: The infection of stock-yards, loading-banks, cars, and markets, and a general diffusion of the plague over the Eastern States. This would imply a national loss, by cattle disease, like that of England, but much more extensive in ratio with our great numbers of stock. Thus England, with her 6,000,000 head of cattle, has lost in deaths alone from lung fever in the course of forty years over $500,000,000. We, therefore, with our 28,000,000, should lose not less than $2,000,000,000 in the same length of time, allowing still a wide margin for the lower average value per head in America. And this terrible drain is for deaths alone, without counting all the expenses of deteriorated health in the survivors, of produce lost, of loss of progeny, of loss of fodder no longer safe to feed to cattle, of diminished harvests for lack of cultivation and manure, of quarantine and separate attendants whenever new stock is brought on a farm, of cleansing and disinfection of sheds and buildings, &c., which become absolutely essential in the circumstances.

We do not include the expense of supervising the trade, examining and quarantining the stock at the frontier of every State, and of the disinfection of cars, loading-banks, stock-yards, and markets. If such were resorted to, after an extensive infection of our Western herds by lung fever, the cattle trade would be virtually stopped. Thus a safe quarantine for store cattle could not be less than three weeks, and a registration and supervision for five weeks more on the farms to which they are taken, would be absolutely essential. Thus the quarantine yards and sheds would be continual centers of infection, and would require to be very extensive, thoroughly isolated from each other, and constantly and perfectly disinfected, the air as well as the solids, to prevent the infection of newly-arrived stock. Such an incubus upon the trade would amount to a virtual prohibition. In rinderpest, sheep-pox, and aphthous fever, quarantine is a comparatively simple and available expedient, as the disease shows itself within a week; but, in lung fever, with the germs lying unsuspected in the system for one or two months, a protective quarantine is practically impossible wherever an active cattle trade is carried on. Hence in the countries of Central and Western Europe, through which the active traffic from the East is carried on, a complete control is usually maintained over rinderpest and sheep-pox, while the peoples have resigned themselves to the prevalence of lung fever as an unavoidable infliction. The same holds in Great Britain. Twice within eleven years has she crushed out invasions of rinderpest, and repeatedly has the same thing been accomplished for sheep-pox; but the lung fever is accepted as a necessary evil, between which and her large importations of continental cattle she must make a deliberate choice.

Happily, in these United States we are as yet under no such compulsion. The lung fever on American soil is still confined to the Eastern States and to inclosed farms, from which it is quite possible to eradicate it thoroughly. Of this possibility we have abundant evidence, alike in the Old World and the New. In several countries of Western Europe, through which there is no continuous cattle traffic between nations on opposite sides. this disease has been killed out and permanently excluded by an intelligent veterinary sanitary supervision. Sweden imported the disease in Ayrshire stock in 1847, but at once circumscribed the infected herds and places, slaughtered the diseased, disinfected all with which they had come in contact, and promptly extinguished the outbreak. Denmark, invaded the same year from a similar source, and on several subsequent occasions from Holland and England, as often quenched the poison by analogous measures. Oldenburg, Schleswig, and Norway, successively invaded by the importation of infected Ayrshires, in 1858, 1859, and 1860, respectively, enjoyed a similar happy riddance, through the application of the same system of suppression. Switzerland, long slandered as the native home of the lung plague, has at last awoke to the truth of the statement of the immortal Haller, made more than a century ago, that this disease only occurs "when an animal has been brought from an infected district"; and, by the judicious use of suppressive measures, has permanently rid the country of the pestilence, and demonstrated that their Alpine air is as clear and wholesome for beast as for man.

In America, Massachusetts and Connecticut have furnished examples equally striking. The former imported the disease in Dutch cattle in May, 1859. In April, 1860. when it had gained nearly a year's headway, an act was passed, and a commission appointed, with full power to extirpate it. After the slaughter of 932 cattle, it was believed that this had been achieved; but new centers of infection were discovered in the two succeeding years, and it was not until 1865 that the commonwealth was purged of the poison. Since that year the lung fever has been unknown in Massachusetts. Connecticut has had a similar experience. Her proximity to New York City and Long Island has brought upon her a series of invasions; but, profiting by the experience of her neighbor, she has, on each occasion, grappled successfully with the enemy, and driven him from her midst.

What has been done by the Scandinavian nations, by Oldenburg and Switzerland. by Massachusetts and Connecticut, can be done by all of our Eastern States. On this point the teaching of history is as unequivocal as on the certainty of the irreparable results if our open Western stock ranges were infected. The one indispensable prerequisite to success is the vigorous and simultaneous action of the various infected States, and its persistent maintenance until the last infected beast has disappeared and the last contaminated place or thing has been purified. It matters little whether controlled by State or national government, if vigor and uniformity of action can be secured; but, as such combined and unflagging work is necessary, it could be best controlled by an intelligent central authority. The United States Government is as much called upon to defend her possessions against an enemy like this—so implacable, so relentless, and so certain, if not repelled, to lay us under an incubus which will increase with the coming centuries, and dwarf the prosperity to which we are entitled—as against the less insidious one who attacks us openly with fire and sword. Let the national Congress consider this matter well. Let every stock-owner press it upon his Representative as a matter that cannot be safely ignored even for a single day. Let boards of agriculture, farmers' clubs and conventions, granges, and all citizens who value the future well-being of the nation, unite in a strong representation on the subject. If the present Congress should neglect it, let citizens make it a test question to every future candidate for their suffrages, and elect only such as are pledged to carry suppressive measures into effect. The danger threatens all classes alike, though the first sufferers will be the stock-owners; for every tax upon production necessarily enhances the value of the product; and, as agricultural progress must be seriously retarded, the tax will not fall upon meat alone, but upon every product of the farm. Nothing can excuse a continued neglect of this subject, the dangers surrounding which increase from day to day, and the final results of which, if once it reaches our Western and Southern States and Territories, can only be computed by the prospective increase of our population and our herds of cattle. For this is not like an evil preying on our currency, banking, trade, or manufactures, the full extent of which may be, in a great measure, seen from the beginning, and the repair of which may be at any time inaugurated by legislative enactment. The animal plague only increases its devastations as we increase the numbers of our herds, and threatens soon to acquire an extension to which no legislation can oppose a check, and a prevalence in the face of which the most desperate efforts of the nation will prove of no avail. Thus our cattle are increasing at the rate of 13,500,000 every ten years, so that by the end of this century they may be exactly doubled, with a prospective loss, if our Western and Southern ranges are infected, of $130,000,000 yearly in deaths alone.

The choice is now in our power. So far as we know, our stock-raising States and Territories are still unaffected. We can still successfully meet and expel the invader; next year it may be too late.

[From the National Live Stock Journal of November, 1878.]

OUR GOVERNMENT AND THE ENGLISH CONTAGIOUS DISEASES ACT.

By an Associated Press dispatch from Washington we learn that "The Secretary of State has been officially notified of the passage of an act by the British Parliament entitled 'The Contagious Diseases (Animal) Act, 1878,' under which, except in the case of countries specially exempted by the Privy Council, in whole or in part, from the operations of the act, all animals landed from abroad in any part of the United Kingdom will, after the 1st of January next, be slaughtered at the port of debarkation. The British Government has also notified Secretary Evarts that, in case the United States desire to be exempted from the operations of the act, the lords will require a statement of the laws which regulate the importation of animals into this country, and the method adopted to prevent the spreading of any contagious disease when it exists in any part of the United States. Secretary Evarts has sent a copy of the act of the British Parliament to the Secretary of the Treasury, in order that he may furnish the desired information preliminary to any action being taken to have the animals shipped from the United States into the United Kingdom exempted."

We think it will puzzle the Secretary of the Treasury to find any methods that have been adopted by our general government "to prevent the spreading of any contagious disease when it exists in any part of the United States"; and if he will take the trouble to investigate the matter pretty thoroughly, he will find that all the regulations that have from time to time been ordered by his department to prevent the introduction of contagious and infectious diseases into the United States from foreign countries are practically worthless. When this fact comes to be reported to the British Government, it is not unlikely that the exemption which the United States now enjoys from the operation of the act will be revoked, notwithstanding our present comparative freedom from any diseases likely to be transmitted by exportation to England. When this condition of things is brought about, and the business of exporting fat cattle, sheep, and swine from this country to England—which has, within the past few years, grown to such enormous proportions and exercised so powerful an influence upon prices in this country—comes to a sudden halt, we shall expect such a pressure to be brought to bear upon Congress as will compel the passage of some such act as that introduced into the House last May by Hon. J. S. Jones, of Ohio, to which reference was made in these columns in June last.

But is it wise in us to await unfavorable action on the part of the British Government before taking such steps as will preclude all probability of this country being included in the prohibition? Clearly, the interest is too large, and the effect of adverse action on the part of the Government of Great Britain upon our farming community would be too disastrous, to justify us in taking any chances in the matter. The regulations now provided by law against the importation of plagues and infectious diseases from abroad are confessedly worthless; and as for the stamping out of such diseases when they do make their appearance, we have absolutely no law that is general in its operation. A few of the States have attempted it on their own account, but most of them have no laws at all upon the subject, and none can be effectual without the sanction of our general government; for Congress alone has the power to regulate commerce with foreign nations and between the several States.

It is imperative that early and efficient action be taken by our Congress upon this matter, if we would not have our present lucrative trade in fat cattle and sheep with England seriously crippled. Members of Congress are now at home among the people, and such a pressure ought to be brought to bear upon them as will compel them to act upon this question as soon as they reassemble at Washington.

In addition to the foregoing, I inclose you copies of the laws passed by the legislatures of Massachusetts and New York for the suppression and extirpation of the disease during its prevalence in those States, and the rules adopted and enforced by the British Government for the extirpation of this and other contagious diseases among farm animals in its Indian possessions.*

All of which is respectfully submitted.

WM. G. LeDUC,
Commissioner of Agriculture.

Hon. A. S. PADDOCK,
Chairman Senate Committee on Agriculture, Washington, D. C.

Since the publication of the above letter (Senate Mis. Doc. No. 71, Forty-fifth Congress, third session), many additional facts in relation to the prevalence of this disease, and the measures taken to suppress, and, if possible, eradicate it in the various localities in which it has been found to exist, have come into the possession of the department. To the information contained in the following letter from the pen of Dr. James Law, which appeared in the New York Tribune of February 25, is due in part the active measures instituted by the authorities of that State for the suppression of this destructive malady:

A REVIEW OF THE DISEASE.

To the EDITOR OF THE TRIBUNE:

SIR: The excitement about the cattle disease has had its proverbial course of nine days, and there are already signs of reaction. From every side we begin to hear statements that the danger has been exaggerated, that the disease only exists in three or four herds, that it is seen only sporadically—not epidemically; that the English live-stock trade must be speedily re-established; and that, in short, the whole thing has been a gigantic mistake. Should this spirit prevail so as to prevent a uniform and concerted action by the different infected States to crush out this baneful exotic, it will rob the country of her best, and perhaps her only chance, of securing and maintaining the European live-stock market.

If the object of this *laisser faire* argument is to soothe the minds of our European cousins, and persuade them that this disease is less dangerous than that of Europe, they may as well save their labor. Europe has learned by centuries of sad experience the true nature of the contagious pleuro-pneumonia of cattle. Europeans now realize that wherever there is one animal suffering from this disease, there is a standing menace to the whole cattle of the country. They know that where they allow the disease to exist at all it decimates their herds yearly. They know that wherever they have boldly grappled with the enemy, crushed out every remnant of the malady and its virus, and jealously guarded their frontiers against its further importation, they have permanently cleared their folds of a disastrous pestilence. They see that wherever the disease has appeared in Western Europe, or in the western or southern hemispheres, it has only been where a diseased animal or its virulent products have carried the seeds into such a land. They know that so long as they allow the free importation of cattle from an infected country, all their efforts to crush it out of their home stock will be absolutely futile.

Turning to England, which has been the main agent in drawing public attention to the matter, she was absolutely ignorant of this disease until forty years ago, and in Youatt's and other veterinary works published prior to this date we find the most unsatisfactory accounts of this and other plagues known only on the continent. But from 1839, when it was first, in the present century, brought to the British Isles, and above all since 1842, which brought the free-trade act and the free importation of continental stock, Great Britain has suffered more from this than from all other animal plagues put together. It was estimated that in the first quarter of a century after its introduction this plague cost England $450,000,000 in deaths alone. The additional losses from deterioration and lack of live stock, and from the infection of forage, &c., which

---

For these acts and the rules alluded to in this paragraph, see appendix.

could no longer be put to their most profitable uses, have never been computed, but must enormously swell the sum total.

England had a hard lesson to learn, and she has been forty years in learning it, but we may depend upon it she has now learned it most thoroughly, and can no more forget it nor treat it with indifference while the present generation survives. Many years ago I was engaged, with other veterinarians who had acquainted themselves with the continental experience and literature, in enforcing on Great Britain the truth that to deal with this disease economically they must kill out the poison within their own borders, and exclude all stock from infected countries. Then, as now, we found many alleging that the disease was native to the soil, and occurred sporadically, not epidemically. Then, as now, we found men bearing the name of veterinarians, who had fallen so far behind the age as to support these allegations, being either criminally ignorant, or so morally oblique that they preferred the wrong because the popular side. So long as it can be shown that this disease never invades a new country, but as imported in the animal body or in some of its products, so long will all claims for its spontaneous generation, its sporadic appearance, or its development from certain local conditions, like swill-feeding, be put out of court.

THE DISEASE PROPAGATED BY CONTAGION.

The history of the malady in all time, and in all countries and hemispheres, east, west, north, and south, testifies with one voice that out of the steppes of Eastern Europe and Asia it is propagated by contagion alone. The unreasoning and misleading talk about "no epidemic" is, therefore, in the highest degree reprehensible. The affection is not an epidemic in the sense of being due to some generally diffused influence, which acts alike upon all the stock of the country, and strikes them down indiscriminately, and without regard to proximity or contact. Were this the case, our efforts to permanently extirpate it were vain. But its spread is always and only proportionate to the facilities for contagion and infection. And the present comparative immunity of America is only due to the fact that the plague reached here at that seaport toward which the greater part of the cattle traffic of the country tends, and from which few animals are removed inland. Given in the United States the same free movement of cattle from our infected points to all points inland as was till recently seen in Great Britain, and there would speedily follow the same general infection of the country. This is sufficiently illustrated in our past American experience. Massachusetts imported the disease from Europe, and although it was met by repressive measures as soon as recognized, it cost the commonwealth two years and $70,000 to extirpate it. It was imported into Brooklyn, and though it had to fight its way against the uniform current of cattle traffic eastward and northward, it has extended to New Jersey, Pennsylvania, Maryland, Virginia, and the District of Columbia.

My recent observations in this neighborhood are in perfect harmony with the above. The stables at Blissville, holding 800 to 900 cattle, fatting and milking, the property of different owners, who could purchase when they chose in the surrounding infected locality, could not fail to become a prominent hot-bed of the disease. Had such stables, with all their drawbacks of overcrowding, filth, and swill-feed, been thoroughly disinfected, filled with healthy Western stock and sedulously secluded from all neighboring cattle and visitors, they would not have become infected with contagious pleuro-pneumonia. Again, at Fifteenth street, Brooklyn, I found that all, or nearly all, the dairies in the vicinity had recently suffered from the disease, and that this infected center was within two blocks of Prospect Park, where the herd of Jerseys had been subjected to its ravages in August and September. At New Lots, Kings County, where I found seven infected herds in a very limited area, the testimony of the owners was to the effect that the disease only appeared and spread through their herds as they bought new cows from jobbers. At Roslyn, Queens County, I found two infected herds; the first contaminated by two cows bought from a New York jobber, and the second by two cows bought from the first. In New York City I found one infected herd, caused by a cow purchased from the same jobber whose cows took the disease to Roslyn. The Connecticut herd which I examined at Morrisania was infected by two cows purchased from a New York jobber, and the same man, according to his own sworn testimony, was proceeding to resell members of the same infected herd into other dairies when his career was cut short by the action of the metropolitan board of health. Nor were the results in such cases but the infection of one or two in a herd; where the diseased cow was introduced a general infection was the usual consequence. All that I could learn about the progress of the disease in this and former years was to the same effect. The malady never appeared apart from the introduction of strange animals, and when introduced the general infection of the herd was the consequence.

RAPID SPREAD OF THE PLAGUE.

The disease is not widely prevalent, because it extends its ravages only by contagion and infection, and the conditions of the American cattle trade have been strongly opposed to this. But the disease has not only held its own for thirty-six years, but has slowly gained against every obstacle until it numbers its victims in six different States. It is not wanting in virulence, but will, when it has a fair opportunity, sweep with remorseless force over the entire land. To this it is daily tending. From Brooklyn it has laboriously crept onward as far as Maryland and Virginia, and unless extirpated it will continue its baleful course until, reaching our open pasturages of the West and South, it will poison the sources of our cattle trade, descend upon our Eastern States with every cattle-train, infect the rolling stock on all our great railroad trunks, and bid defiance to all control. Wherever it has met with similar conditions it has proved thus intractable. In the steppes of Eastern Europe it has held perennial sway despite the best directed efforts of the Russian Government, and on the open pastures of Australia it still prevails, notwithstanding the most persistent and almost ruinous efforts for its extermination. So will it prove should we neglect the present opportunity and allow it to spread until it reaches our unfenced ranges of Texas, Kansas, Colorado, Wyoming, &c.

We are advised to employ inoculation. But what is inoculation? If successful, the production of the disease artificially, with its prominent lesions, in a less vital organ. In every stable where cattle are successfully inoculated the poison is produced in unlimited quantity. It is diffused through the air. It lodges in the dry parts of the building, in the fodder, etc., and is preserved for months and years. Unless these buildings are subsequently disinfected, they are deadly to the first susceptible animal that enters them. Finally, the immunity obtained by inoculation is not permanent, but lasts at the most for about two years. Inoculation, therefore, is a ruinous recourse, unless a country is already generally infected. It is itself a prolific means of spreading the poison. It cannot be effectual, unless the whole bovine race of the country are operated on and all the calves as soon as dropped; and so long as it is practiced, the stables must be considered infected, and the stock coming from such infected centers must be held to be dangerous to the animals. No country in Europe has practiced inoculation to so great an extent as Holland, and no country in Europe is to-day more extensively ravaged by this disease. England has tried inoculation to a very large extent, and England has been reluctantly compelled to abandon it. Australia has fallen back upon it as a dernier resort, and she has found that it only lessens the losses, while it has failed to exterminate the disease.

THE INFECTION MUST BE STAMPED OUT.

The day may come when we, too, may wisely follow Australia in adopting a general inoculation as a palliative of the disease. But this can only be if we criminally neglect the plague until it reaches our Western stock-ranges and bids defiance to all efforts at its extinction. To follow such a course at the present time would be ruinous, indeed, and those who counsel it cannot understand the problem we have to deal with. As already remarked, England, engaged in extirpating the disease from her own herds, will never offer us an unrestricted trade in cattle so long as we harbor this insidious enemy. A maintenance of infection by continued inoculation of our herds assuredly means the indefinite suspension of our foreign live-cattle trade; and nothing will secure the resumption of this trade short of the entire extermination of the malady.

Certificates of soundness of the cattle shipped are not worth the paper they are written on. No one would knowingly export sick animals to Europe, and no one is capable of detecting the existence of this disease during its lengthened period of incubation. We need not shut our eyes to this fact, for assuredly the English, who have had a far longer and harder experience of the disease, will not. Those who, knowing the character of the malady, counsel any measures short of its speedy and absolute extinction, are the true enemies of the live-stock interests and of the country. If their words should prevail, the future generations of Americans, seeing their country more ravaged than even the States of Europe, and by plagues exotic to her soil, will look back with regret to the time when it had been possible for their fathers to have averted such a baleful legacy.

It is still possible for us as a nation to do what has been done by Norway, Sweden, Denmark, Holstein, Oldenburg, Switzerland, Massachusetts, and Connecticut, and what is now being attempted in England, to stamp out this plague, which as an exotic should never have gained a footing on our shores. If the governors and legislatures of the States now infected and if Congress do their duty, they will follow the lead of Governor Robinson, of New York, and spare no effort nor expense until this plague has been banished to the Old World, whence it came. And if every citizen will do his duty he will cause such power to be exerted on these State and national authorities as will forbid any further neglect of this matter. No one having a full acquaintance with the subject can afford to remain silent in face of the existing facts, and this feeling alone has impelled me to pen the above remarks. New York may act alone, but, if so, she must either establish a long quarantine at her border or she will soon again import the disease from New Jersey. New Jersey may act independently, but she must be left in constant danger of infection from Pennsylvania and Maryland. So with the other States. The only path of safety is to wage a war of extermination simultaneously in all the infected States; and should the State legislatures and Congress fail to meet the need, they will prove recreant to their trust, and entail a great evil upon this continent.

Yours, &c.,

JAMES LAW.

ITHACA, N. Y., *February* 21, 1879.

In pursuance of the provisions of an act passed by the legislature of New York in the year 1878, entitled "An act in relation to infectious and contagious diseases of animals," on the 12th day of February last Governor Robinson appointed General Marsena R. Patrick his assistant, and directed him to take active measures for the suppression and extirpation of the disease in Kings and Queens Counties of that State. The following instructions were issued to General Patrick by the governor:

It has been made known to me that the infectious and contagious disease among neat cattle, called pleuro-pneumonia, has been brought into and exists in various places in the counties of Kings and Queens of this State. You are therefore directed, as such assistant, to prohibit the movement of cattle within said counties, except on license from yourself after skilled examination under your direction. You are also directed to compel all owners of cattle, their agents, employés, or servants, and all veterinary surgeons, to report forthwith to you all cases of disease by them suspected to be contagious. When such notification is received, you are directed to have the cases examined, and to cause all such animals as are found to be infected with the said disease destroyed and buried with slashed hides. You are directed, further, to quarantine all cattle which have been exposed to the infection of said disease, or are located in an infected place; but you may, in your discretion, permit such animals to be slaughtered on the premises and the carcasses to be disposed of as meat if, upon examination, they shall be found fit for such use. You will forbid and prevent all persons not employed in the care of the cattle there kept from entering any infected premises. You will likewise prevent all animals and fowls from entering such premises. You will prevent all persons so employed in the care of animals from going into stables, or yards, or premises where cattle are kept, other than those in which they are employed. You will cause the clothing of all persons engaged in the care, slaughter, or rendering of diseased or exposed cattle, or in any employment which brings them in contact with such diseased animals, to be disinfected before they leave the premises where such animals are. You will prevent the manure, forage, and litter upon infected premises from being removed therefrom; and you will cause such disposition to be made thereof as will, in your judgment, best prevent the spread of infection. You will cause all buildings, yards, and premises in which said disease exists, or has existed, to be thoroughly disinfected.

You are further directed, whenever the slaughter of diseased or infected animals is found necessary, to certify the value of the animal or animals so slaughtered at the time of slaughter, taking account of their condition and circumstances, and to deliver to their owner or owners, when requested, a duplicate of such certificate. Whenever any owner of such cattle, or his agent or servant, has willfully or knowingly withheld, or allowed to be withheld, notice of the existence of disease upon his premises or among his cattle, you will not make such certificate. You are further directed to take such measures as you deem necessary to disinfect all cars, or vehicles, or movable articles by which contagion is liable to be transmitted. You are also to take such measures as will secure a registry of cattle introduced into any premises in which disease has existed, and to keep such cattle under supervision for the period of three months after the removal of the last diseased animal and the subsequent disinfection of such premises. You are further authorized and empowered to incur such expenses in carrying out the provisions of the foregoing order as may, in your judgment, be necessary, and to see to it that the bills for such expenses be transmitted to this department only through yourself, after you have examined and approved them, in writing.

L. ROBINSON.

By the governor.

General Patrick at once established his headquarters at the Brooklyn board of health, and called to his assistance Professor Law and many other eminent veterinarians. Active measures were immediately instituted for a suppression of the disease, which will no doubt be continued with the same energy until it is extirpated.

Further legislation having been found necessary for the speedy and complete eradication of this malady, an additional act was promptly passed by the legislature of New York, on the 15th day of April. (For provisions of this act, see appendix.)

During the latter part of February last, and shortly after the commencement of this investigation of the condition of the dairy stock in the vicinity of Brooklyn, Dr. Law was summoned before the Senate Committee on Agriculture, which was then engaged in taking testimony in regard to the prevalence of pleuro-pneumonia among cattle in this country. At the request of this committee he submitted the following written statement:

PLEURO-PNEUMONIA IN NEW YORK AND ELSEWHERE.

STATEMENT OF DR. JAMES LAW.

INFECTION AND INFECTED PLACES AROUND NEW YORK.

Up to the time of my leaving New York we had found in that neighborhood thirteen centers of the contagious pleuro-pneumonia, embracing over twenty separate herds, and more than one thousand animals. At one place alone (Blissville), we are now killing the sick at the rate of twenty head and upward per day. We are further doing all we can to encourage the slaughter under our own supervision of the animals that are in such infected stables, but which do not yet show signs of illness. These are being disposed of at the rate of from thirty to seventy per diem.

Healthy animals slaughtered in this way are sold as human food and their hides disinfected. All infected places are placarded as such, and placed in quarantine, within which neither man, beast, nor bird is allowed to enter or pass out, save the necessary attendants, who are disinfected and forbidden to go near other cattle.

In the infected counties no movement of cattle is allowed save under special permit given after examination. All are compelled, under penalty, to report to General Patrick the existence of cases of contagious disease as well as all suspicious cases. Finally, all sick cattle killed to stay the progress of the malady are paid for by the State, according to appraisement, which shall in no case exceed one-half the original value of the animal. This point I consider all essential to encourage the owners of sick stock to report them, and at the same time to avoid the risk of artificial or careless infection of unmarketable animals for the purpose of selling them to the State.

The minor details of our action I need not record.

CATTLE KEPT AT THE BLISSVILLE SWILL-STABLES.

It having been testified before the committee that the cows in the stables of Gaff, Fleischmann & Co. were there for dairy purposes only, I think it requisite to correct the statement. The stables were filled not only with cows, but also steers and bulls. The stock belonged to many different parties, but mostly to dealers and butchers who hired their board. The owners of the stock had on their part, as a rule, no interest in the milk, which went to third parties as payment for the care-taking of the animals. The healthy cattle fattened rapidly and were sold for beef, and as there was a constant change of stock the contagion had an ample field among the newly-come and susceptible animals, and had a chance of extension to other places and herds with every beast removed, fat or otherwise.

I have had testimony that the fat stock frequently went out of Long Island, but have no *personal* knowledge of this. Now any such movement is prevented, and the consequent danger is at an end.

NATURE OF THE CONTAGIOUS PLEURO-PNEUMONIA.

When speaking of this disease we should strike out of our vocabulary such words as *epidemic* and *sporadic*. *Out of Eastern Europe or Asia the malady is absolutely unknown, save as propagated by contagion or infection.*

Wherever, out of these regions, it has made an inroad, it can always be traced to the importation of a sick or convalescent animal, or of some product of such an animal. Many such instances could be drawn from the records of its existence on the continent of Europe, but, manifestly, those cases are more satisfactory which refer to the extension of the disease to distant islands and continents. During the European wars at the beginning of the century, this malady, like the rinderpest, prevailed all over Europe, wherever the armies marched, and the eastern or steppe cattle were drawn for their support.   But the British Isles remained perfectly exempt until 1839, when the pleuro-pneumonia reached Ireland by some cattle sent by the British consul at the Hague.

It spread from this center, reached England some time in 1841, and since the passage of the free-trade act of 1842 has been kept up by continual arrivals of infected continental stock.

Yet it only reached where the railroads penetrated, and seemed to respect the Highlands of Scotland, where the native black cattle only are bred, and into which outside stock are never brought.

The *United States* knew no such contagious disease until the importation of an infected Dutch cow in Brooklyn, in 1843, and this, together with one or two other importations, have furnished the material for its extension over seven different States.

Australia, with her thousands of herds, was respected until 1858, when an English cow conveyed the poison which has since ravaged her herds without intermission.

The *Cape of Good Hope* remained clear until 1854, when an English cow carried the infection which still prevails in the cape herds.

The same truth is shown negatively by the fact that every country and State that has vigorously stamped out the first arrivals of disease, and taken measures to prevent further importation, has rid its territory of the pestilence.   Among them may be named Norway, Sweden, Denmark, Schleswig, Oldenburg, Switzerland, Massachusetts, and Connecticut.   Some of the countries have been again infected in connection with the Danish and Franco-German wars, which, for the time, destroyed all safeguards, but until such a contingency arrived their herds were preserved in health.

The fact that in our country and in Western Europe this disease is propagated only by contagion, is the grand central truth round which all our thoughts of the malady should revolve, and upon which we should base every measure adopted for its extinction.   If the affection could arise spontaneously, from any faulty conditions of hygiene in our own land, then farewell to all hope of permanently ridding our herds of the plague.   But all history testifies to the contrary, and we can foretell with as much confidence as we can the rising of to-morrow's sun, that if we could once extinguish the products of the imported poison, we need fear no more contagious pleuro-pneumonia until it is again imported from an infected land.

There is no such thing as a *sporadic* case of contagious pleuro-pneumonia, and no epidemic case in the sense that it is due to some condition of life apart from the presence of the virus in the country.   Every case in this country, as in Western Europe, the Cape of Good Hope, and Australia, is the result of direct or indirect contagion, and of that alone.

It is true that affections of the chest will occur in all future time as they occur in other animals, and in man himself, as they occurred in cattle before the importation of the contagious germ, but such cases have not been in the past and will not be in the future the cause of the propagation of the disease from animal to animal, or in other words of the development of a contagium.

Extirpate from the country this exotic contagium and we can supply unassailable beef to the world.

DANGER TO THE COUNTRY OF THE POSSIBLE INFECTION OF WESTERN HERDS.

For ten years I have been publicly warning the country of the danger of allowing this disease to extend to our Western States and Territories.   (See especially National Live Stock Journal, March, 1878, and Transactions of New York State Agricultural Society, 1877-'78.)   Infection of the Western herds means speedy infection of all the cattle cars of the railways, yards, loading-banks, &c., and the starting of a constant stream of infected animals towards our Eastern States and markets.

This means a uniform infection of the country and losses of thousands of millions of dollars in a short space of time.

Worse than this, should the malady extend to our unfenced cattle-ranges it will be practically unmanageable.   Such has been the experience on the open steppes of Russia and the cattle-ranges of Australia, where the most costly efforts at the extinction of the disease have proved futile and the poor palliation of inoculation has been established.   (See National Live Stock Journal, March, 1878.)

DANGER OF INOCULATION.

The public advocacy of inoculation demands a word on this subject.

Successful inoculation in favorable conditions leads to the loss of but two or three per cent. of animals operated on.   The survivors are protected from contagious pleuro-pneumonia for a variable period averaging two years.   But every inoculated animal is infected, the places where inoculated animals are kept are infected, all their products are infected, and there must be the most thorough system of disinfection for all such places and things before immunity can be gained.   Every new animal introduced and every calf born must be inoculated. It becomes evident, therefore, that to the stock of the country at large inoculation produces all the dangers of an equal extension of the disease in the ordinary way.

Inoculation, therefore, is ruinous to any attempt at extinguishing the poison. It has been tried in Holland more extensively than in any country in Europe, and Holland is to-day the most plague-ravaged country on the continent. It has been followed extensively in Great Britain, but she has been reluctantly compelled to abandon it in favor of a system of absolute extinction. It has been practiced widely around New York, yet this district is probably now the most prolific center of the disease in America.

Australia has fallen back upon it as a *dernier resort*, after a fruitless attempt to expel the malady from the open pastures. We, too, must one day come to this wretched palliation, if we neglect to stamp out the disease while still confined to our eastern and inclosed farms, and allow it to reach our western open prairies.

DANGER OF MEDICINAL TREATMENT.

As with inoculation, so with the maintenance of sick animals alive for treatment. The production and diffusion of the poison is in exact ratio to the period during which the animal is allowed to survive after illness has been detected. To treat the sick, therefore, is almost equivalent to propagating the disease, because on a large scale and in all sorts of stables it is impossible to keep up a constant disinfection of the air and other diseased products.

Wherever extinction of the poison is attempted, treatment of the disease must be forbidden under heavy penalties.

IMPORTANCE OF UNITED STATES ACTION.

The isolated action by individual States is eminently unsatisfactory. In New York we are working at the extermination of the disease, but after we have accomplished this we can only preserve our immunity by subjecting all New Jersey cattle to a quarantine of one or two months at our frontier. If New Jersey on her part kills it out, she must quarantine against Pennsylvania, Pennsylvania against Maryland and Virginia, and so on as far as the disease is found to extend. Isolated action will be incomparably more expensive, tardy, and uncertain than a uniform movement under one central head, and everything ought to give way to secure such a desirable result The question involves tens of millions of dollars of our foreign commerce annually, and the trade has been steadily increasing, so that it is surely a matter in which the central government can properly act.

SUGGESTION OF MEASURES FOR THE EXTINCTION OF THE DISEASE.

1st. Appoint a veterinary sanitary staff to act with the Commissioner of Agriculture in stamping out the contagion.

2d. Make it incumbent on all stock-owners and their representatives, and on all veterinarians, to report all suspicious cases to the Commissioner under a penalty.

3d. Let the sanitary staff promptly investigate all such cases and take measures accordingly.

4th. Let every infected county be proclaimed and placarded, and let all movements of cattle within such county be forbidden excepting by special license.

5th. Let all sick animals in an infected herd be at once slaughtered, their hides slashed, and the carcasses deeply buried; and in case the owner has not withheld notice of the existence of the disease let him obtain an order on the treasury for a suitable indemnity, which should in no case exceed one-half the value of the animals; failure to notify should entail loss of the indemnity.

6th. Let all cattle found in infected places be likewise slaughtered, their hides disinfected, and their beef allowed to pass into consumption as food, if fit for this purpose. For such animals, indemnity should be allowed to the extent of not more than two-thirds of the value, after deducting salvage obtained from meat and hides.

7th. Let all infected stables, all manure, and all movable objects that have come in contact with diseased cattle, be subjected to an exhaustive disinfection, and let all cattle afterward placed in such buildings be sequestered in quarantine under the supervision of the veterinary sanitary authorities until at least three months after the removal of the last sick animal and the disinfection of the premises.

8th. Let all railroad cars, ships, boats, wagons, and other movable objects that have become infected be cleansed and disinfected under the direction of the veterinary sanitary staff before they are again used for the transportation of cattle.

The advice to slaughter the exposed as well as the sick cattle I think very important, as it enables us to stamp out the disease quickly and to disinfect once for all, and obviates the necessity for a long-continued and expensive supervision in the case of every infected herd. If such exposed animals are placed in quarantine, as we are still compelled to do by a defeat of the law in New York, we find that every three weeks or a month a new case develops, necessitating continued visitation, professional examination, and slaughter, and repeated and expensive disinfection, without taking into account the enormously enhanced danger of the extension of the disease to other herds.

One other question will not brook an hour's delay. The testimony concerning the two ship-loads of cattle slaughtered at Liverpool may be misleading, but unless a gigantic blunder has been committed it implies that the disease has already reached one or more isolated spots in the West. This was inevitable sooner or later if the disease was not crushed out in the East, and I have constantly uttered warnings on the subject. If it has already taken place it should be treated at once, for the evidence implies that not only has the malady gained a footing in the West, but that the owners of the infected stock are acting unfairly by the country, and selling off their infected stock to make what salvage they can. There are, then, not only of infected cars, stock-yards, &c., but of the sale of lean stock to different localities in the West, whence we shall have new streams of infection, until our unfenced ranges suffer. No delay should occur in ascertaining the facts of the case. If there has been a mistake it will relieve the country to know it, whereas if there is even one center of infection in the West it should be stamped out promptly at any cost.

REPORT ON THE STOCK-YARDS AT THE PORT OF NEW YORK.

In investigating the existence and status of lung fever in the cattle of Long Island and Manhattan Island, I met with several outbreaks in which the disease was traced to cows sold into the herds in question by Patrick McCabe, a New York jobber. Three such instances may be named: First, Mr. Wheelock, farmer, at Roslyn, Queens County, purchased two cows of McCabe in August, which communicated the disease to the whole herd of eighteen head, and to that of a farm about two miles distant, to which two of his (Wheelock's) cows were taken. Second, Mr. Brazzel, Eighty-first street, New York, got a cow from McCabe the week after Christmas which conveyed the disease to his herd. Third, Mrs. Stur, Fiftieth street, New York, had a cow from McCabe about ten days ago on trial. This cow had been sick ever since her arrival, and when I saw her on Saturday was in a condition of advanced pleuro-pneumonia. I had further information, from a man in the trade who has a high reputation for honor, that the cattle that had passed through the hands of this McCabe had been for two years the most prolific source of disease in the dairies of Brooklyn and Long Island.

Accordingly on Saturday last, in company with Dr. Lautard, I went to examine his (McCabe's) premises and stock, when we were much surprised to find that he kept them in the New York public stock-yards at Sixtieth street, and I could not discover that he had any other place. The clerk found in charge of the office at the yards assured us that he constantly kept his cows there, and only removed them as he found purchasers. He did not think he could have any other place for keeping cows.

At the time of our visit he had a number of cows in the yards. At these yards the cows of all the dealers are usually placed in the sheep-house for warmth, but this is immediately adjacent to the inclosures for the other stock, and all alike must enter and leave by the same roads and gates or wharves. Further, when the sheep-house is crowded the cows are turned out into the open cattle inclosures in the yard. Cows are received in these yards indiscriminately from near as well as remote places, including among the former Westchester, Rockland, and Orange Counties, which, according to the best evidence I can obtain, are infected. (I have not yet verified the last fact by personal observation.) No precaution is taken to prevent the proximity or contact of these cows with the other stock.

There seems, therefore, no alternative; we must consider the New York stock-yards at Sixtieth street as infected, and that stock shipped from these yards to Europe will be liable to develop the disease after landing if kept alive long enough to allow of the completion of the period of incubation. That the evil results have been seen mainly in the cows is explained by the fact that they are allowed time after leaving the yards for the completion of the period of incubation (one to two months), whereas the fat cattle even if sent to Europe are slaughtered before this time has elapsed.

JERSEY CITY STOCK-YARDS.

In these as in the New York stock-yards there is the entire absence of any means of separating cows brought from near and infected neighborhoods and stock brought from the West or other uninfected localities. The cow-stable is at the north side of the yards and can only be reached by cattle that have passed through among the inclosures for the other stock. The stable itself is furnished with open gates, not doors, facing the inclosures for other stock and separated from them only by a narrrow wagon road, perhaps fifteen feet wide.

Mr. Fowler, whom I found in charge of the yard, was violently denunciatory of the mere idea that this disease existed anywhere, and of all who would mention such a subject, and could with difficulty be persuaded to give any information regarding the yards, the stock, its proximity in the inclosures, and its disposal. He admitted, however, that they got four or five cows per week, and on rare occasions one or two carloads; that they mostly came from Eastern Pennsylvania, and that they remained in the yards until they were sold to parties in New York City, Brooklyn, Jersey City, Staten Island, &c. I may here state that on the occasion of my visit, late on Saturday night, the cow-stable contained eighteen cows and eight calves waiting for sale; so that, according to Mr. Fowler, I must have hit upon the very exceptional case of an arrival of two car-loads.

I further drew from Mr. Fowler that the fat stock for exportation were taken from any part of the yards, wherever suitable animals could be found, and carried by boats to the ocean-going steamers. There was no attempt made to keep such animals apart from such as might possibly come from infected districts in New Jersey and adjacent States, nor from the inclosures where such cattle had formerly been, *as indeed why should there be, seeing the whole story of the disease was a fabrication?*

As bearing on the question of the probable infection of these yards, I shall add that the malady is well known to exist in Alexandria, Va. I have had the most circumstantial reports of its existence around Washington. According to Dr. Corliss, it prevails to some extent around Newark, N. J. Last year it made havoc in the town of Clinton, and the year before near Burlington, N. J. Further, in making inquiry among the farmers at New Lots, Kings County, New York, whose herds are now infected, I found that they had repeatedly traced the disease to Jersey cows brought into their herds. There is, therefore, the strongest circumstantial evidence that both the Jersey and New York stock-yards, the two points from which cattle are shipped to Europe, are infected places, and that the apparent absence of disease in American cattle when landed in England is due to the fact that they have not yet had time to pass through the long incubation period of the disease.

ABSURDITY OF A CERTIFICATE OF SOUNDNESS.

The professional examination at the yards of animals destined for exportation can never be better than a farce. The most accomplished veterinarian has no means of detecting the presence of the specific poison until the period of incubation has passed, and as this lasts for from three weeks to two months, the evidence of infection contracted in the New York stock-yards cannot possibly be recognized until long after the animals have landed in England. The great mass of our Western cattle is sound so far as the contagious pleuro-pneumonia is concerned, and if infected, it is presumably only after they have been sent East. The disease, therefore, can only be in the incubation stage so long as they remain on our shores, and in this stage no man can recognize it, though it only wants time for its development. Any examination in such a case must be the most empty of forms, and must be prejudicial rather than beneficial, inasmuch as it leads to the certifying of the soundness of animals that may be, and often probably are, infected. It is quite manifest that in the case of cattle that may have been infected in the New York or New Jersey stock yards, an examination a fortnight later on their landing in Liverpool would be almost as great a farce as the examination prior to shipment at New York. Hence the soundness of the English position in ordering the slaughter at the quays of all cattle from an infected country.

THE COURSE OF SAFETY.

If we can be assured that there is not yet an infected center in the great stock-raising regions of the West, the cattle from there might be safely shipped to England under the following regulations:

1st. Let the Western cattle-trains be made up of cars that have never been used for the local cattle traffic in the eastern parts of the Atlantic States or of such as have been thoroughly cleansed and disinfected before use.

2d. Let all such trains be from the West *through*, and let these take on board no live nor dead cattle, nor other unmanufactured products of cattle, east of given points on the respective lines, such points to be designated as soon as we know conclusively how far the pleuro-pneumonia has extended westward. Let such trains pass to designated stock-yards on the quays at least one-fourth of a mile apart from all other stock-yards, or cattle stables, or pastures.

3d. Let such yards be rigidly closed against all visitors, no one being admitted except the necessary attendants, and no one being employed as such who has recently been in charge of other cattle in the East.

4th. Should it be necessary to sell any such stock for home consumption, they must

16 SW

be driven by their attendants to other yards or pastures at a distance, or to the other stock-yards, where buyers may see them. The attendants on the foreign stock-yards may drive such animals into the common stock-yards, but must not, on any account, enter themselves.

5th. The cattle intended for export must be transferred to the ocean-going ships direct, or carried to them on boats that have never been used for conveying other cattle, or that have been subjected to the most thorough disinfection subsequent to such use.

6th. It should be shown that the ocean-going vessels, in which the export cattle are shipped, have not carried, and do not now carry, any hides or other unmanufactured products of cattle; or, if they have previously carried such articles, that they have been thoroughly disinfected since.

PLEURO-PNEUMONIA—THE LUNG PLAGUE—CONTAGIOUS LUNG DISEASE OF CATTLE.

Pleuro-pneumonia is a malignant contagious fever to which, as far as known, cattle only are liable, and in them is accompanied by inflammation and other diseased conditions of the lungs and their membranes, together with great prostration of the entire system.

It proceeds from a poisoned condition of the blood. How, when, or where this poison was first generated it is impossible to tell. Nor is it less difficult to determine its specific nature. So far as reliable information has yet reached, it is never generated spontaneously, but depends entirely on the introduction of a virus or contagion into the system of a healthy animal. A single animal so infected infects the herd: the herd, subdivided and scattered, infects other herds until in time large areas of country have been visited and devastated by the fearful scourge.

Beginning, as we have reason to believe, in the far-off East, and at a remote date, its course has been westward until, crossing the Atlantic in the system of stock imported from European states, it has at length found lodgment here.

The earliest symptoms of the disease are not always easily detected, there being no intensity of inflammation at first, and the period of incubation varying often from eight or nine days to three or four months. The knowledge of the existence of the disease in adjoining States or farms, or even in remote sections from which cattle have been introduced, should serve to put every one on guard and lead to frequent thermometric trials even with cattle apparently in perfect health. While such trials would not, perhaps, in every case determine infallibly the existence or non-existence of the disease, yet in a very large majority of cases—possibly in nine out of ten, and particularly if other symptoms were present—they would lead to a right conclusion. The trial is made by inserting the thermometer in the rectum. If a rise of temperature to $103°-106°$ Fahrenheit is observed we may be reasonably sure that the disease exists, at least in an incipient state.

Its further development is indicated by fits of shivering, often so slight and transient as to escape the notice of all save the practiced eye; by a dull, staring coat, with (frequently) a rigid skin: by a harsh, dry cough, the more apparent when the animal is made to move briskly: by irregular chewing of the cud; constipated bowels; excrement dry; urine diminished, but with high color; and, in the case of cows, by a falling off in the quantity of milk.

At an early stage of pleuro-pneumonia there is a harsh sound or roar produced by the passage of air through the wind-pipe and its subdivisions, which may sometimes be heard at some distance from the sick animal. Occasionally the air rushing through the bronchial tube (made rigid by a mass of hardened lung) produces a very decided whistling noise. A somewhat watery discharge from the nose, increased in the act of coughing, is noticed early in the disease, and driving sick cattle in the earliest stage produces much thirst, and there is sometimes a ropy saliva discharged from the mouth, while the muzzle is hot and dry.

As the malady progresses the pulse rises to seventy, eighty, and even a hundred beats per minute; the respirations to thirty-five and forty per minute, and are labored and audible, while each expiration is accompanied in most cases by a short distinctive grunt or groan, the more marked whenever pressure is applied to the ribs over the lungs.

At this stage the cough increases, the gait becomes more languid, the eyes more prominent and fixed; the countenance assumes an uneasy, pained expression, and a disposition is manifested by the sick to separate from the well. When the animal stands the elbows are turned out, the fore limbs extended, the hind feet drawn forward under the body, the head and neck stretched out, and the back arched, while the nostrils are more or less convulsively expanded at each inspiration. When lying, to which there is a tendency, the animal rests, especially in the latter stages of the disease, on its brisket, or on the affected side, leaving the ribs on the healthy side as much freedom of motion as possible.

With a still further advance in the disease, the pulse becomes more frequent (often rising to 120 per minute) and the heart-beats, at first subdued, are now marked and palpitating; the tongue becomes foul and covered with fur, and the breath has a nauseous smell. Listlessness, grunting, grinding of the teeth, diminished secretions, and weakness rapidly increase; the breathing is more frequent and labored; the animal gasps for breath; the spasmodic action of the nostrils is more marked, the groan more audible; the temperature is irregular, the tendency being to coldness of the horns and extremities. These conditions are followed by a mattery or watery discharge from the eyes and nose, rapid loss of flesh, hide bound, and either obstinate constipation or else a violent watery diarrhœa of fœtid matter associated often with a considerable discharge of clear-colored urine.

Percussion over the lungs will, in the beginning, often reveal the disease when not otherwise apparent. With some practice and a little care almost any one can distinguish the sick from healthy cattle by listening to the sides of the chest. In the earlier stages of the disease percussion gives out a clear or resonant sound, followed, as the malady increases, by a dull, heavy one, easily distinguished from the sound caused by the lungs in health.

Where one lung only is affected, partial, sometimes complete, restoration may result; but acute pleuro-pneumonia, in which both lungs are affected, we may safely assert is never terminated except by death.

As stated above, the period of incubation of this disease varies from eight or nine days to three or four months; the usual average period being from twenty-five to forty days. The acute stage of the disorder varies from seven to twenty-one days. Convalescence extends over a period of from one to three months, during the greater part of which time the convalescent animal is often capable of infecting healthy cattle.

As a rule, in mild outbreaks, the mortality attains twenty-five per cent., and in severe epidemics sixty, seventy, or even one hundred per cent.

In England, the lung disease has more than doubled the ordinary mortality of the country, entailing a loss of many millions of dollars.

While various remedies for this insidious disease have from time to time been recommended and tried, not one of them, nor all of them combined, have proved a specific against its destructive effects; and as a means to be relied on for the protection of the stock of the country, they are worse than useless. As a rule, the malady baffles the skill of the most learned veterinary practitioners, frequently attaining its greatest mortality where most they have combated it.

Nevertheless, as there may be circumstances under which partial relief might be afforded by timely remedial agents, it is deemed expedient to give in this place the treatment which, in general, has been found most efficacious.

The course most obviously to be pursued, when the slightest symptom of the disease is observed, or where the slightest cause for suspicion exists. is to apply the thermometer, to separate at once every suspected animal from the rest, to use disinfectants, to adopt a low diet, and to watch carefully for further developments. The weight of testimony is against bleeding. If constipation is detected it should be removed by a moderate dose of salts. Slight diarrhœa need not be checked; but when violent use a mixture of gallic acid (or its equivalent) and gruel, one-half ounce of the former to one-half pint of the latter; or else, one-half ounce powdered alum to one quart of milk. Sometimes there is considerable swelling or bloating of the stomach, which may be removed by carbonate of ammonia—one ounce in a moderate quantity of gruel, repeated if necessary. To lower the temperature and ease the breathing give acid sulphite of soda, one ounce, twice a day. In an advanced stage of the disease administer one or two ounces of whiskey or of oil of turpentine every three or four hours. If no relief is observed employ copious warm-water injections, and give two or three times a day an ounce of carbonate of ammonia in a quart of linseed-tea. Although out of place in the acute stage of the disease, blisters, setons, rowels, and cauterization may be applied in some cases to advantage after the fever has abated. Several preparations of carbolic acid have been tried with more or less success. Perhaps the best is—

Pure carbolic acid, 1 drachm;
Water,   1 pint;

given at a dose, three times a day.

Convalescence begun, restoration to health will be hastened by giving a teaspoonful of sulphate of iron in the food at each meal. The herd itself from which the sick have been removed should be placed, as a possibly preventive measure, on daily doses of the same preparation. (sulphate of iron,) allowing about half a drachm to a drachm *per capita*, mixed with an equal amount of coriander seeds, given in meal or bran, the better to disguise the iron.

A post-mortem examination of the chest generally reveals layers of yellowish, friable, false membrane (covering-skin) stretching across and around the sack containing the heart. With them is found a yellowish, clotted fluid, highly charged with albumen and shreds of solid deposit. Diseased portions of one or both lungs are found adhering to the membrane of the ribs and diaphragm, from which there is more or less difficulty in detaching them. The membrane covering the lung, usually smooth and glistening, is rough and mottled with a number of more or less marked pimples or warts.

The fluid around one or both lungs varies from a few ounces to several gallons. At times it is tolerably clear when warm, and gelatinizes on cooling; at others it is difficult to separate it from the shreds of lymph and false membranes in which it is held. Pus-cells often abound in it, and it assumes in some cases the character of pus, from which an intolerable stench sometimes proceeds.

On removing the lungs the essential appearances of the disease in all cases will be found quite uniform, although differing considerably in extent.

In recent and mild cases in which only one lung is affected, the surface of the lung may be smooth; parts of it collapsed, as in health, with the normal pink color preserved. The affected part is swollen, hard, and mottled. On cutting into this, the older diseased portions present a very peculiar marbled appearance. The substance of the lobules is solid and of a dark red color, and the tissue between the lobules is of a yellowish-red, more or less spotted with red points, but sometimes of almost pure yellowish-white color. The more recent deposits are distinguished mainly by a lighter red color of the thickened lobules.

At a more advanced stage of the disease the lung will be found harder and of darker color, its tissues having lost a portion of the marbled appearance, the blood-vessels obstructed, and showing how nourishment had been cut off from the lungs, while the older, darker, and more solid portion of the latter have become detached, so that they remain as foreign bodies imbedded in the cavities of the diseased tissue. The admission of air into these cavities, by dissolution of the lung tissue, produces the cavernous sounds which the ear can detect in the living animal.

On taking a warm diseased lung, severing the still healthy portions, making incisions into the parts solidified, and suspending them so that they may drain, a large amount of yellowish serum, of a translucent character, and varying greatly in weight, is obtained. The quantity of this serum, and of the solidified deposit in a diseased lung, is so large that, from a normal weight of four or five pounds, a lung attains ten, twenty, forty, or even fifty pounds.

The condition of the air-passages will be found to vary from one of perfect freedom in the healthy portions of the lungs to a state in which the mucous surface is coated with false membrane, or solid exudations of lymph in the diseased parts. These passages are sometimes found nearly filled, throughout their whole extent, with a deposit similar to that usually found on the surface of the diseased lung.

The heart's sack is sometimes found to be thickened by deposits around it, and not unfrequently to contain an excess of serum. The heart itself is contracted and pale, containing a little dark blood.

The organs of digestion at different stages manifest a state of dryness. The third stomach, which is so constantly packed with dry food in febrile diseases, is in the same condition in pleuro-pneumonia. In advanced cases there is found a more or less diffuse redness, and even effusion of blood in the large intestines, with fluid, fetid, and sometimes slightly blood stained excrement, such as is discharged in life.

Such briefly, and in language free from technicalities, are the description, cause, symptoms, treatment, and post-mortem appearances of pleuro-pneumonia as gathered from previous publications of this Department and other recognized authorities.

APPENDIX.

MASSACHUSETTS.

LUNG FEVER OR PLEURO-PNEUMONIA OF CATTLE.

The following act, for the suppression and extirpation of the disease called pleuro-pneumonia among cattle, was passed by the Massachusetts legislature April 4, 1860:

AN ACT to provide for the extirpation of the disease called pleuro-pneumonia among cattle.

Be it enacted, &c., as follows:

SECTION 1. The governor is hereby authorized to appoint three commissioners, who shall visit without delay the several places in this commonwealth where the disease among cattle called pleuro-pneumonia may be known or suspected to exist, and shall have full power to cause all cattle belonging to the herds in which the disease has appeared, or may appear, or which have belonged to such herds since the disease may be known to have existed therein, to be forthwith killed and buried, and the premises where such cattle have been kept cleansed and purified; and to make such order in relation to the further use and occupation of such premises as may seem to them to be necessary to prevent the extension of the disease.

SEC. 2. The commissioners shall cause all cattle in the aforesaid herds not appearing to be affected by the disease to be appraised before being killed at what would have been their fair market value if the disease had not existed; and the value of the cattle thus appraised shall be allowed and paid out of the treasury of the commonwealth to the owner or owners thereof.

SEC. 3. Any person who shall knowingly disregard any lawful order or direction of said commissioners, or who shall sell or otherwise dispose of an animal which he knows, or has good reason to suspect, has been exposed to the aforesaid disease, shall forfeit a sum not exceeding five hundred dollars.

SEC. 4. The commissioners shall make a full report to the secretary of the board of agriculture of their proceedings and of the result of their observations and inquiries relative to the nature and character of the disease.

SEC. 5. The commissioners shall duly certify all allowances made under the second section of this act, and other expenses incurred by them, or under their direction, in the execution of their service, to the governor and council; and the governor is hereby authorized to draw his warrant therefor upon the treasury.

SEC. 6. This act shall take effect from its passage, and continue in force for the term of one year thereafter, and no longer.

[Approved April 4, 1860.]

On the 12th of June, 1860, the following additional acts were passed:

AN ACT concerning contagious disease among cattle.

SECTION 1. The selectmen of towns, and the mayor and aldermen of cities, in case of the existence in this commonwealth of the disease called pleuro-pneumonia, or any other contagious disease among cattle, shall cause the cattle in their respective towns and cities which are infected, or which have been exposed to infection, to be secured or collected in some suitable place or places within such city or town, and kept isolated; and, when taken from the possession of their owners, to be maintained, one-fifth of the expense thereof to be paid by the town or city wherein the animal is kept, and four-fifths at the expense of the commonwealth, such isolation to continue so long as the existence of such disease or other circumstances renders the same necessary.

SEC. 2. Said selectmen and mayor and aldermen, when any such animal is adjudged by veterinary surgeon, or physician by them selected, to be infected with the disease called pleuro-pneumonia, or any other contagious disease, may, in their discretion, order such diseased animal to be forthwith killed and buried at the expense of such town or city.

SEC. 3. Such selectmen and mayor and aldermen shall cause all cattle which they shall so order to be killed to be appraised by three competent and disinterested men, under oath, at the value thereof at the time of the appraisal, and the amount of the appraisal shall be paid as provided in the first section.

SEC. 4. Said selectmen and mayor and aldermen are hereby authorized to prohibit the departure of cattle from any inclosure or to exclude cattle therefrom.

SEC. 5. Said selectmen and mayor and aldermen may make regulations in writing to regulate or prohibit the passage from, to, or through their respective cities or towns, or from place to place within the same, of any neat cattle, and may arrest and detain, at the cost of the owners thereof, all cattle found passing in violation of such regulations, and may take all other necessary measures for the enforcement of such prohibition, and also for preventing the spread of any such disease among the cattle in their respective towns and cities and the immediate vicinity thereof.

SEC. 6. The regulations made by selectmen and mayor and aldermen in pursuance of the foregoing section shall be recorded upon the records of their towns and cities respectively, and shall be published in such towns and cities in such manner as may be provided in such regulations.

SEC. 7. Said selectmen and mayor and aldermen are authorized to cause all cattle infected with such disease, or which have been exposed thereto, to be forthwith branded upon the rump with the letter P, so as to distinguish the animal from other cattle; and no cattle so branded shall be sold or disposed of except with the knowledge and consent of such selectmen and mayor and aldermen. Any person, without such knowledge and consent, selling and disposing of an animal known to be effected with such disease, or known to have been exposed thereto within one year from such sale or disposal, shall be punished by fine not exceeding five hundred dollars, or by imprisonment not exceeding one year.

SEC. 8. Any person disobeying the orders of the selectmen or mayor and aldermen, made in conformity with the fourth section, or driving or transporting any neat cattle contrary to the regulations made, recorded, and published as aforesaid, shall be punished by fine not exceeding five hundred dollars, or by imprisonment not exceeding one year.

SEC. 9. Whoever knows, or has reason to suspect, the existence of any such disease among the cattle in his possession or under his care, shall forthwith give notice to the selectmen of the town or mayor and aldermen of the city where such cattle may be kept, and for failure so to do shall be punished by a fine not exceeding five hundred dollars, or by imprisonment not exceeding one year.

SEC. 10. Any town or city whose officers shall neglect or refuse to carry into effect the provisions of section one, two, three, four, five, six, and seven, shall forfeit a sum not exceeding five hundred dollars for each day's neglect.

SEC. 11. All appraisals made under the provisions of this act shall be in writing, and signed by the appraisers, and the same shall be certified to the governor and council, and to the treasurer of the several towns and cities wherein the cattle appraised belong, by the selectmen and mayors and aldermen respectively.

SEC. 12. The selectmen of the towns and mayor and aldermen of the cities are hereby authorized, when in their judgment it shall be necessary to carry into effect the purposes of this act, to take and hold possession, for a term not exceeding one year, within their respective towns and cities, of any lands, without buildings other than barns thereon, upon which it may be necessary to enclose and isolate any cattle, and they shall cause the damages sustained by the owners in consequence of such taking and holding to be appraised by the assessors of the town or city wherein the lands so taken are situated, and they shall further cause a description of such land, setting forth the boundaries thereof, and the area as nearly as may be estimated, together with said appraisal by the assessors, to be entered upon the records of the town or city. The amount of said appraisal shall be paid as provided in the first section, in such sums and at such times as the selectmen or mayor and aldermen respectively may order. If the owner of any land so taken shall be dissatisfied with the appraisal of said assessors, he may, by action of contract, recover of the town or city wherein the lands lie, a fair compensation for the damages sustained by him: but no cost shall be taxed unless the damages recovered in such action, exclusive of interest, exceed the appraisal of the assessors. And the commonwealth shall reimburse any town or city four-fifths of any sum recovered of such town or city in any such action.

AN ACT in addition to an act concerning contagious diseases among cattle.

SECTION 1. In addition to the commissioners appointed under the provisions of chapter one hundred and ninety-two of the acts of the year one thousand eight hundred and sixty, the governor, by and with the advice and consent of the council, is hereby authorized to appoint two additional persons to constitute, with those now in office, a board of commissioners upon the subject of pleuro-pneumonia, or any other contagious disease now existing among the cattle of the commonwealth.

SEC. 2. When said commissioners shall make and publish any regulations concerning the extirpation, cure, or treatment of cattle infected with, or which have been exposed to the disease of pleuro-pneumonia, or other contagious disease, such regulations shall supersede the regulations made by selectmen of towns and mayors and aldermen of cities, upon the same subject-matter, and the operation of the regulations made by such selectmen and mayors and aldermen shall be suspended during the time those made by the commissioners as aforesaid shall be in force. And said selectmen and mayors and aldermen shall carry out and enforce all orders and directions of said commissioners, to them directed, as they shall from time to time issue.

SEC. 3. In addition to the power and authority conferred on the selectmen of towns and mayors and aldermen of cities, by the act to which this is in addition, and which are herein conferred upon said commissioners, the same commissioners shall have power to provide for the establishment of a hospital or quarantine in some suitable place or places, with proper accommodations of buildings, land, &c., wherein may be detained any cattle by them selected, so that said cattle so infected and exposed may be there treated by such scientific practitioners of the healing art as may be there appointed to treat the same. And for this purpose said commissioners may take any lands and buildings in the manner provided in the twelfth section of the act to which this is an addition.

SEC. 4. The governor, by and with the advice and consent of the council, is hereby authorized to appoint three competent persons to be a board of examiners to examine into the disease called pleuro-pneumonia, and who shall attend at the hospital at quarantine established by the commissioners mentioned in the foregoing section, and there treat and experiment upon such number of cattle, both sound and infected, as will enable them to study the symptoms and laws of the disease, and ascertain, so far as they can, the best mode of treating cattle in view of the prevention and cure of the disease, and who shall keep a full record of their proceedings, and make a report thereon to the governor and council, when their investigation shall have been concluded: *Provided*, That the expense of said board of examiners shall not exceed ten thousand dollars.

SEC. 5. The selectmen of the several towns, and the mayors and aldermen of the several cities, shall, within twenty-four hours after they shall have notice that any cattle in their respective towns and cities are infected with, or have been exposed to, any such disease, give notice in writing to said commissioners of the same.

SEC. 6. The commissioners are authorized to make all necessary regulations for the treatment, care, and extirpation of said disease, and may direct the selectmen of towns and mayors and aldermen of cities to enforce and carry into effect all such regulations as may, from time to time, be made for that end; and any such officer refusing or neglecting to enforce and carry out any regulation of the commissioners, shall be punished by fine not exceeding five hundred dollars for every such offense.

SEC. 7. The commissioners may, when in their judgment the public good shall require it, cause to be killed and buried any cattle which are infected with, or which have been exposed to said disease, and said commissioners shall cause said cattle to be appraised in the same manner provided in the act to which this is an addition; and the appraised value of such cattle shall be paid, one-fifth by the towns in which said cattle are kept, and the remainder by the commonwealth.

SEC. 8. Whoever shall drive or transport any cattle from any portion of the commonwealth east of the Connecticut River to any part west of said river before the first day of April next without consent of the commissioners, shall be punished by fine not exceeding five hundred dollars, or by imprisonment in the county jail not exceeding one year.

SEC. 9. Whoever shall drive or transport any cattle from any portion of the commonwealth into any other State before the first day of April next, without the consent of the commissioners, shall be punished by fine not exceeding five hundred dollars, or by imprisonment in the county jail not exceeding one year.

SEC. 10. If any person fails to comply with any regulations made, or with any order given, by the commissioners, he shall be punished by fine not exceeding five hundred dollars, or by imprisonment not exceeding one year.

SEC. 11. Prosecutions under the two preceding sections may be prosecuted in any county in this commonwealth.

SEC. 12. All appraisals made under this act shall be in writing and signed by the appraisers and certified by the commissioners, and shall be by them transmitted to the governor and council, and to the treasurers of the several cities and towns wherein the cattle appraised were kept.

SEC. 13. The provisions of chapter one hundred and ninety-two of the acts of one thousand eight hundred and sixty [except so far as they authorize the appointment of commissioners] are hereby repealed, but this repeal shall not affect the validity of the proceedings heretofore lawfully had under the provisions of said chapter.

SEC. 14. The commissioners and examiners shall keep a full record of their doings, and make report of the same to the next legislature, on or before the 10th day of January next, unless sooner required by the governor; and the said record, or an abstract of the same, shall be printed in the annual volume of Transactions of the State Board of Agriculture.

SEC. 15. The governor, with the advice and consent of the council, shall have power to terminate the commission and board of examiners whenever, in his judgment, the public safety may permit.

STATE OF NEW YORK.

AN ACT to prevent the introduction and spread of the disease known as rinderpest, and for the protection of the flocks and herds of sheep and cattle in the State of New York from this and other infectious and contagious diseases. Passed April 20, 1866.

Be it enacted by the senate and assembly of New York:

SECTION 1. It shall be the duty of the health officer of the port of New York, in addition to the duties now imposed on him by existing law, to examine and inquire whether any animals are brought in any vessels arriving at said port in violation of any regulation of law passed by the Congress of the United States prohibiting the importation of such animals.

2. Whenever any animal is brought as a ship's cow, with no intention of landing the same or of violating any such law or regulation of Congress as aforesaid, the same shall be carefully examined and kept in quarantine for the space of at least twenty-one days, and if any symptoms of the infection or incubation of the disease commonly known as the rinderpest or any other infectious or contagious disease shall present themselves, it shall be the duty of the said health officer immediately to cause the said animal or animals to be slaughtered, and their remains boxed with a sufficient quantity of quicklime, sulphate of iron, or other disinfectant, and with sufficient weights placed in said box to prevent the same from floating, and to be cast into the waters of the said port. It shall also be his duty to cleanse and disinfect by suitable agencies the berth or section of the ship in which said animal or animals were lying or slaughtered, and also to cause the clothing and persons of all taking care of the same or engaged in slaughter and burial to be cleansed and disinfected.

3. William Kelley, of Dutchess County, Marsena R. Patrick, of Ontario County, and Lewis F. Allen, of Erie County, are hereby appointed as commissioners under this act, and with powers and duties as hereinafter enumerated.

4. In the event of any such disease as the rinderpest or infectious disease of cattle or sheep breaking out or being suspected to exist in any locality in this State, it shall be the duty of all persons owning or having any interest whatever in the said cattle, immediately to notify the said commissioners or any one of them of the existence of such disease; whereupon the said commissioners shall establish a sanitary cordon around such locality. And thereupon it shall be the duty of the said commissioners to appoint an assistant commissioner for such district with all powers conferred by this act on the said commissioners or their agents or appointees, which said assistant commissioner shall immediately proceed to the place or places where such disease is reported to exist, and cause the said animal or animals to be separated from all connection or proximity with or to all other animals of the ruminant order, and take such other precautionary measures as shall be deemed necessary; and if in his opinion the said disease shall be incurable or threaten to spread to other animals, to cause the same immediately to be slaughtered, their remains to be deeply buried, and all places in which the said animals have been confined or kept to be cleansed and disinfected by any of the agencies above mentioned; and also to cause the same to be carefully locked or barred so as to prevent all access to the same by any animals of a like kind for a period of at least one month. Any animal thus slaughtered shall be appraised under the supervision of said commissioners, and one-half of the value of said animal shall be paid by the State to the owner thereof.

5. It shall be the duty of the said assistant commissioner, immediately on his being notified of his appointment, or at any time thereafter of the breaking out of the said disease in any place contiguous to the same and within the county in which he resides, to give public notices of the same in at least one newspaper printed or published in the said county, and to cause notice to be posted up in at least five conspicuous places in said neighborhood, and it shall be his duty to enjoin, in said notice and otherwise, all persons concerned in the care or supervision of neat cattle or sheep not to come within one hundred feet of the said locality without the special permission of the said assistant commissioner.

6. It shall be the duty of the commissioners appointed under this act, whenever they are advised that any such disease has made its appearance within the limits of the State, to publish in the State paper and in at least one paper published in any county where such disease exists, a statement of the methods approved by the New York Agricultural Society for the treatment of cattle affected therewith, for the isolation of the same, for the disinfection of the premises or building in which said cattle are found affected as aforesaid, and for the prevention of the spread of the same through any agencies of whatever kind.

7. The commissioners aforesaid, and all such assistants as they may appoint, whenever in their judgment or discretion it shall appear in any case that the disease is not likely to yield to any remedial treatment, or whenever it shall seem that the cost or worth of any such remedial treatment shall be greater than the value of any animal or animals so affected, or whenever in any case such disease shall assume such form of malignity as shall threaten its spread to premises, either contagious or infectious or

otherwise, are hereby empowered to cause the said animals to be slaughtered forthwith and buried, as above provided, and to do all such things as are mentioned in the fourth section of this act.

8. The said commissioners or their assistants are hereby empowered to enter upon and take possession of all premises or parts thereof where cattle so affected as aforesaid are found, and to cause the said cattle to be confined in suitable inclosures or buildings for any time requisite in the judgment of the said commissioners or their assistants, and prior to the slaughter and burial of the said animals and the full and complete disinfecting and cleansing of such premises; and all persons, whether owners of or interested in such cattle or otherwise, who shall resist, impede, or hinder the said commissioners or their assistants in the execution of their duties under this act shall be deemed guilty, and on conviction of the same, of a misdemeanor, and shall be punishable with fine not exceeding one thousand dollars, or imprisonment not exceeding the term of six months, or of both, in the discretion of the court before which they shall be adjudged guilty as aforesaid.

9. The commissioners shall have power to establish all such quarantine or other regulations as they may deem necessary to prevent the spread of the disease, or its transit in railroad cars, by vessels, or by driving along the public highways; and it shall be proper for the governor of the State, by public proclamation as aforesaid, to enjoin all persons concerned or engaged in the traffic or transit of cattle or sheep, not to enter upon any such places, or take therefrom any such animal, or to pass through any such locality, and within such distances from the same as in the said proclamation may be prescribed.

10. The sum of one thousand dollars, or so much thereof as may be necessary, is hereby appropriated to pay to the said commissioners for their services, while actually engaged in the duties enjoined upon them in this act, at the rate of five dollars per day to each, and such further sums as may cause them actual expenditures in traveling to and from the places they may be called upon to inspect or visit, and in the printing or publishing of all regulations or notices mentioned in this act. And the further sum of fifteen thousand dollars, or so much thereof as may be necessary, is hereby appropriated out of any money in the treasury not otherwise appropriated, to pay for animals slaughtered by the provisions of this act, and the comptroller is hereby directed to pay for the same on the warrant of the said commissioners.

11. The assistant commissioners are to receive for each and every day while actually engaged in duties provided by this act the sum of three dollars per day, and all actual expenses and disbursements paid or incurred in the discharge of their duties as aforesaid, which said sums shall be a charge upon the county for which he is appointed, and shall, when duly audited by the board of supervisors of the said county, be paid by the county treasurer.

12. The slaughtering of animals for beef after having been exposed to the contagion, or supposed to have been so exposed, may be permitted by the commissioners, or prohibited by them, as they may judge proper.

13. This act shall take effect immediately, and shall continue in force for one year.

AN ACT in relation to infectious and contagious diseases of animals. Passed April 15, 1878; three-fifths being present.

The people of the State of New York, represented in the senate and assembly, do enact as follows:

SECTION 1. Whenever any infectious or contagious disease affecting domestic animals shall be brought into or shall break out in this State, it shall be the duty of the governor to take measures to suppress the same promptly, and to prevent the same from spreading.

§ 2. For such purpose the governor shall have power—

To issue his proclamation, stating that infectious or contagious disease exists in any county or counties of the State, and warning all persons to seclude all animals in their possession that are affected with such disease or have been exposed to the infection or contagion thereof, and ordering all persons to take such precautions against the spreading of such disease as the nature thereof may in his judgment render necessary or expedient.

To order that any premises, farm or farms where such disease exists or has existed be put in quarantine, so that no domestic animal be removed from or brought to the premises or places so quarantined, and to prescribe such regulations as he may judge necessary or expedient to prevent infection or contagion being communicated in any way from the places so quarantined.

To call upon all sheriffs and deputy sheriffs to carry out and enforce the provisions of such proclamations, orders, and regulations; and it shall be the duty of all sheriffs and deputy sheriffs to obey and observe all orders and instructions which they may receive from the governor in the premises.

To employ such and so many medical and veterinary practitioners and such other persons as he may from time to time deem necessary to assist him in performing his duty as set forth in the first section of this act, and to fix their compensation.

To order all or any animals coming into the State to be detained at any place or places for the purpose of inspection and examination.

To prescribe regulations for the destruction of animals affected with infectious or contagious disease, and for the proper disposition of their hides and carcasses, and of all objects which might convey infection or contagion, provided that no animal shall be destroyed unless first examined by a medical or veterinary practitioner in the employ of the governor, as aforesaid.

To prescribe regulations for the disinfection of all premises, buildings, and railway cars, and of all objects from or by which infection or contagion may take place or be conveyed.

To alter and modify from time to time, as he may deem expedient, the terms of all such proclamations, orders, and regulations, and to cancel or withdraw the same at any time.

§ 3. Any person transgressing the terms of any proclamation, order or regulation issued or prescribed by the governor under authority of this act shall be guilty of a misdemeanor.

§ 4. All expenses incurred by the governor in carrying out the provisions of this act and in performing the duty hereby devolved upon him, shall be audited by the comptroller as extraordinary expenses of the executive department, and shall be paid out of any moneys in the treasury not otherwise appropriated.

THE IMPORTATION OF CATTLE PROHIBITED.

The following is an official copy of the act passed by Congress to prohibit the importation of cattle in 1865:

"AN ACT to prevent the spread of foreign diseases among the cattle of the United States.

Be it enacted by the Senate and House of Representatives of the United States of America in Congress assembled, That the importation of cattle be, and hereby is, prohibited. And it shall be the duty of the Secretary of the Treasury to make such regulations as will give this law full and immediate effect, and to send copies of them to the proper officers in this country and to all officers or agents of the United States in foreign countries.

SECTION 2. *And be it further enacted,* That when the President shall give thirty days' notice by proclamation that no further danger is to be apprehended from the spread of foreign infectious or contagious diseases among cattle, this law shall be of no force, and cattle may be imported in the same way as before its passage.

Passed the House of Representatives December 11, 1865.

Attest:

EDWARD McPHERSON, *Clerk.*

THE BRITISH GOVERNMENT.

The following is an abstract of the rules and regulations adopted by the British Government to prevent the spread of the rinderpest and pleuro-pneumonia among cattle, and foot and mouth disease among sheep in its Indian possessions:

1. When cattle or sheep are purchased at a fair, they should always be treated as having been probably exposed to contagion.

2. When cattle or sheep are being removed from one locality to another, they should not be allowed to mix with other cattle or sheep *en route,* and should never be kept overnight in or near quarters previously occupied, as such quarters are often contaminated by having recently been occupied by diseased animals.

3. When cattle or sheep are purchased in a fair or elsewhere, they should, on being brought to the purchaser's premises, be kept by themselves, and not allowed to mix with the old cattle of the farm, at pasture, or watering time, or any other time. They should be kept by themselves in complete isolation for one month or six weeks, in order to have proof afforded whether they are affected with a contagious disease or not.

4. When cattle are traveling, or are moved from one district to another, they are

liable to be exposed to contagion and contract disease; therefore, on their arrival at home, they should be carefully inspected, and if they have passed through an infected district, they should be kept by themselves for some time. (See Rules 20 and 21.)

5. When diseases of a contagious nature, or supposed to be of a contagious nature, appear among cattle, the first important duty is to separate the sick from the healthy animals.

6. Carefully inspect all the animals, and remove to the hospital any showing the slightest symptoms of disease.

7. Divide the healthy cattle into several lots, making each lot as small in number as space will permit. Picket the cattle in such lots a good distance apart and to windward of the sick cattle. Frequently inspect each lot, and remove at once any animal in the least unwell. By steadily adopting this plan, the disease will be found in a few days to exist only among one or two lots, and by at once removing to the hospital any becoming sick, the disease will speedily be arrested in spreading through the herd. Each lot should be kept isolated from other cattle for a period from four to six weeks.

8. The hospital to contain the diseased cattle should be inclosed by a strong fence and isolated. The attendants and the sick cattle must not be permitted to leave the isolated area. Food and water may be taken to the attendants and cattle, but no forage, water, litter, clothing, or anything else should be taken from the hospital. Dogs should not be allowed to go to and from the hospital, as they may carry contagium to places where healthy stock may be.

9. The dry litter, &c., of the hospital should be burnt inside the hospital area, and the moist dung and discharges, &c., should be frequently removed from the stalls and buried in pits dug in the hospital premises. These pits should be six feet or more deep, and should be filled with the wet litter, dung, &c., of the hospital up to within two feet of the surrounding ground surface, and then quicklime and good fresh earth should be used to fill up the remaining two feet.

10. The stalls, walls, &c., and ground of the hospital should be scrupulously cleaned by frequent sweepings and washings, and after every cleansing disinfectants, lime, ashes, or even dry earth, should be plentifully scattered over the floors and ground, and the wood-work and walls should be first washed and then whitewashed.

11. The hospital should be well ventilated; sulphur fumigation should be daily carried out for an hour or so in the hospital building, and at this time the doors and windows may be closed and the ventilators only kept partly open.

12. The constant burning of sufficient litter, opposite the doors or the windward side of the building, at seasons when flies are numerous and troublesome to cattle.

13. The sick cattle should be kept scrupulously clean, and have thin gruel and fresh green grass in its season for diet. The healthy cattle should also be kept on laxative food, as cattle fed on hard dry food have the disease in a more severe form than those fed on laxative fodder.

14. When these contagious diseases have prevailed among cattle or sheep, they should not be allowed to pasture, or to be kept with unaffected herds, until a month or six weeks have expired after the last case of disease occurring among the affected lot.

15. Animals that recover should be well washed with warm water and soap prior to being removed from the hospital, and, if obtainable, carbolic acid should be added to the warm water in the proportion of one wineglassful of the acid to a gallon of warm water.

16. Carcasses of stock that die of rinderpest, black-quarter, and other forms of anthrax fever, and pleuro-pneumonia, should be buried and covered with at least four feet of earth.

17. The hides of cattle that die of these contagious diseases should be either well scored or slashed with a knife, thus destroying their value, and should be then buried with the carcasses.

18. The surface of earth floors of stalls and ground on which cattle affected with contagious diseases have been kept should be removed and buried, and the earth below should be well dug up and turned over, and the floor remade with fresh earth. Brick and stone floors may be scraped, washed, and disinfected with quicklime or carbolic acid.

19. Poles of carts and harness, or saddlery, &c., used by animals affected with contagious diseases, should be washed and disinfected.

20. The periods of incubation of rinderpest, black-quarter, and other forms of anthrax fever all believed to be within twenty-eight days; so a month has been named as the time for an animal supposed to have been exposed to the contagium of these diseases to be kept isolated.

21. The period of incubation of pleuro-pneumonia varies from two to six weeks, but has been found, as a rule, to be about forty days; so, when cattle have been exposed to the contagium of this disease, they should be kept isolated for forty-five days.

A STRANGE CATTLE DISEASE.

Mr. W. W. Lenoir, of Shull's Mills, Watauga County, North Carolina, gives the following account of a strange disease which has prevailed among cattle in that State for several years past:

SIR: Your letter, directed to my former residence in Haywood county, North Carolina, reached me after long delay, but ought to have been answered sooner. I retained no copy of the letter written by me in 1872, to which you refer, in relation to the strange disease among cattle which has been of late years in the Northwest and North, incorrectly called the Texas fever, but which was known throughout a large portion of the South for many years before the independence and annexation of Texas by the vague name of the distemper, and is still so named.

It evidently prevailed first near the coast, and a dim outline of its history and progress, and of the imperfect knowledge and erroneous theories which prevailed concerning it, can be traced in this State, and probably in other Southern States, in the legislation concerning it.

In North Carolina we have a broad belt of land adjoining the coast, stretching entirely across the State from Virginia to South Carolina, which is almost a level plain, and which extends far enough inland to include many counties and parts of counties. This belt is composed of large bodies of exceedingly rich alluvial swamp-lands, which are rarely dry enough for cultivation without artificial drainage, and which lie along the streams, and are separated from each other by bodies of level, sandy, dry land which form the remainder of this level belt.

These swamp-lands are covered with dense forests of cypress, juniper, oak, and quite a variety of other kinds of trees, many of them of immense size. The dry sandy lands between the swamps are covered almost exclusively with forests of pine trees.

Above this level belt lies another broad belt, which also sweeps entirely across the State, and is called the sand hills. The alluvial lands along the streams extend through and above the sand hills, and have a similar forest growth, but are narrower, and form sometimes swamps and sometimes rich alluvial bottoms dry enough for cultivation in grain without ditching. The uplands of the sand-hill region are composed of innumerable hillocks, and low flat ridges, and narrow plains of very sandy land, the forest growth of which is almost exclusively pine.

Above the sand-hills and extending from them to the Piedmont region, another broad belt runs across the State, which may be called the midland belt of North Carolina. This is an undulating region, composed of clay upland, interspersed with fine alluvial bottoms along the streams. This belt is almost destitute of pine, except in the old fields, of which there are far too many—lands which have been once in cultivation, and have now grown up in thickets of what are called old-field pines. The principal native forest growth of this belt of the State is oak, with an abundant mixture, however, of hickory, poplar, walnut, dogwood, sourwood, gum, and a variety of other trees.

Above this midland belt of the State comes the Piedmont region, extending to the foot-hills and lower portions of the southeastern slopes of the Blue Ridge, and including the secondary ranges southeast of the Blue Ridge, called Lauratown, Brushy, and South Mountains, &c.; and the fine Piedmont valleys of the Dan, Yadkin, Catawba, Broad, and other rivers; which lie between the smaller mountain ranges and the Blue Ridge.

The Piedmont region is marked by a surface becoming by degrees more and more undulatory, broken, and at length mountainous: by the presence still of alluvial bottoms along the streams; by a greater variety of soil as well as surface of the uplands, portions of which are here found to be somewhat sandy; by a greater variety of forest trees, and by the partial reappearance of pines, which are now found scattered over the uplands among the other trees, not in excess, but in ample abundance.

Finally, we have the mountain region, including the summit of the Blue Ridge, which in North Carolina forms the water-shed between the Atlantic and Mississippi waters, and extending from it to the Alleghany range, which forms the State line between North Carolina and Tennessee. This highly elevated mountain belt has a cool, moist, temperate, healthful climate, and a delightfully varied surface of lovely valleys and rich mountain sides. Its agricultural resources are wonderfully varied and extensive. It is a land eminently suitable for permanent pastures and meadows; and when its immense forests are subdued and its lines of transportation opened up, it will soon become the finest grazing, stock-raising, and dairying land in the United States.

Please excuse this slight outline of the State, which is interesting in itself and has some bearing on the subject.

An early statute on the subject of the "distemper," enacted in North Carolina many years ago, prohibited the driving of cattle from the pine lands in the eastern portion of the State to the oak lands in the middle portion of the State. This marked the progress the disease was making at that date, and indicated the belief that its cause existed in and was confined to the pine lands in the eastern portion of the State. But the disease has slowly crept across the midland belt and into and nearly across the Piedmont belt of this State. A recent North Carolina statute, passed I think in 1876–'77, prohibits the driving of cattle from below the Blue Ridge into Watauga county, which is on and west of the Blue Ridge, in the mountain region of the State.

I regret that I have not the means, in this secluded locality, of giving you exact dates and fuller references. I think that when I wrote the letter in 1872, to which you refer, the "distemper" had just reached Morganton, in Burke county, North Carolina, within a few miles of the foot-hills of the Blue Ridge. I am glad to be able to state that its progress, as it approaches the Blue Ridge and reaches higher elevation, seems to be slower; and strengthens the belief generally held here, that it will not get a permanent foot-hold in the cool climate of our mountains west of the Blue Ridge.

Some of the facts connected with the progress and contagious character of this disease are so strange as to challenge credulity: and yet so important and so easily verified, that it is still more strange that they are so little known, and have been subjected to so little careful and systematic investigation.

The progress of the disease over the region which it infests may be compared to that of the disease called ringworm on a surface of the human body. There is a slowly advancing angry external border around the infected region, in which border the disease is violently active, killing a large proportion of the cattle where it first makes its appearance, on many of the farms all, or nearly all. This angry border advances at an irregular rate, and presents an irregular outline, pausing in places for several years; and then perhaps advancing suddenly and destructively several miles in a single season. I think I have observed in Caldwell and Burke counties, North Carolina, and it is probably the case elsewhere, that it sometimes makes more rapid progress along the deep valleys than above them. I have not observed that it advances along the leading thoroughfares of travel and traffic, except as they conform with such valleys. In Wilkes county, North Carolina, it has made a long pause on the south bank of the Yadkin River. But I fear it is about to get a permanent foothold on the north bank. James Gwyn, Elkin, Surry county, North Carolina, who lives on the north bank of the Yadkin in Wilkes county, has lost cattle twice from it, and can inform you of its progress, in his neighborhood.

But though thus irregular in its outline and progress, this angry border which surrounds the infected region has, at all times, a tolerably definite location, and is designated among us as the "distemper" line. The region within, over which the disease has already passed, is said to be within or below the "distemper" line. The region beyond it to which the distemper has not permanently reached, is said to be above or without the "distemper" line.

The disease is most fatal in autumn, disappears after white frost, and does not reappear until warm weather in the late spring or summer.

The country below the distemper line, like the surface within the border of the ringworm, seems to be comparatively free from the disease. The cattle have become acclimated, and there are only occasional cases of the "distemper." But whenever cattle from above the distemper line are driven below it after warm weather is well advanced, or in winter, and suffered to remain there till then, there is a strong probability that they will take the "distemper" and die of it.

When cattle from below the distemper line are driven above it in winter, they may remain there permanently without any probability that they will suffer from or propagate the disease. And if cattle from below the "distemper" line, and acclimated there, are driven above the distemper line after warm weather has set in, they will thrive and fatten, and show no outward appearance of the disease. But they impart the disease in its most destructive character, especially when they have been heated by hard driving or work, to the healthy cattle around them. Cattle only passing over the road which they have traveled, it may be several days before, if it has not rained in the mean time, will take the disease and die. As cattle are very apt to smell the dung of other cattle in passing over it, it seems probable that in such cases the germs of the disease are inhaled from the dung.

Still more wonderful than this, when taken in connection with it, is the fact that the cattle thus taking the disease from apparently healthy cattle, and dying of it in its acute form, may die surrounded by healthy cattle of their own neighborhood to which they will not impart the disease. However violent such accidental outbreaks of the disease may be at the time, it never gains a permanent foothold when carried in this way far above the slowly advancing "distemper" line.

I am not skilled nor well-read in the diseases of cattle, or in other diseases. In the only book I have which treats of the diseases of cattle, a slip-shod American rehash and abridgment of a standard English work, the disease called in England red water resembles our so-called "distemper" more than any other disease described in it.

The organized and widely extended inquiries of your department might determine some very interesting and important questions concerning this disease. Has it causes which give it a spontaneous origin in certain localities in the Southern States? Can those causes be removed? What are the best methods of preventing the spread of it beyond those localities, and of suppressing it where it has already a permanent foot-hold beyond them? What is the best treatment of the animals attacked by it, &c.

A widespread belief exists that it is caused by ticks. I am sure that this is an error. But it is worth investigating for the sake of exploding it. The ticks which often prey in disgusting numbers on the cattle at the South, both above and below the "distemper line," and may well aggravate the distemper, or any other disease, are worth investigating on their own account. Cattle may be kept free from them by the regular addition of brimstone to their food or salt.

Hoping that this very meager and imperfect statement may aid you in directing a more minute and accurate investigation of this disease, which has been so fatal to Southern cattle and has so depressed their value,

I am, very respectfully, yours,

W. W. LENOIR.

Hon. WM. G. LE DUC,
    Commissioner of Agriculture.

RINDERPEST OR CATTLE PLAGUE.

The following letter, addressed to the Commissioner of Agriculture, gives the symptoms and *post mortem* appearances of the destructive disease known as rinderpest or cattle plague:

SIR: At your request I will give as fully as possible the symptoms and *post mortem* appearances of a fatal disease in cattle, known as rinderpest or cattle plague.

The disease I allude to made only one great invasion in Great Britain during the present century (in the years 1865 and 1866), and swept away many thousands of cattle, the money loss from its ravages being between ten and twelve million pounds sterling. At that time (1865 and 1866) I was in practice in a large agricultural district of England (Berkshire), and as soon as the disease visited that county I was appointed by the government one of the cattle-plague inspectors, and therefore had ample opportunities of examining large numbers of animals affected with the disease, and availed myself of the chance of making many *post mortems.*

The disease is purely contagious, and therefore preventable. It is a specific, malignant fever, indigenous to the Asiatic steppes of Russia, runs a definite course, and generally terminates fatally. It is essentially a disease of the bovine family, but may be communicated to the sheep, goat, deer, &c. It has a period of incubation varying from four to ten days; during this period the animal gives no indication of being affected.

Symptoms.—Primary fever, as indicated by a rise in the temperature; a remarkably dull and dispirited condition of the animal, which will stand with its head hanging down, ears drawn back, and coat staring, refusing all food or even water. Rumination is suspended; if made to move it shows great prostration of strength, and frequently staggers as if about to fall. The skin is hot in places, and remarkably so between the limbs; an eruption on, and a peculiar appearance and condition of the mucous membrane of the mouth is seen; it is red and furred, presenting raw-looking spots, especially on the inner side of the upper lip and along the roof. The breath is fetid, and the mucous membrane of the vagina alters to a dark-red color. These signs are rarely absent. Tears early trickle from the eyes, which are red and expressive of suffering, and a watery discharge flows from the nostrils. There is a continuous increase of these secretions, which become more or less purulent in the advanced stage of the malady; rigors and twitching of the superficial muscles, failing pulse, oppressed breathing, sores on the skin, with discharges from the same. Emphysema of the tissues of the neck and back; the extremities are cold at the commencement of the disease, and in the latter stages the increased heat of the body gives place to a remarkable coldness along the course of the spine. Secretion of milk is arrested very suddenly, the animal grinds its teeth, arches the back, moans, and shows signs of great uneasiness.

At first the bowels are constipated, but soon violent purging commences, leading to dysentery, the evacuations being slimy, liquid, and sometimes of a dirty-yellow color, tinged with blood, of a fetid character, with much straining. The urine is scanty and dark in color. The buccal membrane becomes covered with a yellowish-white material, which can be easily stripped off, showing an ulcerated surface under it. The animal now stands with great difficulty, gets quite drowsy and unconscious; the breathing short, quick, and more painful.

The animal will sometimes sink as early as twelve hours from the commencement of the attack, but in many cases the disease will be protracted to the fifth or sixth, and occasionally to the eighth or ninth day. As death approaches the mucous membranes acquire a leaden hue, with dark-colored spots on their surface. Tympanitis sets in, and the discharges from the bowels are involuntary.

The mortality in Great Britain was very great. The disease is highly contagious, and will not yield to medical treatment. Vaccination and inoculation were tried, but all seemed only to spread the pestilence.

Post mortem appearances will differ according to the part of the organism chiefly affected, and especially according to the time of duration of the malady. In many cases, the roof of the mouth will be found covered with a dirty-yellow exudation upon an ulcerated surface, the lining of the larynx, pharynx, and all the mucous membranes of the mouth is of a deep red color, and often covered with a layer partaking of the characters of lymph and pus combined, varying from the finest film to a quarter of an inch in substance. The lungs are often covered with a soft membranous exudation; emphysema of them is also very commonly found, but not always. On opening the abdominal cavity, the omentum is frequently found to present patches of redness; the intestines are altered in color, from the condition of the mucous membrane being partially seen through their walls. On cutting the rumen (or paunch) a quantity of undigested food is generally found, but nothing more than a tinge of redness in patches can be found here. The reticulum (honey-comb) does not show any signs of the disease; the omasum (manifolds) affords, in the majority of cases, very characteristic indications of the effects of the malady, its folds being inflamed in patches, or ulcerated in patches, even showing large perforations from sloughing, with claret-colored edges. The contents of this stomach are dry and caked.

The fourth, or true digestive stomach, the abomasum (rennet) is inflamed and shows specific lesions of the disease. The contents are nearly always fluid, and often mixed with blood; the mucous membrane is not only intensely red, but is studded with superficial erosions; the membrane can be easily removed from the submucous tissues, in some cases showing deep sloughs or ulcers. This condition is more marked near the pyloric region, being of a claret color.

The intestines show similar morbid changes, particularly the jejunum and the ileum, also the cæcum, which shows a peculiar mottled appearance from the accumulations, in the follicles, of a dirty-white or yellowish secretion. The liver is mostly unaffected, but the gall-bladder is remarkably full. The lining membrane of the vagina is of a dark red color and semi-detached condition.

I have given above all the early symptoms of this disease, together with the *post mortem* appearances, and I am sure you will agree with me that it differs materially from any other disease of the cow or sheep.

Sheep will take the disease from cattle; in order to test this, experiments were tried at the Royal Veterinary College, London. Sheep took the disease from cattle and died, showing the same *post-mortem* appearances. Cattle also took it from the sheep and died. Afterward Professor Simonds found a large number of sheep in England affected, from having been in company with or near diseased cattle.

Professor Law, of Cornell University, says: "Treatment of this plague should be legally prohibited under all circumstances, all the attempts of the different schools of medicine, and of empiricism have only increased its ravages; while nations and districts that have vigorously stamped it out, and excluded it, have saved their property."

I trust we have no cases of this terrible scourge in this country, and that the reports in some of the Philadelphia newspapers of the past week, of its prevalence in the vicinity of Washington, may prove to be unfounded, as I have cause to believe they are. But if ever introduced into this country, the victims and all other cattle with which they had been in contact, should be promptly destroyed and buried deeply, and the places and things with which they have come in contact be disinfected in the most perfect manner.

I have the honor to be, sir, your most obedient servant,

JOHN W. GADSDEN,
M. R. C. V. S., England.

PHILADELPHIA, *October 25, 1878.*

GLANDERS.

EXPLANATION OF ILLUSTRATIONS.

[These illustrations are photographic copies of the plates accompanying Professor Gerlach's treatise on glanders, published in the *Jahresbericht der Koeniglichen Thierarzneischule zu Hannover*, 1868. The same illustrate the morbid anatomy of glanders.]

PLATE I.—FIG. I. Development of glanders-cells of connective-tissue corpuscles in the mucous membrane of the septum. Enlargement 300.
1. Spindle-shaped cells, with a large oval nucleus.
2. The same, more swelled; nucleus larger; a second nucleus developing.
3. Cells like No. 2, but with ends blunted; more granulated and approaching decay.
4. Round cells of different size, with a large nucleus; the largest ones have a dark, granulated nucleus; beneath free nuclei and granulated detritus.
FIG. II. Microscopic cut from gray-yellowish glanders; nodules of the mucous membrane of the septum, in which (cut) can be seen spindle-shaped cells in different stages of development to round cells with a fibrous intercellular substance. Enlargement 300. At *a* the spindle-shaped cells and at *b* the round cells prevailing.
FIG. III. Development of glanders-cells of epithelium elements in the pulmonal nodules. Enlargement 300.
1. Normal cylinder-cell with a nucleus.
2. Cylinder-cell with a second nucleus developing.
3. Cylinder-cell with two and three developed nuclei.
4. Bag-shaped rudiments of cylinder-cells filled with young round cells.
5. Giant-cells with young round cells.
6. Small and large round cells with a large, dark, and granulated nucleus.

PLATE II.—FIG. IV. Lower end of the septum with glanders-nodules and ulcers. (Natural size.)
1. Various gray glanders-nodules.
2. A group of glanders-nodules with a round hole in the middle. (Incipient glanders-ulcers.)
3. A solitary glanders-ulcer.
4. Confluent glanders-ulcers with elevated borders and dirty bottom.
FIG. V. Transversal cuts through the gray nodules in the mucous membrane of the septum. (Natural size.)
a. Gray nodule in the midst of the tissue of the mucous membrane; the upper layer of the mucous membrane raised.
b. Gray nodule in the upper layer of the mucous membrane, visible on the surface.
FIG. VI. A piece of the lower border of a lung, cut surface. (Natural size.)
1. Miliary tubercles.
2. Tubercle of the size of a pea.
3. A large glanders-nodule developing.
FIG. VII. Also a piece of the lower border of a lung, cut surface. (Natural size.)
1. Miliary nodules surrounded by a red crust.
2. Large gray glanders-nodule (glanders excrescence) growing yet in one direction.

GLANDERS.

Fig.I

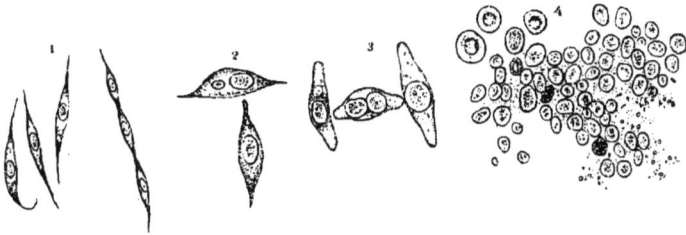

Fig.I.

Development of glanders-cells of connective-tissue corpuscles in the mucous
membrane of the septum.

Fig.II.

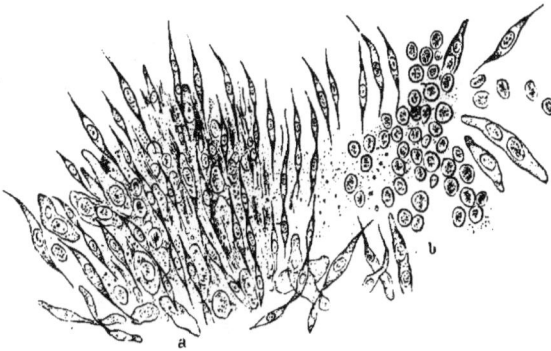

Fig. II.

Microscopic cut from gray-yellowish glanders.

Fig.III.

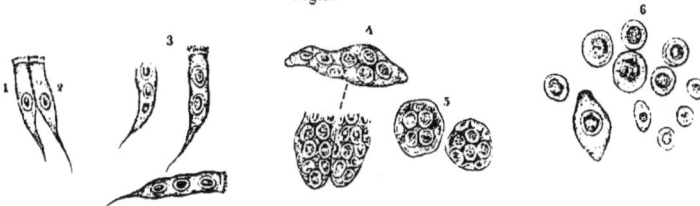

Fig. III.

Development of glanders-cells of epithelium elements in the pulmonal nodules.

A.Hoen & Co.Lithocaustic.Baltimore.

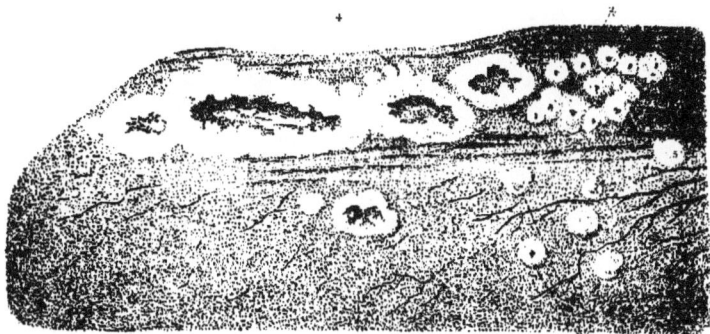

Fig IV
Lower end of the septum with glanders nodules and ulcers (natural size).

Fig V
Transversal cuts through the gray nodules in the mucous membrane of the septum
(natural size).

Fig VI
Piece of the lower border of a lung, cut surface (natural size)

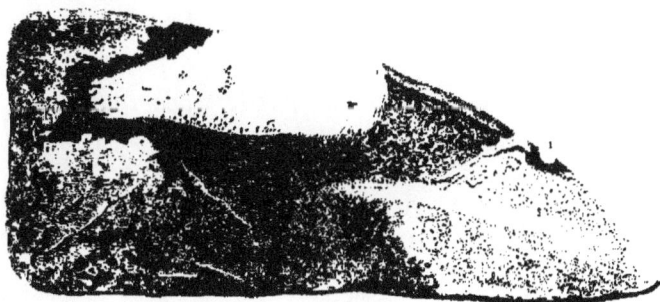

Fig VII
Also a piece of the lower border of a lung, cut surface (natural size).

A Hoen & Co.Lithocaustic Baltimore

GLANDERS.

By Dr. H. J. DETMERS, V. S., Chicago, Ill.

DEFINITION.—Glanders is a contagious disease *sui generis* of animals belonging to the genus *equus*. It has usually a chronic course, can be communicated by means of its contagion to several other species of animals and to human beings, and must be considered incurable if fully developed. The principal seat of the morbid process is usually in the mucous membrane of the nasal cavities. Three main symptoms, viz., discharges from the nose, swelling of the submaxillary lymphatic glands, and particularly ulcers of a peculiar, chancrous character in the mucous membrane of the septum of the nose, characterize glanders, and are, therefore, of the greatest diagnostic value. Wherever these three symptoms, or only two of them, are present and fully developed, there the diagnosis is secured. But unfortunately this is not always the case; sometimes two, and even all three, principal symptoms may be wanting, and still the horse may be affected with glanders. In such a case the seat of the morbid process is not in the nasal cavities, but further on in the respiratory passages, or even in the lungs. Several such cases have come to my observation, and have also been described by others, especially by Professor Gerlach. In still other cases, in which the disease might be called "external glanders," but is better known by the name of "*farcy,*" the morbid process has its principal, or even its exclusive, seat in the subcutaneous connective tissue and in the skin or cutis. The late Professor Gerlach, in his treatise on Glanders, published in the "*Jahresbericht der Koenigeichen Thierarznuschule zu Hannover,*" 1868, discriminates, in consequence of these differences, three distinct forms: Nasal or common glanders, pulmonal glanders, and farcy. As such a division of glanders proper into nasal and pulmonal glanders—farcy is described by every author under a separate head—facilitates considerably the diagnosis, and explains also at once why just those symptoms which are usually looked upon as most characteristic remain sometimes imperfectly developed, or entirely unobserved, it will be convenient to adopt Gerlach's classification.

1. NASAL GLANDERS.—This form is that which is most common, best known, and characterized by the three principal symptoms which have been mentioned.

(*a.*) *The discharge from the nose,* although the most conspicuous of those three symptoms, is really the one which is the least characteristic, or of the least diagnostic value, because several other diseases of the respiratory organs are also attended with discharges from the nose, which are more or less similar. It is true, the discharge in glanders possesses some properties which, if considered as a total, are characteristic and are not found combined in any other disease; but the difficulty is one or another of these qualities is not always sufficiently developed. Consequently, if the other two principal symptoms, the swelling of the lymphatic glands and the ulcers in the nose, are absent or not observed, the discharges

from the nose are seldom characteristic enough to serve as the sole basis
of a reliable diagnosis. The same are frequently one-sided, and, according to most authors, oftener from the left than from the right nostril.
According to my experience they are nearly, if not quite, as often from
the right as from the left nasal cavity, and, at any rate, just as often
from both nostrils as from one only, but always more abundant from one,
either right or left, than from the other. At the beginning the discharges are usually thin, almost watery, frequently greenish, or somewhat similar in color to grass juice; afterward the same appear to be
composed of two different fluids, one yellowish and watery and the other
whitish and mucus. Still later the discharges become thicker, more
sticky, exhibit frequently a mixture of different colors, are sometimes
greenish, sometimes dirty white or grayish, contain not seldom streaks
of blood, and, in advanced stages especially, particles of bone or cartilage. They have a great tendency to adhere to the borders of the nostrils and to dry there to dirty yellow-brownish crusts. As to quantity,
the nasal discharges in glanders are seldom very copious, at least not
as copious as in many other diseases—strangles, for instance. The quantity, however, varies. Sometimes, especially when the weather is warm
and dry, the discharges may be very insignificant or be absent altogether,
and, at other times, particularly if the weather is rough, wet, and cold,
will increase in quantity and become comparatively abundant. Several
authors have attached special importance to one or another of the various properties as something characteristic, by which the nasal discharges
in glanders can be distinguished from those of other diseases, but, in reality, none of those properties are constant enough, or belong exclusively
to glanders, to be alone of great diagnostic value. Solleysel and Kersting considered the stickiness as such a characteristic, but the discharges
in strangles are frequently just as sticky. Pinter and Vilét relied upon
the specific gravity; they found that the nasal discharges of glanders,
which consist partly of matter and partly of mucus, sink to a certain
extent in water, while the mucus discharges of distemper swim on the
surface. This test is of some value, but is not decisive, because matter
is sometimes admixed also to the nasal discharges of other diseases.
Others have laid stress upon the one-sidedness of the discharge, but the
latter is just as often from both nostrils as only from one, and a one-sided
discharge belongs also to some other diseases; is, for instance, observed
in a catarrhal inflammation of one of the frontal or maxillary sinuses,
if cavities in one of the three last molars of the upper jaw have effected a
fistulous opening into the maxillary sinus, if a polypus has developed
in one of the nasal cavities, &c. Professor Gerlach considers the greenish color as a very important characteristic, but that, too, is not reliable,
because it is not constant, is usually observed only at the beginning,
and belongs frequently, also, to the nasal discharges of catarrh, strangles,
and influenza, if the patients are kept on green food or in a pasture. The
nasal discharge constitutes a characteristic symptom of glanders only, if
all its essential properties are present (sufficiently developed), and are
considered as a whole. If the other principal symptoms (swelling of the
lymphatic glands and ulcers in the nasal cavity) are absent or remain
unobserved, some minor symptoms, which may happen to be present,
and the absence of all such symptoms which are peculiar to other diseases, make frequently a diagnosis possible.

(b.) A *distinctly limited swelling of the submaxillary lymphatic glands*
constitutes the second essential symptom, which is more characteristic
of glanders, and of greater diagnostic value than the discharge from the
nose. The swelling corresponds to the discharge; that is, if the latter

is one-sided, for instance, from the left nostril only, the glands of the corresponding left side of the head are affected, and if the discharge is from both nostrils the glands of both sides are swelled, but always those of that side the most on which the discharge is most copious. The swelling does not exhibit any conspicuous sign of inflammation, and is usually not painful, except at the beginning or after a sudden increase of the morbid process. It is always distinctly limited, and the swelled gland is always hard and usually of the shape and size of a peanut; may occasionally, however, be found as large as a hen's egg. Large inflammatory swellings without distinct limits do not belong to glanders. At first the swelled glands are more or less movable beneath the skin, but afterwards, in an advanced stage of the disease, the same frequently appear to be attached more or less firmly to the bone and are immovable. The swelling, unless irritated by external causes, never dissolves in suppuration like the inflammatory swellings common in distemper, and is absent only if the lymphatic glands have been extirpated, if the lymphatics have become obliterated, or if the morbid process in the mucous membrane of the respiratory passages is situated too high to be within the province of those lymphatics which are connected with the submaxillary glands, for the swelling is caused solely by a deposit of deleterious matter which has been absorbed by the lymphatics. Professor Gerlach looks upon every horse as probably affected with glanders which shows a distinctly limited, hard, knotty, and painless swelling of the submaxillary lymphatic glands. I will not contradict a man of his experience and learning, and admit that such a swelling constitutes a very suspicious and characteristic indication of glanders, especially if some other symptoms of that disease are also present; but I am obliged to remark that I have seen horses not affected with glanders in which those glands were swelled to the size of a peanut, and were hard, without pain, and movable.

(c.) *Ulcers of a peculiar, chancrous character* on the mucous membrane of the nose, and especially of the septum or cartilaginous partition between the nasal cavities, constitute by far the most characteristic symptom, and, in fact, the only one which makes the diagnosis a certainty, even if all other symptoms should be absent or imperfectly developed. Still, such is never the case; if there are ulcers in the nose, then there is also a discharge of matter mixed with mucus from the corresponding nostril. In some cases these ulcers are present, but are situated too high to be seen unless the horse is examined in bright sunlight and the rays of the sun are reflected by a mirror into the cavity of the nose. The seat of the ulcers is usually on the septum and near the nasal bone. Their size and shape vary (Fig. IV). Some ulcers are small, isolated, almost round; others are large, of an irregular shape, and of uneven depth. All produce matter, have elevated, corroded borders, a dirty, steatomatous-looking bottom, and are never covered with a scab. At first small gray specks or elevated gray spots (glanders-nodules), varying in size from that of a pin's head to that of a pea, make their appearance (Fig. IV, 1 and 2, and Fig. V, a and b). These nodules soon decay and form ulcers. Gradually the ulcers increase in size and depth (Fig. IV, 3); their borders become more elevated and corroded; the process of decay goes on; and if two or more small ulcers are close together, they become confluent, unite, and constitute one large, irregularly-shaped ulcer (Fig. IV, 4), which continues to increase in size and depth. Decay and destruction work their way deeper and deeper, even into the cartilage, and if ulcers happen to be existing in both cavities, or on both sides of the septum, it occurs not seldom that the latter becomes perforated. I observed several such cases, one especially in Lee Centre, Lee county, Illinois, in 1866, in which the hole in the lower or anterior part of the septum was fully as large as a silver half-dollar. The borders of the same appeared irregular, corroded, much swelled or elevated over the surface of the septum, and coated with a dirty-looking, discolored, and blood-streaked glanders-matter. The disease, in that case, was far advanced, and the animal about ready to die.

Sometimes it happens that a glanders-ulcer shows a tendency to heal; it loses its chancrous character; granulation makes its appearance; a scurf or scab is formed; a healing takes place, and a fibrous, whitish-colored, somewhat puckered or star-shaped scar is left behind.

Some authors have attached considerable diagnostic importance to a bluish or lead-gray color of the nasal mucous membrane, and to bluish or lead-gray spots, which usually make their appearance before it comes to ulceration. Such a bluish color, however, is not a constant symptom—in some cases only small red specks can be seen on an otherwise rather pale mucous membrane, and is not characteristic either, because it is observed also in catarrhal diseases, and in horses driven against the wind in cold weather.

(d.) *Minor symptoms.*—The three principal symptoms just described are usually accompanied by some others of minor diagnostic value, but under certain circumstances very important, especially if one or another of the principal symptoms should happen to be imperfectly developed. As such minor symptoms, may be mentioned, first, an accumulation of a glassy, whitish-gray mucus in the inner canthus or corner of the eye of the diseased side of the head. It is a symptom which usually makes its appearance at the beginning of the disease; second, a lusterless, dry, and dirty-looking, or so-called "dead" coat of hair; third, more or less difficulty in breathing; fourth, a peculiar short and dry cough, somewhat similar to the well-known cough of a horse affected with heaves. These last three symptoms, of which the cough is the most characteristic, make their appearance only after the morbid process has made considerable progress. In some cases the plain outbreak of the disease, or the appearance of plain and unmistakable symptoms, is preceded by a swelling of the inguinal, the axillary, and other lymphatic glands.

The difficulty of breathing, and the peculiar and somewhat characteristic cough, though only minor symptoms in common or nasal glanders, rise to great diagnostic importance if the morbid process has its principal seat in the lungs instead of the mucous membrane of the nasal cavities—if, in other words, the animal is affected with that form of the disease which Professor Gerlach has called "*pulmonal glanders.*"

It happens sometimes that a horse is affected with glanders and communicates the disease to other healthy animals, but does not itself show any of the three principal symptoms characteristic of that disease; has no discharge from the nose, no swelled glands, or ulcers in the nasal cavities. The late Professor Spinola, in his lectures on veterinary pathology at Berlin, related such a case to his students, which will serve as an illustration. It is substantially as follows: In a village near Berlin glanders broke out in a stable in which several horses were kept. A veterinary surgeon was called, who made an investigation and condemned every horse that showed any symptoms of the disease, and every animal condemned was immediately killed. The horses apparently not affected were kept for several weeks under police control, and from time to time inspected, but finally released. Among them was one old sorrel horse which had the heaves, and which had been brought into the stable a short time before the first case of glanders made its appearance. This sorrel horse soon after was sold to a man in another village, and came into a stable containing also quite a number of horses. In that stable, too, glanders broke out. A veterinary surgeon (another one) was called, and every horse showing symptoms of glanders was condemned and immediately destroyed. The old sorrel horse, however, which was known to have "the heaves," was again released after some length of time, together with those which had remained exempted, and was sold once more, this time to a man who kept over 30 horses (I have forgotten the exact number) in his stable a few miles from the city. In this last stable glanders likewise made its appearance after some lapse of time, but in that case Professor Spinola was called. He, too, after a careful investigation, condemned every horse that showed any symptom of glanders, and insisted upon condemning also the old sorrel horse, whose history was then unknown to him, notwithstanding that no symptoms of disease, except such as are usual attendants of heaves, could be observed. The owner hesitated to consent to the loss of a horse apparently not affected with glanders, but Spinola insisted upon the condemnation. The *post mortem* examination revealed that the old horse, which had the "heaves," was affected with pulmonal glanders in a very high degree; and Spinola, after learning the history of the old sorrel, was convinced that the latter had caused the outbreak of the disease in all three stables. Professor Gerlach, in his valuable treatise, cites several cases, which to relate would lead too far. Some cases, though not so strking as that related above, have also come under my own observation. In pulmonal glanders the morbid process has its principal seat in the lungs, and may remain limited to the latter for months, and even for one or two years; and during that time, or as long as the morbid process is confined to the lungs, no prominent symptoms may make their appearance except such as are usual attendants upon heaves—some difficulty of breathing, and a peculiar short, weak, and dull cough, which must be heard, but is not easily described. Finally, however, but not before the disease has made considerable progress, the difficulty of breathing increases, more or less discharge from the nose makes its appearance, emaciation sets in, the natural glossiness of the coat of hair disappears and becomes rough, stands on end, and exhibits a so-called dead and dirty-looking appearance. The skin, too, loses some of its natural elasticity, and the animal becomes "hide-bound."

The morbid changes are revealed only at the *post mortem* examination. Smaller and larger glanders-nodules (usually called tubercles) present themselves in different stages of development and subsequent decay in the tissue of the lungs. Some of them constitute formations rich in glanders-cells (see illustrations), and others, especially if the disease is of long standing, as decayed, cheesy, dried, and shrunk substances and glanders-tumors of a sarcomatous and fibroid character. In some of the oldest ones even a deposit of lime-salts may have taken place. I remember one case, which occurred in Germany, a few miles from my residence, about twenty years ago, when I first commenced to practice. I was called to examine a horse suffering from some pulmonal disorder. The symptoms were those of pulmonal glanders in an advanced stage of development; even nasal discharges had made their appearance. I diagnosticated glanders, but being young and without much experience, declined to take the responsibility of condemning the horse, because the laws of Germany are very strict in that respect, and provide that every horse affected with glanders be destroyed immediately. I therefore reported the case, not to the proper executive authorities, but to the veterinary surgeon-general, who, at my solicitation, came immediately and examined the animal. He did not pronounce it a clear case of glanders, but doubted, at least hesitated. The owner, however, consented voluntarily to have the horse killed. The *post mortem* examination revealed pulmonal glanders in a very advanced stage. A similar case, of which I shall have to give a brief account in another chapter, I had an opportunity to observe in 1866, near Dixon, Lee county, Illinois.

As the principal symptoms of pulmonal glanders are essentially, for some length of time at least, only such as are also observed in common cases of heaves (one of the most frequent disorders of horses), the diagnosis must frequently be based, as a lawyer would say, upon circumstantial evidence.

A horse must be suspected of being affected with glanders, first, if the peculiar, weak, and dry cough constitutes, compared with the difficulty of breathing, the predominating symptom; if the animal becomes more and more emaciated and hide-bound, and if the appearance of the coat of hair is such as to indicate the presence of a cachectic disease. Second, if it is known that the animal in question has been exposed to the contagion. Third, if other horses have become affected with glanders or farcy, after having been together with the animal that shows those symptoms. Fourth, if a horse apparently affected with heaves has previously exhibited other symptoms, more or less characteristic or suspicious, of glanders. Fifth, if other symptoms, such as are observed in so-called "nasal gleet," or incipient nasal glanders, make their appearance.

3. FARCY, OR EXTERNAL GLANDERS.

The name "farcy" is given to such cases of glanders in which the morbid process has its seat in and immediately beneath the skin, and in which nodules, boils (glanders-buboes), and ulcers of a very infectious and chancrous character make their appearance in the subcutaneous tissue, and in the skin itself. Glanders-nodules and lenticular ulcers in the tissue of the skin, boils beneath the skin, smaller and larger open ulcers penetrating the same, a strand-shaped swelling of the subcutaneous lymphatics, swelled lymphatic glands, and œdemata, the latter especially in the legs and on the head, constitute the most essential symptoms.

Professor Gerlach discriminates two forms: Subcutaneous glanders or common farcy, and exanthematous glanders or skin farcy.

(*a.*) *Subcutaneous glanders or common farcy.*—The morbid process in this rather frequent disease has its principal seat in the subcutaneous connective tissue, and in the lymphatic system of the skin and between the skin and the muscles, but especially on the inner side of the hind legs, on the hips, on the neck, between the fore legs, and on all such places where the skin is thin and fine. At first distinctly limited swellings of an inflammatory character (incipient boils or glanders-buboes) make their appearance in the subcutaneous tissue. These swellings or boils soon commence to dissolve, or to decay, from within; the ulceration begins in the center, but the matter, being very corrosive, soon works its way into the skin, the boil finally opens, and presents a farcy-ulcer with a steatomatous bottom, and elevated, corroded, and inflamed borders. At the same time, or even before the formation of the first ulcer has become completed, deleterious matter is absorbed by the nearest lymphatics, and deposited in the lymphatic glands. The former, in consequence, swell to hard and plainly visible cords or strands, and the latter to painful and distinctly limited tumors. The partial or total

closing of the lymphatic vessels and glands thus effected interferes
with, and even prevents, a performance of their functions, or stops
the absorption of lymph, and œdematous swellings, more or less extensive, are the necessary consequence. The same make their appearance especially if the seat of the morbid process is on the inside
of a leg, and if either the inguinal or axillary glands are swelled and
closed by a deposit of deleterious matter. The more extensive and complete the swelling and closing of the lymphatic vessels and glands, or
the more lymphatics are affected, the more extensive is also the œdema.
Lameness, usually caused by such an œdema, is also a frequent attendant.

The roundish boils or tumors increase in size from that of a hazel-nut
to that of a hen's egg. At first, when such a boil is making its appearance, it is not fastened to the skin; the latter can yet be moved a little
in every direction over the boil, but soon the neoplastic process and the
subsequent decay will extend to the tissue of the skin, and boil and skin
will become firmly united before the ulcer breaks and discharges its extremely infectious and corrosive contents, consisting of decaying glanders-cells or matter, and lymph.

(b.) *Exanthematous glanders or skin farcy.*—In this form of glanders
or farcy the principal seat of the morbid process is in the tissue and
in the lymphatics of the skin or cutis. It is a rare form in horses, but
the only one in which external glanders or farcy makes its appearance
in a human being. Distinctly limited swellings (nodules and tumors) of
the size of a pea to that of a hazel-nut, either isolated, or united and resembling a string of beads, make their appearance in the tissue of the
skin. These swellings soon break, and then present round ulcers with
elevated and corroded borders. The discharge consists of a mixture of
matter, composed mainly of decayed glanders-cells and lymph. In
other, though rather rare cases, the swellings are very small and numerous,
and present themselves as small nodules, some of which are so small as
to be scarcely visible, while others are about as large as common peas.
These small swellings, too, are soon changed to ulcers, which are usually
flat, lenticular, and constantly suppurating. If close together the same
become frequently confluent. Only one case of skin-farcy has ever come
under my observation. It was about five years ago, at Manhattan,
Kans. Numerous small ulcers were crowded closely together on the
nose and the muzzle of the horse, which was also affected with nasal
glanders.

On the human skin, not being covered with hair, the whole process
can be observed much better than on the skin of a horse. Professor
Virchow's description of skin-farcy in men may, therefore, find a place.
Virchow says:

At first these spots are much reddened, but very small, almost like flea-bites; then
papular swellings are formed; the surface of those swellings rises gradually rather in
the shape of a round and solid elevation than of a pustule, and assumes a yellowish
color, which gives it a pustulous appearance. If the epidermis is removed from such
a flat or roundish papule or nodule, which is not depressed in the center, but surrounded by a swelled and reddened court, a puriform, moderately consistent yellowish
fluid is formed, which contains but few organized constituents, and consists mainly of
the decayed elements of the formerly solid nodule. The fluid, therefore, is not lodged
in a pustulous elevation of the epidermis, but in a small hole in the corium, which
penetrates the latter as if it had been made with a punch. After some time the fluid
(matter) becomes colored by hemorrhagic admixtures; still later its color is changed to
bluish red, and finally small brown or blackish crusts or scabs are formed. Such eruptions appear sometimes in enormous numbers on the whole body.—(Gerlach's Treatise.)

Nasal gleet.—This is a name which I have accepted only with great re-
luctance, because it signifies no definite disease, and is used frequently,

as I shall hereafter have an opportunity to show, to cover ignorance, fraud, and crime. It can be retained only if applied exclusively to such cases of disease (usually occult or incipient glanders) in which the horse has a suspicious-looking discharge from the nose, but shows no other characteristic symptoms sufficiently developed to base upon them a sure diagnosis. So, for instance, it may happen that a horse has a chronic discharge of matter and mucus from one or both nostrils, and, perhaps, also a distinctly limited swelling of the submaxillary lymphatic glands, and yet neither the discharge nor the swelling may be sufficiently characteristic to justify the belief that the horse in question is affected with glanders, because the latter is a disease which, for obvious reasons, demands a correct and positive diagnosis. To declare that a horse has glanders is equal to condemning the same to be killed. The term "nasal gleet," therefore, is convenient and admissible, if used exclusively to signify a disorder of the respiratory organs attended with suspicious discharges from the nose, and other symptoms common in glanders, but not yet fully enough developed or sufficiently characteristic, one way or another, to make the existence or absence of glanders a certainty. Such a disorder, of course, must be considered as incipient or occult glanders till every doubt has been removed.

Chronic and acute glanders.—Glanders, as a rule, is a chronic disease. The morbid changes develop slowly. Of the various forms in which the disease is able to make its appearance pulmonal glanders, unless complicated with one of the other forms, or with other inflammatory or feverish diseases, is the most chronic, or takes the longest time to produce conspicuous symptoms and to become fatal. It takes frequently two or three years before the animal succumbs. Nasal glanders is usually not quite so slow in its progress; still it also very often takes half a year longer before the morbid process makes sufficient headway to produce plain, unmistakable symptoms, or before the chancrous ulcers, characteristic of glanders, make their appearance in the mucous membrane of the septum of the nose. Farcy, or external glanders, is usually the least chronic (comes the soonest to a termination) of the various (uncombined) forms of glanders. Plain and unmistakable symptoms (veritable farcy-ulcers) make their appearance almost always within three months and frequently within a week or two after the infection has taken place. In mules and asses, however, the various forms of glanders are usually less chronic, make a more rapid progress, are more destructive, and come sooner to a termination than in horses. The progress of the morbid process depends also to a great extent upon the constitution and the organization of the animal and the mode and manner in which it is kept. Weather and temperature, too, have considerable influence; warm and dry weather usually retards, and cold, wet, and stormy or inclement weather usually accelerates and spreads the morbid process. Most authors discriminate between acute and a chronic form of glanders. From a practical standpoint such a distinction is perfectly admissible, but to separate acute and chronic glanders as two different diseases, as has been done by some (French) authors, must lead, and has led, to very dangerous mistakes and to great confusion. Every form of glanders, as I have said before, is naturally—*es ipso*—more or less chronic in its course, but may become acute, either from the first beginning or at any stage of its development, and sometimes very suddenly, under any of the following conditions:

1. If a complication takes place either with one of the other forms of glanders or with another disease or disorder. Sometimes even a small

wound is sufficient to inaugurate the acute course or a rapid progress of the morbid process.

2. If glanders has been communicated by a direct introduction of glanders-matter into a wound, or a direct contact of the contagion with the blood. The greater the quantity of glanders-matter introduced the more concentrated the contagion inoculated, or the larger the wound the more acute or rapidly progressing and spreading is usually the morbid process of the communicated disease.

3. If the constitution of the animal has been weakened, or if the vitality of its organism has been seriously impaired either by glanders itself or by any other disease, although the course of glanders is naturally slow or chronic from the beginning, it is usually changed to an acute one as soon as the morbid changes have become sufficiently important and extensive to weaken essentially the constitution of the animal, and to cause a profuse infection or spreading of the contagion through the lymphatics in the animal organism. Toward its fatal termination glanders, therefore, always changes its course from chronic to acute. Unlike most other diseases it commences chronic and ends acute.

4. Exposure to wet, cold, and inclement weather, catching cold, hard work, close, dirty, and ill-ventilated stables, unhealthy food, &c.—in short, everything that is calculated to produce an injurious influence upon the organism, or is calculated to impair the health of the animal, has a tendency to accelerate the morbid process, to change the chronic course of glanders to an acute one, and to hasten the outbreak after an infection has taken place.

The morbid process of glanders is accelerated and caused to spread more rapidly if the latter becomes complicated with an inflammation, or with any very feverish or very typhoid disease. The morbid processes of glanders and inflammation increase each other reciprocally. The inflammatory process adopts, to a great extent, the nature and characteristics of glanders, and the morbid process of the latter disease becomes blended with the former, and assumes the attributes of an inflammation. In either case all the symptoms become very violent, and the morbid process progresses and spreads very rapidly, particularly in those tissues which are in a state of inflammation. Ulceration, too, becomes extensive in a short time, and the lymphatics, by absorbing the deleterious matter, seem to spread the contagion and the elements of glanders rapidly through the whole system. If the original disease is glanders, farcy will also make its appearance within a short time; and *vice versa*, existing farcy will soon be complicated with nasal and pulmonal glanders of an inflammatory character. The exudations produced by an inflammation which has assumed the nature of glanders are always very deleterious and corrosive and destroy like a caustic the tissues with which they come in contact. The morbid changes effected by such an inflammation resemble those of a malignant diphtheria. In extreme cases the morbid process may become so violent as to cause the neoplastic process, characteristic of glanders, to be superseded by immediate destruction and mortification. In such a case profuse, diphtheritic ulceration and destruction of tissue take the place of the neoplastic production of glanders-cells and their subsequent decay. The glanders-cells are destroyed (decay or perish) before their formation has been completed, consequently are absent.

That a direct and abundant introduction of glanders-matter into a wound, or a direct contact of the contagion with the blood, is well calculated to produce an acute form of glanders, or sufficient to inaugurate a rapid progress of the morbid process, is probably best illustrated by a case which occurred about eleven years ago, near Dixon, Lee county, Illinois, where I was then practicing. A farmer, Mr. B., came to my office with a horse which he had recently bought, and which was apparently suffering from some pulmonal disorder. The animal was in a moderately good condition and free from fever. The morbid symptoms observed consisted in a slightly laborious breathing, a short, dull, but somewhat loose (not dry) cough, some discharge from one nostril, and a slight swelling of the submaxillary lymphatic glands of the same side of the head. The symptoms, consequently, were the same as are usually observed in pulmonal glanders; but as none of them were sufficiently developed or presented sufficiently characteristic properties to indicate with certainty the presence of glanders, and as no ulcers—the most important diagnostic symptoms of glanders—could be discovered in the nose. I hesitated to make a definite diagnosis, but informed the owner of my suspicion, and advised him to put the horse, if convenient, to hard work for the purpose of accelerating thereby the morbid process (if glanders), and to return the animal for further examination within a week or so. A few days afterwards the same farmer came again to my office with another horse with a badly torn eyelid and an inflamed eye for treatment. This latter horse, which I will call horse No. 2, had been bitten in the eyelid and had the same torn by the horse with the suspicious symptoms, which I had seen before, and which I will call horse No. 1. In examining the wound, which probably had been made during the night, I found the borders very much swelled, and the wound and the conjunctiva of the eye in a condition which strengthened my suspicions of horse No. 1 being affected with glanders. Still, by means of a few stitches, I united the margins of the wound as well as circumstances permitted. After I had performed the operation I examined the horse as to his general health, but especially as to symptoms of glanders. With the exception of some feverish acceleration of the pulse and the very inflamed condition of the torn eyelid and the conjunctiva, no morbid symptoms could be found. The horse appeared to be in good health and free from any respiratory disorder. The next day I saw both horses, Nos. 2 and 1, on B.'s farm, a few miles from Dixon. Horse No. 2 had high fever; the wound in the eyelid presented considerable swelling and had suppurated; some of the stitches had been torn out; and a lump of grayish and glassy mucus had accumulated in the inner corner or canthus of the eye. These symptoms, though comparatively insignificant under other circumstances, convinced me still more that the torn eyelid would not heal and that horse No. 1 was affected with glanders, and had communicated the contagion to horse No. 2. In the condition of horse No. 1 no essential changes had taken place, except perhaps a slight increase in the discharges from the nose. About a week later horse No. 2 presented plain and unmistakable symptoms of glanders, consisting of lameness, swelling of the inguinal glands, copious discharges from the nose, swelling of the submaxillary glands, and diphtheritic ulceration on the septum. The condition of horse No. 1 was almost unchanged. Both horses were killed the next day. The *post mortem* examination of horse No. 1 revealed, besides the characteristic morbid changes in the lungs, indicative of pulmonal glanders of long standing, only a few small ulcers high up on the septum, while horse No. 2 showed all the essential symptoms of fully-developed acute nasal glanders and of incipient farcy, but scarcely any morbid changes in the lungs. Whether the inoculation with glanders-contagion effected by the biting and tearing of the eyelid constituted the first communication of the contagion to horse No. 2 by horse No. 1, or whether a previous infection had taken place (both horses had been worked together, and had been kept in the same stable a week or two before the eyelid was torn), I was unable to decide, but hold myself convinced that the direct introduction of a comparatively large quantity of the contagion into a fresh wound, and the immediate contact of the same with the blood, constituted the cause of the acute course of the disease, inaugurated by the inflammation in the wound of the eyelid. There can be no doubt of the disease having been communicated by horse No. 1 to horse No. 2, because subsequent inquiries elicited the fact that horse No. 1 had become infected with glanders several months before he came into the possession of Mr. B., by another horse to which the disease had been communicated by a condemned United States Army horse affected with glanders and sold by the government to a farmer, in whose possession he died.

Another case, perhaps not less illustrative, occurred in the same year, also not far from Dixon. I was called upon to examine a mule which showed suspicious symptoms, indicating the presence of glanders, but as no ulcers could be discovered in the nose a definite diagnosis could not be made. This, however, was the more necessary and desirable, as the mule in question had come from another State (Indiana), and had been bought only a few days before. To get out of the difficulty and to force a decision, I inoculated the mule with his own nasal discharges under the sternum behind the fore legs. In a few days a nice farcy-ulcer had developed, the symptoms of glanders proper, too, had made considerable progress, and the chronic course of the disease had been changed to an acute one.

Wherever glanders presents itself as an acute disease, either an uncommonly large quantity of the contagion has been introduced at once and brought in direct contact with the blood, or a complication of some sort has been effected.

The nature of glanders.—The hypothesis in regard to the nature of glanders, and the theories concerning the morbid changes and their relative importance, have differed very widely, and have recently undergone great changes. Although modern investigations have proved beyond a reasonable doubt that all the old hypotheses are erroneous, some of them soon get to have their adherents.

At the end of the last and the beginning of this present century most veterinarians looked upon glanders as a blood disease. Bourgelat (1779), Kersting (1784), and Coleman (1839), supposed that glanders proceeded from a morbid, corrupt, or defective composition of the blood and was the immediate cause of the disease.

Later veterinarians advanced different opinions. Dupuy (1840) called glanders an *affection tuberculeuse*, considered it, together with strangles or distemper, grease-heal, &c., as a tuberculous disease, and denied, like most French veterinarians, the existence of a contagion. Marel (1825) looked upon glanders as the natural consequence of a chronic inflammation of the nasal mucous membranes. Dance and Cruveilhier connected glanders with an inflammation of the lymphatics. Loiset found thrombosis in the lymphatics of the mucous membrane of the nose, and after that a tendency prevailed to consider glanders as a pyæmic disease. This new doctrine culminated in the hypothesis of Tessier, who denied the absorption of matter, substituted a formation of matter (pus) in the blood, and pronounced glanders as one of many diseases in which a tendency to produce matter is primarily existing in the blood. Finally clinical observations were made in France which removed (?) every doubt as to the pyæmic nature of glanders. Renault (*Recueil de méd. vétér.*, 1835, p. 396) published observations, according to which glanders

proceeded from a fistule on the withers, from bruising of the upper eyelid, and from a fistule of the spermatic cord. Dupuy (*Bulletin de
l'Académie de méd.*, 1836, p. 481) observed that glanders proceeded from a
seton on the shoulder. Riss (*Recueil de méd. vétér.*, 1837, p. 602) observed
several cases of glanders which were caused by severe contusions of the
nose. Rey observed that glanders made its appearance after a fracture
of the nasal and maxillary bones. Afterwards Renault and Bouley
(*Recueil de méd. vétér.*, 1840, p. 257) endeavored to corroborate or to affirm
these observations by direct experiments. They injected matter into the
veins of horses, and claimed to have produced glanders-ulcers in the
nose of a horse by such an injection of innocent matter. Rey (*Recueil
de méd. vétér.*, 1867, p. 417) looks upon the experiment of Renault and
Bouley as a singular case, but Professor Hering in Stuttgart (*Repertorium*,
1868, p. 36) does not find it singular at all, and says that he made the
same experiments a long time ago, and had succeeded in producing in
some cases glanders, in other cases suppuration (in the lungs), and in
others no result at all. Such statements are, to say the least, exceedingly queer, particularly if made by such a learned and experienced
man and otherwise so reliable an authority as Professor Hering, because
such observations are, and must be, based upon a mistake either one
way or another. There are three possibilities: Either the matter injected
into the veins must have been taken from a horse affected with glanders
or farcy, the animals experimented on must have been previously
infected with the disease, or exposed in some way to the contagion. A
previous infection must be considered as the most probable solution,
because the horses subjected to such experiments are usually old or
condemned animals bought for anatomical purposes at from two to four
dollars a head, or the disease produced was no glanders at all. A great
many experiments with injections of matter (pus) into the veins of
horses—probably the most that ever have been undertaken—have been
made at about the same time, but independently and at different places,
by Professor Guenther in Hanover (*Nebel u. Vix Zeitschrift*, 2. B.) and
Professor Spinola in Berlin (*Ueber das Vorkommen der Eiterknoten in den
Lungen*, 1839). The same were afterwards repeated at various times by
Professor Gerlach, the late director of the Royal Veterinary School in
Berlin, who died in 1877. Neither of these three very reliable investigators nor anybody else, except Bouley and Hering, has ever succeeded
in producing (?) glanders in a horse by an injection of innocent matter
(pus) into the veins.

All those hypotheses and theories, notwithstanding some of them
were only short-lived, contributed a great deal in creating the confusion in regard to the contagiousness or non-contagiousness of glanders
(*la morve*), which, until recently, has been prevailing among the French
veterinarians. Bouley separated acute glanders and chronic glanders as
two distinct or entirely different diseases, and considered chronic glanders as non-contagious, and acute glanders and farcy as contagious and
pyæmic diseases. Godine (*Elémens d'Hygiène vétérinaire, suivis de re-
cherches sur la morve, etc.*, 1815), went still further, and denied the contagiousness of glanders altogether. Bouley, however, finally admitted
that contagious acute glanders might, under certain circumstances, be
developed from non-contagious chronic glanders. These fallacious doc-
trines of the professors of the Alfort veterinary school, not only caused
great confusion in regard to diagnosis (glanders not being considered as
a disease *sui generis*, was frequently confounded with other diseases),
but also great losses, amounting to millions of dollars, to the people of
France, by preventing a strict condemnation of glandered horses, and
allowing thereby an unlimited spreading of the disease.

The veterinarians of Belgium, too, became infected with the French or rather Alfort confusion, otherwise they never would have stated in their official reports (*Bulletin du conseil supérieur d'agriculture du royaume de Belgique Arme*, 1858, Bruxelles, 1860), that of 810 glandered horses, 136 had been cured. The veterinary school of Lyons, France, has always kept aloof from the errors of the Alfort institution in regard to glanders, and has never denied the contagiousness of that disease.

The German veterinarians, though differing at times considerably in opinion as to the nature of glanders, have never doubted its contagiousness; and German governments have always been very strict in taking the most effective measures against the spreading of that terrible enemy of the equine race by requiring a prompt destruction of every horse reported by a veterinary surgeon as being affected with the disease. As a consequence, glanders has become a rare disease in Germany, and the annual losses are very insignificant.

Most of the older German veterinarians looked upon glanders as a dyscratic disease. Some believed they had found the immediate cause in a qualitative change of the animal albumen; others, in a morbid increase of fibrin. As to the morbid changes, some thought they had discovered something characteristic in a stagnation of lymph in the lymphatics, others in a formation of tubercles, and still others considered glanders as a product of scrofulosis. A few went even so far as to hold glanders to be identical with tuberculosis and scrofulosis. The tuberculosis doctrine originated in France, and gained a good many adherents willing to look upon glanders as an equine tuberculosis. The scrofulosis doctrine was based upon the erroneous supposition that glanders proceeds or develops from strangles or distemper, and that the latter is a scrofulous disease. Erdt (in his *Rotzdyscrasie und ihre verwandten Krankheiten*) declared glanders, as recently as 1863, to be a dyscratic disease, and discriminated a scrofulosis, blennorrhœic, septicamic, carcinomatous, syphylitic, and other forms of glanders, but considered scrofulosis glanders as the generic form. Professor Gerlach, in his valuable treatise from which several of the notes just given have been taken, refutes the theories of Erdt by the following statement, for the correctness of which I can vouch from my own knowledge of the facts:

The breed of the milk-white (white-born) horses of the royal stables of the late Kings of Hanover was kept pure by continuous in-and-in breeding. As a consequence more than half of the number of colts born perished every year of scrofulosis diseases. At the *post-mortem* examinations the mesenterial glands presented every stage of scrofulosis from simple swelling to a cheesy degeneration. Still, never a case of glanders has occurred, neither among the colts nor among the grown horses. This proves that scrofulosis really makes its appearance in colts, and that in exactly the same form as in children, and that it is therefore not justifiable to attribute an entirely different disease of horses to scrofulosis.

For our present better knowledge of the nature and the morbid anatomy of glanders we are indebted especially to the thorough, unbiased, and scientific researches and investigations of Professors Virchow (*Handbuch der speciellen Pathologie*, Bd. 2, and *Die krankhaften Geschwulste*, Bd. 2); Leisering (*Bericht ueber das Veterinairwesen im Koenigreich Sachsen*, 1862 und 1867); Ravitsch (*Virchow's Archiv*, Bd. 23); Rolop, (*Magazin von Gurlt und Hertwig* Bd. 30), and Gerlach (*Jahresbericht der Koenigl. Thierarzneischule zu Hannover*, 1868).

THE MORBID PROCESS.

Glanders commences as a neoplastic process—new morbid formations (glanders-cells) are produced. The mucous membrane of the respiratory passage, the lungs, the subcutaneous tissue and the cutis, and, occasionally, some of the connective tissues of other parts of the body, constitute the primary seat of the morbid changes. The lymphatic vessels and glands become secondarily affected. The neoplastic process, however, does not in every case of glanders occur in all those tissues named; its seat in a certain tissue determines the form of the disease. In common or nasal glanders the morbid changes have their main seat in the mucous membrane of the nasal cavities, and of the maxillary sinuses; in pulmonal glanders the same make their appearance principally in the lungs; and in farcy the neoplastic process is taking place either in the subcutaneous connective tissue (common farcy), or in the cutis itself (skin-farcy). In other tissues, morbid changes, as a general rule, occur only if glanders has become complicated with another disease—an inflammatory process, for instance. The products of the neoplastic process consist of round cells, and of spindle-shaped cells. The latter, usually, undergo further changes; some of them develop to round cells, and others serve as the elements of excessive or morbid growths of connective tissue, which, however, do not pre sent anything characteristic, and must be considered as subordinate products of the neoplastic process. The round cells are in shape and form similar to granulation-cells and matter-corpuscles, but vary in size from that of the latter to two, three, four, five, and in some cases even ten times as large. The youngest round-cells, or those latest produced, present rather delicate outlines, and are the smallest; the oldest ones, which are distinguished by their granulated contents and their dark color, are the largest, and sometimes very large. All have large nuclei, which grow in the same proportion as the cells, and present in the older ones a dark, granulated appearance. (Fig. I, No. 4, and Fig. III, No. 6.)

The formation of these cells constitutes the real formation of all the morbid changes in glanders, and may, therefore, be considered as something characteristic of the disease, and the cells themselves are appropriately designated as glander-cells. These glander-cells have two different sources; they proceed from connective-tissue corpuscles, and also from epithelium-cells.

1. *Development of glanders-cells from connective-tissue corpuscles.*—The latter become proliferous and swell; the nucleus of each cell or corpuscle grows larger; a second and a third nucleus are produced within the walls of the cell, but not by a division of the first one. The other contents of the cell gradually granulate, the appendages or extensions drop off; finally the whole body of the cell decays. The nuclei become free; the nucleus-envelope or membrane expands, and becomes distinct from the interior, and the metamorphosis of a nucleus into a nucleated cell is thus completed. Such a new cell presents at first a very delicate contour and a large and bright nucleus, but, under favorable circumstances, will soon become firmer and grow larger. Under unfavorable conditions no further development will take place. (Fig. I, Nos. 1 and 4.)

2. *Development of glanders-cells from epithelium-cells.*—A process of proliferation makes its appearance in the tesselated and cylindrical epithelium-cells, is plainest, however, in the latter. At first the oval nucleus increases in size; then a second, and finally a third nucleus are formed at a little distance from the upper obtuse end of the first, which is not divided. The formation and growth of these nuclei cause the cylindrical cell to increase in size, or to swell, and to change its original shape till it is transformed to a mere bag filled with nuclei and small round cells. Finally the bag or the old cell-membrane decays and breaks, and the nuclei and young cells are liberated. (Fig. III, Nos. 1

and 4.) Such a production or development of glanders-cells just described can take place in young or undeveloped and incipient epithelium-cells, because round giant-cells filled with nuclei and small round cells are formed frequently in the deeper or youngest strata of the epithelium. (Fig. III, No. 5.)

Wherever such a neoplastic growth is making its appearance the process is always essentially the same. The original nuclei of the primary epithelium-cells and connective tissue-corpuscles increase in size, and new nuclei are formed within the external membrane, or envelope, of the primary cells. These nuclei are transformed into small round cells, which are liberated by the decay of the old mother or brood-cells, and constitute what is called daughter-cells, and grow larger. This growth and development constitutes a characteristic peculiarity of the large round glanders-cells, which distinguishes the same from otherwise similar granulation-cells, matter-corpuscles, and tubercle-cells, because the latter, during their whole existence, remain unchanged at their first stages of development. Although young glanders-cells are small, and large ones old, the difference in size does not depend exclusively upon the age of the cells. Other growth-promoting and growth-retarding influences must be existing, because some cells grow faster than others, and some do not seem to grow at all. Under certain circumstances only small cells can be found, which are not different from common matter corpuscles, and in other cases a great many large ones, sometimes of an extraordinary size, present themselves. If the morbid process is a violent or a very rapid one, the glanders-cells are always small; rapid development and a fluid intercellular substance constitute the agencies which deprive the cells of their ability to grow, or cause them to remain small, and of a somewhat uniform size. Consequently, in all those cases in which the morbid process of glanders is blended from the beginning with more or less inflammation and exudation, the glanders-cells will be small and numerous; and as the imflammatory exudations destroy and dissolve the intercellular substance, the latter and the exudations themselves will constitute a fluid in which the glanders-cells are kept suspended. The glanders-matter thus formed does not present, under the microscope, any characteristic differences from any other matter or pus. A production of glanders-matter and of numerous small glanders-cells is common if the neoplastic process has its seat in the subcutaneous and intermuscular connective tissues consequent in farcy. In all those cases, however, in which glanders presents itself as a chronic disease, free from any complications with inflammatory processes, &c., whatever, in which the formation of the glanders-cells is a gradual and slow one, and in which the intercellular substance is not destroyed and dissolved, the glanders-cells will grow to a certain size, and young cells with delicate contours and large, bright nuclei, older and larger ones, and very large ones with dark-colored nuclei and granulated contents, will present themselves.

The vitality of the neoplastic products of glanders is limited, but differs considerably according to circumstances. The small, rapidly produced, and therefore numerous, cells, suspended in a dissolved intercellular tissue and exudations, are similar in every respect to matter-corpuscles; the same not only do not grow, but shrink and decay very soon. If the intercellular substance does not decay, but retains its original connective properties, the glanders-cells not only grow larger, but also a great deal older, than matter-corpuscles or tubercle-cells. This vitality will be the greater the larger the space or the greater the amount of the connective intercellular substance between the single cells. Their

age, however, probably never exceeds a year or several months, notwithstanding that some glanders-nodules, tubercles, and tumors may
exist, apparently unchanged, a much longer time, because the constituents of the latter, the glanders-cells, change. Old ones decay, and new
ones take their place even if the whole tubercle or tumor remains essentially as it is. It is to be supposed that such a change is taking place,
because every old glanders-tubercle or tumor contains always old and
new cells in different stages of development.

The retrogressive metamorphosis may be called a fatty necrobiosis.
At first small granules (fat granules) make their appearance in the
nuclei; the latter swell or increase in size, and grow darker; granules appear also within the cells, but outside of the nuclei; finally the
envelopes or external membranes of the cells decay and fall to pieces,
and a granulated detritus is left behind. Therefore, after a regressive
metamorphosis has set in, the glanders-nodules or tubercles and tumors are found to contain a granulated detritus, small and large
granulated cells, and free granulated nuclei, if examined under the
microscope. The glanders-cells may thus perish or be destroyed
without any simultaneous decay of the intercellular substance. In
such a case the further changes which are going on in the tissues, in
which the glanders-cells are imbedded, differ according to circumstances. If the glanders-cells are but few, and rather far apart, the
granulated detritus is removed by absorption, and the morbid process
comes to a termination by local healing. In other cases new glanders-
cells are produced, and take the place of the old ones, and the morbid
growth (tubercle or tumor) continues to exist. If the decaying glanders-cells are numerous and lodged close together, the retrogressive
metamorphosis is usually attended with a morbid or excessive growth
or production of intercellular connective tissue; and the absorption of
the detritus in such a case is attended with, and makes room for, a somewhat extensive production of new fibrous (scar) tissue; linear and somewhat prominent, white stripes, usually uniting in a common center, corresponding to the center of the former neoplastic process, make their
appearance and constitute a star-shaped, whitish scar or cicatrix. In
chronic glanders such cicatrices occur very often in the mucous membrane of the septum; the hard, fibroid, and callous swellings, which
are sometimes found in the mucous membrane of the nose, and the
fibroid tumors which occur in the lungs, and which are easily distinguished from the more pulpy glanders. Nodules and tumors are produced in the same way.

Frequently, however, that is, in all such tubercles and tumors in
which the glanders-cells are numerous and separated only by very little
intercellular tissue, the decay or retrogressive metamorphosis of the
glanders-cells involves and causes a simultaneous decay and destruction
of the intercellular substance, and of the tissue in which the morbid
products are imbedded. The continuity is destroyed, and an abscess is
formed. The decay usually, though not necessarily, begins in the center of the indus of cells, and it seems that certain external influences
are able to change or to accelerate the whole process. So, for instance,
a general decay or a formation of ulcers or abscesses does not usually
take place in the mucous membrane of the maxillary cavities, but almost
invariably, or, at any rate, a great deal earlier in such parts of the nasal
mucous membrane, which are exposed to the current of air passing
through the nose at each breath. The irritation caused by the passage
of air probably constitutes the cause of the more frequent occurrence of
glanders-ulcers in the mucous membrane of the septum than in any

other part of the nasal mucous membrane. If glanders has become complicated with inflammation, the whole process, as has already been mentioned, is entirely different. In farcy, too, in which the morbid changes have their seat in the loose subcutaneous connective tissue, the abscesses are formed in a somewhat different way.

The infectiousness of the neoplastic products of glanders constitutes a specific and pathognomonic attribute of the same, which excludes identification with any other otherwise similar neoplastic or morbid products. The same specific agency, or the same virus, which is instrumental in communicating the disease from one animal to another, constitutes also the cause which spreads the morbid process within the organism of the affected animal. The efficiency does not seem to be dependent upon any particular shape or form of the morbid products, but to be inherent in the material, because not only the live glanders-cells, but also the dead or decayed ones, the granulated and cheesy detritus, and the watery transudations are infectious. The immediate changes produced by a local infection within the tissue, or the creeping of the morbid process from cell to cell, can be seen only under the microscope. If the glanders-process is not complicated, that is, if no other disease is existing, the spreading of the morbid process, or the progress of the local infection, is a very slow one, but is accelerated or becomes rapid if a complication sets in. The morbid process, however, spreads not only by means of a direct infection from cell to cell, but also by means of the lymphatics, which absorb infectious elements and deposit the same in the nearest lymphatic glands. That this is the case becomes evident if an animal is inoculated with glanders-virus. The lymphatics proceeding from the inoculation wound soon commences to swell like strands or chords, and undergo not seldom ulcerous decay. The lymphatic glands, too, commence to swell to solid and painful tumors which afterwards become harder and firmer, but less painful. A morbid production of connective tissue causes the firmness of the swelling, and usually renders such a diseased gland impervious to a further passage of the contents (lymph and infectious glanders elements) of the lymphatics, and prevents, therefore, a further spreading of the infection. If, however, a lymphatic gland thus degenerated becomes finally itself a seat of the neoplastic glanders process, or of the production of glanders-cells, the lymphatics which pass from that gland to another one will also absorb infectious material, and cause thereby a further spreading of the infection and of the morbid process. In nasal glanders, a swelling of the submaxillary lymphatic glands (which receive directly through the lymphatic vessels the lymph from the seat of the morbid process), unattended with any affection whatever of the lymphatics beyond them, is a very frequent occurrence. Hence the spreading of the morbid process by means of the lymphatics is also a usually slow one in chronic glanders; several months may elapse before a new source of infection is formed. The spreading, however, will be a comparatively rapid one in all cases of glanders in which a complication with another destructive or acute disease, as an inflammatory process, has taken place. The morbid process is also apt to spread more rapidly through the lymphatics in common farcy in which loose connective tissue constitutes the seat of the disease. The morbid process of glanders, therefore, is infectious; a spreading of the same is not only effected within the tissue by a propagation of the glanders-cells, but also by means of the lymphatics which absorb the virus and carry the same to the nearest lymphatic glands, where the progress of the morbid process stops if the latter are degenerated by an excessive production of connective

tissue, but proceeds further if those glands become the seat of a
neoplastic production of glanders-cells, as is usually the case in farcy,
and always if glanders is complicated with inflammation.   It is evident
that by such a spreading of the virus and absorption of deleterious
glanders-matter some infectious elements, whatever their nature may
be, will finally pass into the blood, and cause in that way a general disorder, or a general dyscratic condition usually called "glanders-dyscrasy."
That virus or infectious elements pass over into the blood, and pervade
the whole animal organism, becomes apparent by the fact that the blood
and the various animal secretions, the sweat for instance, possess contagious properties already at an early stage of the disease, or before the
morbid process has spread much beyond its original seat, and are able
to communicate the glanders from one animal to another.   It may appear to be somewhat strange that the early infectiousness of the blood
and of the various secretions does not effect a general outbreak of the
glanders-process in every suitable part (mucous membranes and connective tissues) of the animal body, and that, notwithstanding the facility
with which the glanders-contagion communicates the disease from one
animal to another, the morbid process remains usually for a long time
confined to certain parts of the organism.   It is, however, not any more
surprising than a healing, or a cessation of the morbid process, of other
equally contagious diseases—pleuro-pneumonia of cattle for instance—
while the organism is yet replete with the contagion, which, in very
small quantities, is able to communicate the morbid process to other
animals.   The truth is, our knowledge concerning the true nature of the
contagious principle of the various contagious diseases is yet too limited.   If the theories of Hallier and others, based upon the discovery
of micrococci, &c., in the blood and in the secretions of animals affected
with contagious diseases should prove to be correct; if, in other words,
those micrococci—in glanders *Malleomyces equestris*, H.—do constitute
the infectious elements, and the real, immediate cause of the morbid
changes, all those strange phenomena may yet find a satisfactory explanation.   If, however, those micrococci should not constitute the contagious, and should not be the cause of the morbid process, but the
product of the same, or if their presence should prove to be a merely
accidental one, it will be difficult to reconcile those facts.   Professor
Gerlach, who discards those theories as unfounded, hints at an exhaustion of predisposition as affording a possible explanation.

THE ANATOMICAL CHANGES.—The morbid products of the glanders-
process make their appearance usually in more or less distinctly limited
nests, or in shape of nodules or tubercles and tumors, which vary considerably in size.   Some of them are as small as the size of a pin's head,
and are called miliary tubercles; others are larger, of the size of a pea;
and still others are quite large, and constitute tumors or glanders-ex-
crescences.   Practically, therefore, a discrimination between glanders-
tubercles or small nests of glanders-cells, and tumors or large ones,
is admissible.   The former, however, must not be looked upon as
identical with genuine tubercles as occurring in tuberculosis.   A
glanders-tubercle is a different thing altogether, only the name has
become too convenient to be abolished.   Glanders-tubercles occur—1,
in the substance and in the subserous tissue of the lungs; 2, in the
mucous membrane of the nasal cavities and of the maxillary sinuses,
but especially in the mucous membrane of the septum; 3, in the swelled
and indurated submaxillary glands; and, 4, in the cutis.   Some authors have considered the presence of small miliary tubercles in the
lungs as the criterion of the presence of glanders, but others have

found that glanders may exist and still no tubercles may be found in the lungs. Professor Roell, in Vienna, found miliary tubercles in only about 66 per cent. of all cases that came under his observation, and Professor Leisering, in Dresden, and Professor Gerlach, in Berlin, searched for them frequently in vain. Glanders-tubercles make their appearance in the lungs only if the morbid process, which has its principal seat usually—I would like to say, normally—in the mucous membrane of the nose, extends to the lungs; or if original nasal glanders has become complicated with pulmonal glanders, which, in the course of time, is a common occurrence. In those cases in which such a complication is existing from the beginning, or in which pulmonal glanders constitutes the primary disease and nasal glanders the complication, miliary tubercles are found in the lungs frequently within a short time after an infection has taken place, sometimes within from one to three weeks. The same are imbedded in the healthy pulmonal tissue, are surrounded by a court of turgid blood-vessels (Fig. VII, No. 1), have each a small blood-vessel of their own, are at first grayish-white and rather soft, consist of more or less uniform and rather small round cells, with nuclei, connected with each other by a delicate intercellulary tissue, and become, when older, enveloped by a fine tissue of connective fibers. The court of turgid or congested vessels around the tubercles disappears after some time, the blood-vessel which enters the tubercle becomes obliterated, and the substance of the latter, receiving no more nutriment, undergoes decay. A necrobiotic process commences, the round cells shrink, the intercellulary substance decays, and the interior of the tubercle is changed to a cheesy substance, in which finally lime-salts are deposited. The whole process is the same as that which is taking place in a true tubercle in tuberculosis, therefore every difference disappears after the retrogressive process has set in. Hence, glanders-tubercles have frequently been identified with veritable or tuberculosis tubercles, and glanders itself has, at times, been looked upon as a tuberculosis of horses, which assumes peculiar forms, different from tuberculosis of other animals; but as real common tuberculosis occurs in horses as an independent disease, the same as in other animals, as the cells of a glanders-tubercle are usually somewhat larger than those of a genuine (tuberculosis) tubercle, and as, finally, each glanders-tubercle possesses a full intercellulary substance, and has a blood-vessel of its own, either of which is wanting in the veritable (tuberculosis) tubercle, there can be no doubt as to glanders and tuberculosis of horses being entirely different diseases. Besides that, in tuberculosis of horses, the single tubercles are usually a great deal larger than the miliary tubercles of glanders, and only the smallest ones (those of the size of a pea) present some similarity to the larger glanders-tubercles. The retrogressive process does not present anything characteristic.

In the mucous membrane of the nose the glanders-tubercles or nodules are always plainest on the septum (Fig. IV, Nos. 1 and 2). They, too, vary in size from that of a pin's head to that of a pea, and project but little over the surface of the membrane, and are therefore sometimes scarcely visible. At a *post mortem* examination, however, the same can be seen and felt more plainly, because then the mucous membrane is less succulent and swelled. Either singly or in groups they are imbedded in the mucous membrane, usually in the upper layer, and are distinguished from the reddened membrane by their gray, grayish-white, or grayish-yellow color. Sometimes these tubercles, or glanders-nodules, are situated deeper, in the middle or lower layer of the mucosa, and therefore less distinctly circumscribed, and indicated only by a slight elevation

above the surface of the membrane. but not by any distinct color. On
a cut, however, the same can be seen very plainly (Fig. V, *a* and *b*).
The substance of the glanders-nodules in the nose is more or less soft,
and consists of round cells, free nuclei, spindle-shaped cells, and a fine
connective intercellular substance. The spindle-shaped cells are lodged
mostly side by side; some of them, the younger ones, are rather thin,
and others are swelled in the middle, and are ripe and near breaking.
The nodules or glanders-tubercles present usually a gray-yellowish color,
if composed principally of round cells, and their color is somewhat indistinct if spindle-shaped cells constitute the prevailing element. The
retrogressive metamorphosis consists in a decaying to a fatty or cheesy
substance. A real shrinking and exsiccation and a deposit of lime-salts
do not occur. Glanders nodules or tubercles in the cutis are a comparatively rare occurrence in horses, but are observed very often in human
beings affected with glanders. As the skin of horses is coated with hair,
only the larger tubercles or nodules will be noticed; the very small ones
usually escape observation till the regressive process has been completed,
and has changed them to small lenticular ulcers. Otherwise the morbid
changes are the same as in the mucous membrane.

Miliary tubercles, finally, can also be found imbedded frequently in the
morbidly increased connective tissue of the indurated submaxillary and
other lymphatic glands. On a cut the same can frequently be pressed
out of the surrounding tissue as small knots or nodules. An exsiccation
is a frequent occurrence, but a deposit of lime-salts has not yet been
observed.

Glanders-tumors, or very large nests of glanders-cells, can be found
fully developed only in the lungs, but are even then not as frequent as
the tubercles. They have their seat usually immediately beneath the
pulmonal pleura, especially toward the lower sharp border of the lungs.
In some cases, however, the same are also found imbedded in the pulmonal tissue, and are then not often numerous. The tumors, or glanders growths, are either distinctly limited, and varying in size from that
of a cherry to that of an apple, or the same are more or less diffuse.
The large tumors seem to be composed of two or more smaller ones which
have increased in size till they have come in contact with each other
and have united. The intermediate pulmonal tissue in such a case has
disappeared. Large tumors thus produced are frequently of an irregular shape. The pulmonal tissue surrounding the gray or grayish-yellow tumors is at first hyperæmic, and the outlines of the latter are more
or less indistinct, but afterwards the same become more defined. On a
cut these tumors present an appearance somewhat similar to bacon. In
some cases the same are more or less firm and solid, like a fibroid growth,
and in others of the consistency of a sarcoma. (Fig. VII, No. 2, presents the grayish-yellow cut surface of a glanders-tumor in natural size,
for the most part distinctly limited from the hyperæmic pulmonal tissue,
but at one end yet encroaching upon the latter, and not yet presenting
a distinct demarcation. Fig. VI, No. 3, is a smaller glanders-tumor in
natural size. presenting yet visible, small, round, primary nodules and
some remnants of pulmonal tissue, indicating plainly that the growth
takes place, not from one but from several centers, and is not effected
by peripheric apposition.) Under the microscope the constituents are
found to be essentially the same as those of the smaller nodules or tubercles. The round cells, however, vary much more in size. Some are
very large and distinguished by their dark and granulated nuclei.
Numerous epithelial mother-cells, containing nuclei and incipient cells,
spindle-shaped cells in different stages of development, some, maybe,

very much swelled or just breaking, and others decayed and discharging their granulated contents and large nuclei, and a connective intercellular substance which gives the whole tumor its continuity and a certain degree of solidity, constitute the principal components. The softer glanders-tumors, similar in consistency to a sarcoma, are composed mainly of round cells, while the firmer or more solid ones consist principally of spindle-shaped cells, and contain comparatively few round cells imbedded in the intercellular substance, which latter is here and there fibrous and solid, and thereby the cause of the greater firmness. The presence of both kinds of cells, spindle-shaped and large, round ones, proves that connective-tissue corpuscles, as well as epithelium elements, contribute to the formation of pulmonal glanders-tumors. The retrogressive metamorphosis proceeds, according to the observations of Gerlach, in two different ways. Sometimes all components of the glanders-tumor, the intercellular substance as well as the glanders-cells, undergo a process of decay which proceeds either from one center—if the tumor is a simple one—or from several centers simultaneously, if the tumor is a complicated one. In the former case the whole tumor is changed to one cavity with cheesy contents, but in the latter two or more larger or smaller cavities, corresponding to the number of the original tubercles or tumors, are produced. The contents of the same present also a cheesy appearance. Sometimes, however, the whole process is different. The round-cells decay and are absorbed, and an excessive growth or production of connective tissue is taking place. The tumor becomes harder and firmer, and assumes finally the characteristics of a fibroid growth, which contains interspersed in its tissue a few round-cells, and may not undergo any further changes for a long time. Such fibroid tumors correspond to the fibroid cicatrices which occur frequently in the mucous membrane of the septum, and are found not seldom if the morbid process has been a very slow or chronic one. If glanders is acute or complicated with other morbid processes which accelerate its progress, such hard and firm fibroid tumors or cicatrices are never formed. On the contrary, the glanders-tumors decay rapidly, often before the same have had time to assume definite shape and form.

Glanders-ulcers or abscesses are produced if the intercellular substance of the tubercles undergoes dissolution. Dissolved intercellular substance and decayed and decaying glanders-cells constitute the matter. The process is about as follows:

Farcy-ulcers in the subcutaneous connective tissue. The development or the growth of a farcy-tumor is always attended with some local inflammation in the surrounding tissues. A violent proliferation begins in the center of the tumor, and numerous small round-cells which can scarcely be discriminated from matter-corpuscles are produced. The inflammatory process furnishes a sufficient quantity of exudation to loosen and to envelope the round-cells almost immediately after the same have been produced. Some white blood-corpuscles may become intermixed, but the same must be regarded as strangers, because a very large majority of the cells suspended in the fluid exudation are the product of the proliferous process. So it may happen that a farcy boil or tumor shows fluctuation, and contains matter within a few days, or is changed to an abscess much sooner than a common boil. The matter of a farcy-ulcer does not exhibit any distinctive difference from other pus except in so far as it possesses infectious qualities. Almost as soon as a farcy-boil has been changed to an abscess, or contains matter, the nearest subcutaneous lymphatics commence to swell to plainly visible chords or strands, and in their course not seldom new boils are formed, which also undergo the same metamorphosis as the first one. Hence it happens very frequently that farcy boils and ulcers make their appearance in rows somewhat resembling strings of beads, which constitutes one of the characteristics of the disease. A little later the nearest lymphatic glands, too, commence to swell and to be changed to hard and more or less painful farcy-buboes. The circulation or the current of lymph in the lymphatics of such a swelled gland or glands becomes interrupted, and in consequence œdematous swellings make their appearance in the parts in which such an interruption has been effected, usually in a leg. The swelling of the lymphatics and of the lymphatic glands, the lymphatic abscesses, and the appearance of œdemata have led to mistakes; an inflammation of the lymphatics has been supposed to constitute the primary and the production of farcy-ulcers a secondary morbid process. Sometimes, it is true, it is rather difficult to find the primary boils or ulcers from which the morbid process has spread. The comparatively rapid dissemination of the glanders-virus through the lymphatics in the loose subcutaneous connective tissue explains why farcy usually spreads sooner over the whole body, and becomes fatal in much less time than either pulmonal or nasal glanders.

The products of the glanders-process, however, do not always present themselves as distinctly limited growths in form of nodules, tubercles, tumors, and boils. The morbid products in certain cases, especially in such in which an inflammatory exudation is taking place in the same parts in which the glanders-process has its seat, become diffuse, and the glanders-cells almost as soon as produced are carried off by the exudation. Gerlach discriminates two forms of diffuse glanders, viz., glanders-catarrh and diffuse production of glanders-cells in the mucous membranes.

1. *Glanders-catarrh.*—If the glanders-process makes its appearance in a mucous membrane, the first morbid changes and symptoms are always those of glanders, blended with a catarrhal affection. Consequently the first stage of nasal glanders may appropriately be called a "glanders-catarrh," and may under favorable circumstances exist almost unchanged for a long time without being attended by any other characteristic symptoms except perhaps some swelling of the submaxillary lymphatic glands (so-called nasal gleet). Afterward, in a more advanced stage of the disease, more characteristic morbid changes make their appearance, but the catarrhal discharge from the nose remains. In glanders-catarrh the secretions of the nasal mucous membrane differ only in so far from those observed in a common catarrh as they present frequently a greenish or green-yellowish color, and contain very soon epithelium-scales and small, round glanders-cells similar to matter-corpuscles. With the appearance of the epithelium *débris*, however, the somewhat characteristic greenish color usually disappears. The glanders-cells have their source in the epithelium-producing layer of the mucosa, and develop from epithelium-cells, but are carried off or washed away by the fluid exudations. Still the discharge itself, although containing glanders-cells, offers no characteristic of great diagnostic value except its infectiousness, which exists from the very beginning. The microscope reveals no essential differences, neither between the nasal discharges in glanders and in catarrh nor between farcy matter and common pus.

2. *Diffuse production of glanders-cells in the mucous membrane.*—The glanders-cells are not produced in certain limited spots or nests, but in diffusion over large parts of the mucous membrane. The latter appears swelled and loosened in its tissue, and contains larger or smaller numbers of round glanders-cells of different size. Afterwards an exuberant

morbid growth of connective tissue makes its appearance, which causes the mucous membrane to become more or less thick and callous. If the glanders-process extends to the frontal and maxillary cavities, the naturally fine mucous membrane, especially of the latter, is usually found coated with a muco-purulent secretion, and presents more or less uneven swelling and degeneration, caused by an exuberant neoplastic production of connective tissue elements. In the nasal cavity, but especially on the septum, the diffuse glanders-process penetrates not seldom the whole mucous membrane, and extends to the submucosa. Callous swellings are formed by an exuberant production of neoplastic elements of connective tissue, and within these swellings appear diffuse center-stations, or nests of round cells, which (latter) gradually undergo decay and are absorbed. Fibrous or scar-tissue, which afterwards shrinks or contracts to a scar or cicatrix, takes their place. So it may happen that scars or cicatrices make their appearance without any ulceration having preceded. These scars or cicatrices usually contain a center, from which several whitish strands of fibrous tissue, produced by the same process, are radiating in different directions. Still not every scar or cicatrix found on the mucous membrane of the septum has been produced in the same way, without any preceding ulceration. Under favorable circumstances a healing even of a glanders-ulcer will now and then be effected, but in such a case the scar left behind is usually less prominent or conspicuous, and is destitute of such long radiating strands of fibrous tissue.

Glanders-ulcers.—The same, if present, constitute the most characteristic and unmistakable morbid change of the whole morbid process, and are found usually in the mucous membrane of the septum, especially toward the nasal bones, but also in the mucous membrane of the conchæ, the nasal ducts, the larynx, and the windpipe, and, in rare cases, in the cutis. Professor Gerlach says he has found ulcers in the mucosa of the throat and windpipe only in acute glanders. I remember one of chronic glanders that occurred in 1869 in Quincy, Ill., in which, at the *post-mortem* examination, numerous ulcers presented themselves in the nasal ducts and in the mucous membrane of the larynx and windpipe, but none on the septum. In that horse the only observable symptom consisted, for a long time, in difficulty of breathing, resembling a kind of roaring when exercised. The *post-mortem* examination, made by myself, revealed glanders in a very advanced stage of development, notwithstanding that the horse, a fine black roadster, was not suspected of being affected with glanders up to within two weeks before he was killed.

Glanders-ulcers are always preceded by glanders-nodules or tubercles in the mucous membrane or skin, respectively, and are the product of a decay of the glanders-cells and a dissolution of the intercellular substance of those nodules or tubercles. The process, however, by which these ulcers are developed is not always the same, but varies somewhat according to the size and situation of the tubercles. If the latter are large, of the size of a pea, and extend deep into the mucous membrane, a depression, which soon changes to a small hole, at first not larger than a pin's head, makes its appearance in the middle of the external surface. This hole, however, soon grows larger (Fig. IV, No. 2), and constitutes within a few days an ulcer corresponding in size to that of the former tubercle (Fig. IV, No. 3). The deeper the latter extends into the mucosa or submucosa the deeper will also be the ulcer.

If the glanders-tubercles are very small and superficial, or, as it sometimes happens, visible only as gray specks or dots, the proceeding is a little different. At first the epithelium is cast off; a small, scarcely visible loss of substance takes place, which gives the incipient ulcer the appearance of a small erosion. In other cases the decayed, superficial part of the tubercle presents itself as a yellowish-gray mass, which remains for a short time coated with epithelium. The decaying tubercle, in such a case, has the appearance of a small pustule. In both cases, finally, small, flat, lenticular ulcers are formed, which, if numerous and close together, as frequently happens (glanders-tubercles, if very small, are usually situated close together in groups), become soon confluent, and present then one large, flat ulcer with an uneven bottom. A few days ago I had an opportunity to observe small lenticular, and one medium-sized confluent ulcer, on the right side of the septum of the nose of a former circus-horse that had been affected with glanders—had had discharges from the nose—for over eight months.

A glanders-ulcer once formed grows in depth and circumference as follows: At the bottom and on the borders of the ulcer, and also in the immediate neighborhood of the same, appear again gray specks and nodules (nests of round cells), which also undergo decay, become confluent with the ulcer, and increase thereby the size and depth of the latter. The bottom of a glanders-ulcer presents a grayish-yellow (bacon-like) appearance, marked with red blotches, and is composed mainly of round glanders-cells, the decay of which adds to the depths of the ulcer. Consequently, as after each decay new round cells make their appearance, a glanders-ulcer is not only able to work its way through the mucous membrane and its connective tissue, but also into and even through the cartilagenous septem and the osseous conchæ. This, however, takes place only in a very advanced stage of the disease, and under the influence of a complication with an inflammatory process. The bottom of a deep ulcer presents usually a dirty appearance, caused by decay or decomposition of tissue and blood (Fig. IV, No. 4). Growth of a glanders-ulcer in circumference is a very common occurrence. The process is usually a rapid one, if the ulcer is composed originally of small lenticular ulcers, so-called erosions, with corroded gray or inflamed and red borders. If two or more of such compound ulcers happen to be in close proximity of each other, the same very often become confluent in a comparatively short time, and present then one large ulcerating surface. In the cutis the ulceration process is exactly the same, and is invariably preceded by a formation of glanders-tubercles. The latter have their seat usually in the skin of the lips and nostrils. seldom in the skin of the legs and of other parts of the body. In the cutis, too, deep ulcers and flat and lenticular ones can be discriminated. In some cases the cutis-ulcers have a special tendency to increase in depth—if the preceding tubercles have been large—while in others a tendency to grow in circumference is prevailing. The latter is the case especially if the tubercles have been small and close together. Both kinds of ulcers, however, like those in the mucous membrane, produce abundant exudation and matter, a peculiarity by which deep glanders-ulcers situated in the skin are easily discriminated from farcy-ulcers or glanders-abscesses. Besides that. the latter are always kettle-shaped. have red and elevated borders, and are situated in the subcutaneous connective tissue, while the former have their seat in the skin.

THE CAUSES AND ORIGIN OF GLANDERS.

As to the causes and origin of glanders, opinions, especially in former times, have differed very widely. A great many veterinarians, particularly in France, and there until quite recently, either denied its contagiousness altogether (La Fosse, sen. and jun., Fromage Defengre, and Dupuy barely admitted the possibility of an infection; Coleman (English), Smith (English). and Rodet considered only acute glanders as a contagious disease, as did Hutrel d'Arboval and many others), or expressed doubt as to the existence of a contagion.—Dutz. Consequently a spontaneous development or the possibility of the same was not questioned except by a few decided contagionists, such as Volpi in Italy, White in England, and, in modern times, Gerlach in Germany. Nearly all German, most of the English, and a great many French veterinarians (it is but just to mention among the latter Solleysel (1669), De Saunier (1734), Bourgelat (1765), Garsault (1770), Vitet (1783), Gohier (1813), Delwart, and Leblanc) admitted that most cases of glanders owe their origin to infection, but did not doubt the possibility of a protopathic, and even of a deutropathic development. Even at the present day an auchtochthonous and a deuteropathic development, too, are looked upon as something possible, or even self-evident and of frequent occurrence, not only by non-professional men, but also by a great many veterinarians of high standing. As causes of auchtochthonous glanders, all possible injurious agencies have been accused, the same as in all other contagious diseases, such as pleuro-pneumonia of cattle, for instance, which latter, as is now more generally admitted, spreads, and is caused exclusively by infection or by means of the contagion. The principal causes of glanders have been considered spoiled, decayed, and insufficient food, or food of a bad quality or unsuitable composition; dirty, crowded, and ill-ventilated stables; overwork, hardships, and exposure of any kind or description; in short, nearly everything that is calculated to have an injurious effect upon the animal organism. A great many horses in every country and in every clime are exposed to some or to all of the injurious influences just enumerated, and there is not the least doubt that these influences are well able to weaken the constitution of an animal, to produce emaciation and debility, and to cause a whole army of more or less dangerous and frequently fatal diseases, but still glanders is not any more frequent among horses thus exposed and suffering than among others which are well kept and well treated in every respect. In every country and in every clime a larger or smaller number of horses are exposed to all those injuries mentioned, are worked to death, starved to death, suffocated to death in foul stable-air, poisoned to death with spoiled food and with impure, stagnant water, and still there are countries in which glanders is an unknown, or, at least, an exceedingly rare disease, while in other countries in which horses, on an average, are not kept any worse, or, may be, are kept much better, glanders is a very frequent disease, and causes annually great losses. As a general rule, which, however, suffers apparent exceptions as I shall show hereafter, glanders is frequent in all those countries in which a great many horses are imported, and rare in all those countries in which more horses are raised than needed, or from which horses are exported. Besides that, nobody has ever succeeded in producing glanders by merely exposing or subjecting a horse that has never been exposed to the influence of glanders-contagion to any or to all the injurious agencies and influences which have been mentioned as being accused as the causes of protopathic glanders. In the West, where I have lived and practiced during the last thirteen years, glanders, as I have been informed by reliable persons, used to be an almost unknown disease before the civil war, but has been spread by condemned army horses during and immediately after the war, and is now frequent and can be found everywhere.

Among asses and mules glanders is comparatively not as frequent a disease as among horses, notwithstanding that the former have more predisposition, are easier and sooner infected, and succumb quicker. If a protopathic development were possible. or frequently taking place, one would suppose that it would occur especially in those animals (asses and mules) which possess the greatest predisposition, or, in which, if affected, the morbid process is always the most rapid and the most violent. Besides that, asses and mules particularly, are, as a general rule, more exposed to bad treatment and to all those calamities which have been looked upon as probable causes of glanders, than horses. That glanders is not so frequent among asses and mules as among horses, is simply due to the fact that the former are less numerous and usually less exposed to the contagion, because less used on the road and for traveling purposes, than horses. An exception, perhaps, may be made with the American army, or with any other army in which mules are extensively employed, and in them, I suppose, cases of glanders are just as frequent, and perhaps more frequent among the mules than among the horses.

In modern times, most veterinary writers, it seems, have abandoned the possibility of an autochthonous or idiopathic origin of glanders, but the deuteropathic development is yet upheld by a great many. The diseases supposed to terminate in glanders are especially strangles or distemper, influenza, catarrhal affections of the respiratory mucous membranes, and ulceration in various parts of the animal body. To enumerate all the cases recorded in the veterinary literature in which glanders is said or believed to have developed from other diseases, or been produced by an absorption of matter, would lead too far, for the same are very numerous. As to the different theories that have been advanced, I have to refer to what has been said in the first part of this treatise. To show, however, now easily mistakes may be made, I may be allowed to relate a case that occurred last summer in Chicago. Several horses, constituting the stock of a bankrupt circus, all animals in a very fine condition, were put up for keeping by the authorities in charge, in a certain livery and boarding stable. In the same stable influenza prevailed, and nearly every horse, excepting those circus-horses, became affected with influenza in its so-called catarrhal rheumatic form. Deaths did not occur, but some horses became affected severely. After the circus-horses had been in the livery-stable for several weeks they were sold by the United States marshal, and the day after the sale it was found that one of them, a fine black gelding, was affected with plainly developed nasal glanders, and had communicated the disease already to his stall-mate, which exhibited sufficient symptoms, a slight discharge from the right nostril and a characteristic swelling of the right submaxillary lymphatic gland, to warrant the diagnostication of glanders. After the discovery had been made, it leaked out that the black gelding had been "running from the nose" for over eight months. When the sale took place, some of the livery and boarding horses had not yet fully recovered from their influenza. Now, if one or more of the same should have become infected with glanders, and if the merely accidental discovery of the existence of that disease in one of the circus-horses had not been made. the cry would have been raised immediately that glanders had developed from influenza. Further comments, I think, are unnecessary. It may suffice to suggest that a great many apparent developments of glanders from other diseases may have taken place in a similar way. There also can be no doubt that a great many cases of occult glanders (so-called nasal gleet) have been looked upon and treated as distemper, catarrh, influenza, &c., and afterwards. when plain symptoms of glanders made their appearance, it was more convenient all around to suppose that glanders had proceeded from the disease first diagnosticated, than to admit a diagnostic mistake. So with farcy. It undoubtedly has happened a great many times that the first symptoms of farcy have been mistaken for an inflammation of the lymphatics, and as farcy in its further course becomes frequently complicated with glanders, it is easy to conclude that an inflammation of the lymphatics constitutes a primary disease of glanders. Under certain circumstances I admit it is rather difficult to discriminate at once an inflammation of the lymphatics and subsequent ulceration or formation of abscesses from genuine farcy, and so mistakes, undoubtedly, have occurred.

Besides all that the diseases looked upon as the possible progenitors of glanders are similar to the latter only in regard to a few external symptoms but entirely different as far as the morbid process is concerned. They lack altogether during their whole course, from first beginning to their final termination, the specific characteristics of glanders, and a conversion of any one of them into the latter disease must be looked upon as just as impossible as it is to change a cow to a horse, or a goat to a hog. Still, this does not exclude the possibility of an animal affected with one of those disorders, or with any other disease, becoming infected with glanders or farcy. On the contrary, a diseased condition of the respiratory mucous membranes seems to facilitate an infection if an exposure to glanders contagion is taking place. At any rate the morbid process of glanders is always much more violent and makes a more rapid progress in a diseased organism than in one that is otherwise perfectly healthy. To get at the bottom of the facts and to guard against mistakes, it will be necessary never to lose sight of the specific characteristics of the glanders process.

Notwithstanding all those cases of apparent deuteropathic development of glanders which can be found in the veterinary literature of nearly every country, I am not afraid to say I do not believe that a case of real deuteropathic glanders, one that can stand a thorough and unbiased investigation, has ever occurred. Gerlach, in his treatise, repeatedly mentioned, says, on page 115, "A genuine development (protopathic and deuteropathic) must be considered as not proved."

Glanders, as well as pleuro-pneumonia, Russian cattle-plague, and scab and mange, will cease to exist if a propagation by means of infection is made impossible. If, for instance, within the limits of the United States all animals affected with glanders were destroyed at once, and at the same time every place where glanders-contagion may be existing were thoroughly disinfected, and if any importation of glandered horses or of the contagion were successfully prohibited or prevented, glanders would at once become extinct, and would never make its appearance again within the limits of the United States, unless imported again from other countries. It is a disease that can be eradicated.

I said before that glanders is most frequent in those countries in which numerous horses are imported from other countries. This is an undeniable fact except in regard to those commonwealths in which good veterinary schools provide a sufficient number of thoroughly educated veterinary surgeons, and in which stringent laws enforce the immediate destruction of every animal affected with glanders, prohibit veterinary quackery, and do not allow anybody to keep or to treat a glandered animal unless he is a qualified veterinary surgeon, and gives sufficient bonds to pay possible damages.

I know very well that I shall be contradicted, but mere denials, or questions asking where glanders originally comes from, if a spontaneous development does not take place, will not do. Such questions, of course, I cannot answer. When Gerlach first pronounced pleuro-pneumonia of cattle a pure contagion, that is, a disease propagated exclusively by means of infection, Professor Spinola asked pertly if Gerlach had imported pleuro-pneumonia from the moon, but failed utterly—and everybody else, too—to show a solitary case of an unmistakable and well-authenticated spontaneous development. If any one can show me a case of spontaneous glanders, not caused by infection, or give satisfactory and unmistakable proof that a protopathic or deuteropathic development of glanders has occurred, I will take back what I have said, but not before.

The contagion.—The contagion must be considered as the exclusive cause of glanders. When I lived in Dixon, Lee county, Illinois, from the fall of 1865 to September, 1868, I had an opportunity of observing numerous cases of glanders. A friend of mine, D. W. McKinney, dealer in horses and proprietor of a livery-stable, knew nearly every horse in the whole county, and taking special interest in those cases of glanders, assisted me in inquiring into the history of every horse affected. As a result, every case, without exception, was traced back to an infection by condemned United States army horses that had been sold to the farmers.

The contagious principle is developed during the very first stages of the disease, and even before plain symptoms have made their appearance. It exists most concentrated in the immediate products of the morbid process, but especially in the discharges from the nose, and in the contents of the glanders and farcy ulcers. It is present also in all the secretions and excretions of the affected animals, as has been proved by numerous direct experiments. Professor Gerlach, in order to ascertain if the contagion is contained not only in the fluid animal humors and excretions, and in the fluid and solid products of the morbid process, but also in the pulmonal exhalation and in the perspiration, has made several interesting experiments, and has found that an inoculation of a healthy horse with artificially condensed exhalation and perspiration of a glandered animal produces the disease. He has, however, not succeeded in communicating glanders by injecting defibrinated blood of glandered horses (100 and 200 grains respectively) into the veins of healthy animals. Still, the contagiousness of the blood has been established long ago by Abildgardt and Viboeg in Copenhagen.

The experiments of Gerlach and of others, and numerous clinical observations, too, have proved beyond a doubt that the contagion contained in the exhalation and perspiration clings, though only in small quantities, to the aqueous vapors exhaled by the respiratory organs and perspired by the skin. The contagious principle, therefore, is volatile only in a limited degree, and to produce an infection by means of the exhalation and perspiration at a distance of several feet requires usually some length of time. So it happens very often that a horse occupying with a glandered horse the same stable, but not the same stall, remains exempted. The more forcible and accelerated the breathing and the more abundant the perspiration of the horse affected with glanders, the greater, it seems, is the danger of an infection of healthy horses that are near, or occupy the same stable.

Another question not easily answered, and yet an object for investigation, may be asked; that is, Do organic forms constitute the contagion; is the contagious principle bound on, or inseparable from, organic forms; or is its action merely a chemical one? On this question the opinions of the best authorities differ. Professor Gerlach, in his successful experiments with condensed exhalation and perspiration, found no organic forms whatever in the perfectly limpid drops; further, he found no organic forms in the very infectious caseous substances taken from the mucous membrane of a horse affected with diphtheritic glanders. He, therefore, has come to the conclusion that the glanders-contagion does not consist in, nor is bound on, organic forms, and that the action of the contagious principle must be a chemical one. On the other hand Hallier and others have found organic growth (micrococci) in the humors of glandered horses and in the products of the morbid process of glanders, and are inclined to consider those micrococci as the agency which causes the disease, produces the morbid changes, and effects a communication of the glandered process to other healthy animals. If Hallier and others are right, a great many mysterious phenomena observed in glanders find an explanation, but if Gerlach's observations are correct, Hallier's theories necessarily fall to the ground. Gerlach says : " Hallier finds everywhere fungi," and Chauveau finds everywhere cells. Still, notwithstanding my high regard for Gerlach and the thoroughness of his investigations, I think the finds of Hallier and of other investigators cannot be discarded ; positive evidence is always of more value than negative proof. Haeckel (History of Creation, vol. 1, Protista.) and Klebs (*Archiv fuer experimental-Pathologie*, 1873), separate the microscopic organisms found in glanders and in other contagious diseases from the class "fungus," and consider them as a separate class, belonging neither to the animal nor to the vegetable kingdom. Whatever may be the truth as to the real nature of the contagious principle, future investigations must reveal. I myself have had no opportunity to make thorough microscopical investigations of the morbid products of glanders, and can, therefore, not advance any definite opinion of my own. Mere speculations cannot bring any facts to light; thorough and patient observations are necessary.

The glanders-contagion, whatever its nature may be, communicates glanders and farcy not only to the animals belonging to the genus equus, but also to other animals and to man. Numerous cases are reported every year in the periodical veterinary literature. The only domesticated animal that seems to be exempted or to be destitute of any predisposition is the ox.

Glandered horses, as soon as the disease has been diagnosticated, are usually removed to the cow-stable, or to pens or places where cattle are kept, and still no case, as far as I have been able to learn, is on record in which an ox or a cow has contracted the disease. Sheep are easily infected. Goats, too, possess sufficient predisposition. Ercolani described a case in "*Il medico veterinario*," 1861, and Wirth succeeded in communicating glanders to a male goat by means of inoculation (*Archiv fuer Thierheilkunde*, Bd. 6, Heft 1, 1844). Hogs seem to possess but little predisposition, and cases of dogs becoming infected and dying of glanders have been communicated by Nordstroom (*Tidskrift for Veterinairer*, etc., 1862) and Langeron (*Revue vétérinaire*, etc., Toulouse, série I, 1876). Several cases are on record in which wild animals, lions especially, have become infected with glanders by being fed with meat of glandered horses. According to the experiments of Viborg and Bingheim, the flesh of a horse affected with glanders can be eaten without danger of infection if properly cooked or fried.

One important phenomenon must be mentioned, and that is, that glanders always becomes a frequent disease after any great war. Such was the case in our own country after the great civil war, as I have mentioned before, and also in Germany and France, but especially in the latter country, after the war of 1870–'71. Cases of glanders will also be frequent during the next few years in the Turkish Empire, and in those

Turkish provinces which have become independent, when separated from the Ottoman territories. The cause of this frequency is an obvious one. It consists in the abundant opportunity of infection. One horse affected with (occult) glanders in either of the hostile armies can, for obvious reasons, communicate the disease with the greatest facility to a large number of animals. The fact of glanders becoming frequent after each large war has been used very frequently as an argument in favor of a protopathic development, but if it is looked upon in a proper light it proves, if anything, the exclusive spreading of the disease by means of the contagion.

Prevention and treatment.—As to a medical treatment, there is scarcely a remedy known in the whole materia medica that has not been used against glanders, but, so far at least, with very poor success. It is true a great many *pretended* cures are on record. But if the slow or chronic progress of the morbid process, its frequent remissions in warm and dry weather, exacerbations in rough, cold, and inclement weather and in a foul atmosphere, and the great confusion that has prevailed in regard to the true nature of glanders are taken into consideration, it is no wonder that mistakes and deceptions have occurred. Some of the cases that are said to have been cured have no glanders at all, and in others the pretended cures have been only temporary—a mere remission. *Confirmed glanders must be considered as incurable;* and it would, therefore, be for the benefit of every one if our general government (Congress) would enact a law which should make it a criminal offense to keep and to use a horse, or any other animal, known to be affected with glanders. Any attempt to cure should also be strictly forbidden, because a prompt and immediate destruction of every animal affected with glanders, a disease which spreads only by means of its contagion, constitutes the best, surest, and cheapest, and in fact the only prevention.

A case of recent occurrence will serve to illustrate how glanders spreads, and how much cheaper it is to destroy a glandered horse at once than to permit the same to communicate the disease to healthy animals. It will also show the necessity of a stringent law making the sale of an animal known to be affected with a contagious disease a criminal offense.

Last fall Mr. George T. . . . ., Pottawatomie county, Kansas, bought a horse of a Mr. Ch. . . , Manhattan, Riley county, Kansas, and pastured and stabled the same with his other horses, about twenty-four or twenty-five in number. The horse in question, when bought, had some discharge from the nose, which, of course, was pronounced to be nothing but the product of catarrh—in common parlance, a cold. In the course of the winter several of Mr. T. . . . .'s horses commenced to have discharges from the nose. Mr. T. . . . . became alarmed, and brought the new horse, whose nasal discharges had increased, and who showed other symptoms of disease, such as a staring coat, emaciation, &c., to me for examination. I found the symptoms to be those of an advanced stage of glanders. Subsequent inquiries revealed some of the previous history of the animal. Mr. Ch. . . had bought the horse from another man, whose name I do not remember, only a few days before he sold the same to Mr. T. . . . ., and had kept the animal, while in his possession, strictly separated from his other horses, because he knew that the same had a chronic discharge from the nose, and had had it for about two years. Is not such a transaction criminal? And still, in the case mentioned, there is no redress to be had. Mr. T. . . . . is a comparatively poor man; his farm is mortgaged, and all the property he may call his own consists in his stock, but especially in his horses. As I moved away from Kansas early in the spring, I have not learned how many of his horses have become affected, but several had contracted the disease before I left. Besides that, his horses had been together quite often with those of his neighbors, on the prairie, before he knew them to be affected with glanders. It is possible that he has lost, or will lose, nearly every animal he has. Mr. Ch. . . . does not own anything; all his property is in his wife's name; consequently Mr. T. . . . ., if he sues for damages, will have to pay lawyers' fees and costs, but cannot recover anything. If there were a United States law which made it a criminal offense to sell animals affected with contagious diseases, or to own and to keep animals which exhibit symptoms of contagious diseases, and to neglect to advise the proper authorities of the fact, such cases as the one related would not occur. If Mr. T. . . . . were not an honest man, he would undoubtedly have kept still, and would have sold his glandered horses to other innocent parties, and contributed in that way in spreading the disease. I could relate numerous similar cases, but think this one will suffice, especially as this article is already too long.

A successful prevention of glanders is possible only if the contagion—which, even if it should not constitute the sole and only cause of the disease, causes at least nine hundred and ninety-nine cases of one thousand—is thoroughly destroyed wherever it may exist or wherever it may be found. Consequently every animal affected with glanders should be killed as soon as the nature of the disease becomes known, and be buried sufficiently deep or be cremated. But as the contagion adheres frequently also to the stables—manger, floor, partition, &c.—that have been occupied, the stable utensils—brush, curry-comb, &c., and the harness, blankets, halters, bridles, saddles, &c.—that have been used or been in contact with glandered horses, it is of great importance to know what will best and most effectually destroy the contagion. Professor Gerlach has made very interesting and valuable experiments, to relate which, however, would lead too far. I will therefore only state the results arrived at. The discharges from the nose, glanders-matter, &c., lose their infectiousness if perfectly dried by being exposed to currents of air or to the rays of the sun; but kept moist, for instance in a damp cellar, wrapped up in a moist rag, or adhering to the corners of the manger, to a damp wall or floor, or to the bedding or the manure, &c., the contagion seems to possess great vitality, and may remain effective for half a year or longer. Putrefaction does not destroy the contagious principle. Chlorine destroys the contagion, and is therefore a very efficient disinfectant, provided the chlorides used come in actual contact with the contagion. A brief exposure of the infectious substances, nasal discharges, glanders matter, &c., to the influence of chlorine in a gaseous state, mixed with the atmosphere, is ineffective. As a remedy to be given internally, chlorine, in shape of chlorine-water, for obvious reasons cannot be used; chemical combinations will be effected before an absorption can take place. The best and surest destroyer of the glanders-contagion is carbolic acid. It may be used not only as a disinfectant or for the purpose of destroying the contagion clinging to the wood-work of the stable &c., but also in incipient cases of farcy, and in cases in which an infection with glanders-matter has just taken place in a wound, for instance, as a local remedy. If applied to the glanders-ulcers on the septum, or to farcy-ulcers, a tendency to heal will make its appearance. As a disinfectant, a solution of carbolic acid in glycerine or alcohol and water (1:1 or 2:20) is perfectly strong enough to be effective. Old straw, hay, and bedding must be burned, and blankets, &c., are best disinfected by exposing the same for some time to a temperature of 212° F., or higher, either in an oven or in boiling hot water.

As to a therapeutic treatment only a few words will be necessary. Some of the most heroic medicines have been used with very doubtful results. So, for instance. Professor Ercolani, in Turin, claims to have had good success with arsenate of strychnine, but others who have made the same experiments have had no success whatever. Lacaze (*Revue Vétér.*, &c., Toulouse, 1876), asserts to have been successful with large doses of alcohol, but he discriminates contagious and noncontagious glanders, and so no comment will be necessary. In former times cantharides were considered as a remedy, but later investigations have proved them to be perfectly worthless. That every kind of mercurial combination and a great many sure-cure nostrums have been used and been advertised as specific remedies, as in every other incurable disease, is too self-evident to need any further mentioning.

The only rational treatment of a horse or other animal affected with glanders consists in a proper and effective application, in the right place, of either half an ounce of lead or five inches of steel; and until such treatment is invariably adopted, or made compulsory, there will be no prospect whatever of freeing this country from this loathsome disease, dangerous even to man, in whom, if once infected, it is just as incurable as in horses.

INDEX.

○

www.ingramcontent.com/pod-product-compliance
Lightning Source LLC
Chambersburg PA
CBHW021500210326
41599CB00012B/1072

* 9 7 8 3 3 3 7 2 4 1 2 0 9 *